D1058609

POLLUTION AND POLICY

POLLUTION AND POLICY

A Case Essay on California and Federal Experience with Motor Vehicle Air Pollution 1940–1975

JAMES E. KRIER
and
EDMUND URSIN

UNIVERSITY OF CALIFORNIA PRESS
Berkeley Los Angeles London

University of California Press
Berkeley and Los Angeles, California

University of California Press, Ltd.
London, England

ISBN: 0-520-03204-7
Library of Congress Catalog Card Number: 76-3881

Printed in the United States of America

1 2 3 4 5 6 7 8 9 0

CONTENTS

PREFACE AND ACKNOWLEDGMENTS

This book owes its origins—and much more—to David Cavers, Fessenden Professor Emeritus of the Harvard Law School, and president of the Council on Law-Related Studies. He first suggested to us the value of a study of California and federal vehicle pollution control policy; he called to our attention the willingness of the Council to consider funding such a project. Our decision to undertake the study, and the Council's decision to support it, followed not long thereafter.

Professor Cavers deserves full acknowledgment for the initial idea of the study, but we must take credit and, unfortunately, any appropriate blame for its final shape. As studies so often do, this one evolved over the course of time into a form quite different from that planned at the outset. In particular, we determined early in our research that a perspective far more historical than the original conception was necessary: the events and policies in question, we sensed, could be best understood only in the context of what preceded them. A good grasp of the present required more than a glimpse of the past. This awareness meant a considerable expansion of our efforts, and therefore far more research and writing time than originally contemplated. In our judgment, it is to the unending credit of Professor Cavers and the Council that they not only granted this time most freely but left us entirely alone in the meanwhile to pursue our work. Professor Cavers in particular apparently harbored a faith that we would one day finish, even in those times when we ourselves were faithless. So for the initial idea, the generous support, and the faith, patience, and tolerance, we acknowledge our gratitude to Professor Cavers and the Council.

From the outset of its funding the Council had hoped for, though it did not insist upon, support from other quarters. Additional support became especially important as the project expanded, and was eventually provided by the United States Environmental Protection Agency. We are particularly indebted to the agency for its generous funding because the support was forthcoming despite skepticism on the part of some of its personnel as to the value of a largely historical study, and inasmuch as the agency, like the Council, exhibited—to a degree that must surely be uncharacteristic of government agencies—a willingness to fit its deadlines to our work, rather than insisting on the opposite approach. The person most responsible for this ungrudging and unusual flexibility is Dr. Marshall Rose, our project officer throughout most of the study. Besides bearing the burden of our unbureaucratic ways, Dr. Rose provided encouragement, advice, and editorial assistance (as to both form and substance). Neither he nor the agency is, of course, responsible for the contents and conclusions of our research.

A number of students at the UCLA Law School were most resourceful in gathering a large volume of primary and secondary written materials and in conducting over one hundred interviews: Richard Clark, Natalie Hoffman, Mary Keller, Randall Kennon, Marc McGuire, Fred Mueller, and Leda Williams. We give our warm thanks for their valuable participation in the study. Three secretaries in the Law School—Betty Dirstine, Dorothy Goldman, and Carol Sartorius—devoted much time and patience, much more than we deserved, to the typing of draft and final manuscripts. They did their work with cheer and skill that contributed to the final product in many ways, and we are most grateful to the three of them. We also thank the many people who consented to interviews, and the many libraries that let us harbor their books and materials over an unconscionably long period of time.

This book being a cooperative effort, we should indicate who is responsible for what. Edmund Ursin conceived much of the original project design, and he took on virtually all the burdens of supervising the research. He also wrote drafts concerning some of the events in the period 1950–1960, which appear primarily as parts of chapters 5 and 6; beyond this, his ideas permeate the balance of the study.

Portions of Parts IV and V of the book are based to some extent on work previously published elsewhere. Chapter 13 draws considerably from a 1974 report prepared by Eugene Leong in partial satisfaction of the requirements of the doctoral program in Environmental Science and Engineering at UCLA. The report reflects Dr. Leong's actual participation in the air pollution policy process in California; we are most grateful to him for lending us his insights. Chapters 14 and 15 include material originally appearing in somewhat different form in three articles by James Krier in the *Natural Resources Journal* (with W. D. Montgomery), the *UCLA Law*

Review, and the *Journal of the Air Pollution Control Association* (with K. Heitner). We express our appreciation for permission to use portions of these articles.

Those parts of the book not mentioned above are primarily the work of Krier, who also prepared the initial and final versions of the entire manuscript, and who takes sole responsibility (but not credit) for form and content.

The three articles by Krier mentioned above were written in the course of work as a consultant to the Environmental Quality Laboratory of the California Institute of Technology. The articles are small instances of the many direct and indirect ways in which association with the Laboratory over a period of several years has contributed to our study. One particular contribution was crucial: despite the kindnesses of our supporters, it is doubtful the project would have been finished even by its much-revised deadline had not the Laboratory generously afforded Krier a summer in which to concentrate full time on completing the initial manuscript. For the luxury of that time, and for many other generosities, we are grateful to the Laboratory, its staff (especially Glen Cass, Betsy Krieg, and Gregory McRae), and its former and present directors, Professors Lester Lees and Norman Brooks, respectively.

A number of persons read the book in manuscript form and offered suggestions that we have tried to heed in revising it for publication. Professors Lawrence Friedman, Joel Primack, Gary Schwartz, and Dan Tarlock deserve special thanks for constructive criticism and moral support, as does our editor Bernice Lifton and also Shirley Warren of the University of California Press. Nettie Lipton and Udo Strutynski of the Press provided much help and many kindnesses. We also thank the University of California; the Foundation for Research in Economics and Education; and the Centre for Socio-Legal Studies, Wolfson College, Oxford, England—the first for giving Krier time (sabbatical leave), the second for giving him financial support (a travel grant to England), and the third for giving him space and facilities (through a Visiting Fellowship) to work on the final manuscript. The friendship and hospitality of colleagues at the Centre are especially appreciated.

Finally, Ursin thanks his wife, Nona Egan Ursin, for encouragement and advice; Krier thanks his wife, Wendy L. Wilkes, for editorial assistance, for helping to organize data, and for helping to organize him.

J. E. K.
E. U.

January 1977
Los Angeles, California

INTRODUCTION

Something happened in Southern California in the early 1940s. In the first year of that decade (perhaps a bit later—published accounts differ slightly), the area experienced a brownish, hazy, irritating, and altogether mysterious new kind of air pollution that was more persistent than, and quite different from, the isolated instances of irksome smoke that had troubled major urban centers from at least the mid-1800s. The new problem was, of course, smog. (The word, a synthesis of "smoke" and "fog," fits the type of air pollution common to places like London, but it is something of a misnomer for the photochemical air pollution of Los Angeles and many other urban areas, pollution that has little to do with smoke or fog. But the word has become so common that we are going to employ it here.)

The smog problem became more serious over the next few years following its onset, and so too did attitudes about it. People clamored. Public officials responded. Studies and controls began. These continue today, and so does the problem—at local, state, and national levels. The situation is especially pressing in the Los Angeles area, where the federal government proposed, among other controls designed to achieve stringent new national air quality standards, regulations that could literally have banned the sale of gasoline for driving during the six smoggiest months of the year (May through October). This book is the story of how the smog problem was discovered, of control policy and how it too was (to some considerable degree) "discovered," and of how and why that policy evolved to such extremes. The book's eventual focus will be on motor vehicle pollution and what has been and might be done about it—done not so much in the technological as the institutional sense, in the sense of how to *think about* organizing and deploying our technology and other techniques of control.

But we come to that only eventually, because a good part of the story deals with the years of discovery and debate that went into identifying the motor vehicle as a central contributor to polluted air.

Let us be more precise about what the book concerns. Primarily, it is an account and an appraisal of the identification of and responses to the problem of motor vehicle air pollution in California and at the federal level over a period spanning some thirty-five years—roughly 1940-1975. The problem is an important one currently receiving a great deal of attention from state and federal lawmakers; the public and the news media; and those in academic disciplines ranging over science, engineering, economics, law, and political science. To our knowledge, however, none of this attention has taken the form of historical study—the method of inquiry that we have adopted to considerable degree.

A good deal of our inquiry focuses on the California experience. It was in California that the important contributions of motor vehicles to air pollution were first discovered, and it was there that the first measures to control those contributions were put into operation. As the direct result of these measures, the state came to occupy a unique position in vehicular emissions control. In 1967 Congress explicitly preempted authority to control emissions from new automobiles (there had been an implicit call for such a policy expressed in the history of federal legislation passed in 1965); thereafter, the states could prescribe no vehicular emission standards, whether more or less demanding than the federal ones, for other than used vehicles. There was, however, one exception: California was permitted, as it is today, to set more stringent standards in certain instances. In part, the exception was intended to enable the state to cope with particularly severe problems in some of its areas (especially Los Angeles), but it was also motivated by California's pioneering experience in the field (an experience, like most pioneering ones, characterized as much by failure and frustration as by grand accomplishment), and by the accompanying purpose to employ California as a testing ground for new approaches to control.[1]

Since 1967, vehicular pollution control policy in California and the rest of the states has come to be more heavily infused with federal inputs, especially as a result of the Clean Air Amendments of 1970.[2] Under that legislation, the federal government established uniform air quality standards for the nation and instructed the states to implement measures that, together with strict federal emission standards for new vehicles and some stationary sources, would achieve them. In some areas this means little more than that state governments must effectively control stationary sources. In other areas, however—those with severe air pollution problems like Los Angeles—attaining the federal air quality standards necessarily entails dramatic improvements in technology and, pending those improvements (or, more probably, in addition to them), increased control of used vehicles by way of technological controls and, more significantly, increased control of all vehicles by way of measures (like the ban on gasoline sales mentioned above) to curtail the number of vehicle-miles driven. The matter of designing and implementing used-vehicle controls and limitations on miles driven (transportation controls) has been particularly vexing to California and federal air pollution control authorities, and it will command a considerable amount of our attention.

Because California was the first of the states to study and regulate air pollution from automobiles, and because it has been more or less a laboratory engaged in sustained efforts to cope with unusually difficult problems, its years of trial and error, failure and success, should hold instructive lessons for other states. Moreover, because the state's efforts have been a major influence shaping the federal approach to vehicular pollution control,[a] and have also of late been much shaped by the federal approach, study of the California experience should be revealing with respect to federal as well as state policy.

From a broader perspective, the events on which we focus are typical of most environmental problems and perhaps a number of other social problems involving significant interplay among legal, technological, and institutional issues. For this reason, we hope the analysis might help enlighten understanding not only of vehicle pollution control policy but contain as well lessons (often only implicit) for environmental policy generally, and for a wider class of similar policy concerns.

We term our study a "case essay," rather than the familiar "case study," because the latter phrase suggests an approach that is more data bound, more exhaustive, more rigorous, and less impressionistic, than is our own. The essay format permits us a certain license, without pretensions to the contrary, that we have found useful—particularly useful inasmuch as the more we studied the questions at hand, the more we considered them in need of further study beyond our capabilities, our energies, and our time. The work is necessarily a tentative, speculative attempt; hence the essay format.

As we said above, previous work in the area does not reflect primarily historical inquiry, and there are some who believe such an inquiry can be

[a]"Air pollution control precedents established in California were . . . an important part of the institutional context of Federal air pollution control legislation." R. Dyck, "Evolution of Federal Air Pollution Control Policy" (Ph.D. diss., University of Pittsburgh, 1971), p. 213.

of little value. For example, early efforts to interest the United States Environmental Protection Agency in the study presented here met with no success (a situation ultimately reversed). The Clean Air Amendments of 1970 were said to have resulted in "a substantial departure from previous state and federal approaches to air pollution control"; the legislation presented state and federal governments with new responsibilities, and for this reason the agency concluded that "an historical analysis of the formulation of air pollution controls in California has little bearing" and thus was "not considered relevant to . . . current program needs."[3] We can only disagree: the present approach to air quality control might well have taken quite different, and perhaps quite improved, shape had there been a better understanding of the events and approaches that preceded it. That point was made, perhaps a bit too eloquently, by a veteran of California's struggle with air pollution. He said:

> In our Nation's capital there is a relatively small, but beautifully proportioned, building. Over its entrance is engraved Archives of the United States of America. At the sides of this entrance are contemplative seated figures. On one pedestal is engraved "Study the Past"; on the other, "What Is Past Is Prologue."[4]

We believe that, at least in this instance, the past has indeed been prologue in many important respects, and our study attempts to reveal them. We begin in Part I with an effort to put the substance of the study into perspective by providing a framework of facts and concepts that figure in the later discussion. The framework describes the nature of the pollution problem in terms of some of its most important technical, behavioral, and institutional components. It is intended primarily for those with no special knowledge of scientific, engineering, legal, or economic aspects of air pollution and its control; we hope it is neither too abstract for them nor simplistic for the specialists. The purpose is solely to introduce handy touchstones for later reference, and it should be borne in mind that we will eventually return to expand and apply the observations sketched in Part I.

Parts II, III, and IV of the book contain the historical narrative of events over the years 1940 to 1975. Part II concerns the initial period (up to 1960) of discovery, debate, and delay in the policy process; Part III focuses on state and federal control efforts that occurred over the decade 1960–1969. Large portions of the contemporary scheme of control, which began in 1970, are the legacy of these early years and are, in any event, illuminated by considering them in some detail. Part IV describes the development and implementation of the contemporary scheme between 1970 and (roughly) 1975. We should emphasize that events in 1970 and after took directions in some respects quite different from those of the preceding years, and that their culmination represents our present (though quite unstable) pollution policy.

Throughout the historical narrative we have tried to be selective. We describe only what we consider to be significant events, and as to these the amount of detail tends to vary inversely with what has been written in accessible form by others. As it happens, many events over the years in question are especially interesting and important (sometimes even intriguing or amusing), yet oddly enough they have been discussed only scantily, if at all—dribs and drabs and piecemeal accounts in often remote sources. Other developments have been treated as thoroughly as necessary in quite easily obtained literature, so that we can simply summarize and provide references for those who wish to pursue the inquiry.

Despite such efforts at economy, the historical narrative often requires a quite careful reading. It is long simply because the span of years covered is itself long and rich. It is at times busy with details that interrupt the smooth flow of large events, yet those details must be confronted because they are often central parts of the story. It is, on occasion, convoluted, tracing one chain of events over a period of years, then returning to trace another. But this, if not inevitable, is at least convenient. History, especially the history of policy made in a many-layered federal system, does not develop in a nice linear progression of significant event followed by significant event. Rather, at any one point many events are occurring simultaneously, and to make sense of them often requires that we break out related happenings and order them according to a logic that violates the strict laws of time.

To help cope with the difficulties of length, detail, and occasional convolution, we use the familiar convention of organizing chapters in the historical narrative by blocks of time each of which has its own coherence in terms of significant events. Beyond this, we provide a chronological table at the beginning of each of the busier chapters to give a sort of thumbnail sketch of what was happening when. (See tables 5, 8, and 9 at the beginnings of chapters 6, 8, and 10, respectively.) Even then, a firm grip on the lines of the story is difficult without having at least a sketch of the whole picture in mind. That being so, a brief outline here of the high points of Parts II, III, and IV should prove useful.

Essentially, the story is one of an emerging awareness of the contribution of motor vehicles to urban pollution problems, and a gradual move from local to state to predominantly federal controls. The first significant response to the mysterious new pollution problem that had thrust itself into the sunny life of Southern California in the early 1940s came in 1947—state legislation permitting, but not requiring, countywide air pollution control districts. (Local ordinances enacted over the years 1945-1947 had proved inadequate to handle the problem.) Los Angeles County established a district immediately and embarked on a program intended to control pollution but keep industry in business as well. The effort succeeded halfway: by 1949 business was going as usual while pollution conditions were growing worse, in part because both the 1947 legislation and Los Angeles district controls virtually ignored motor vehicles.

That attitude toward motor vehicles had been a hallmark of the decade, despite a few scientific studies that pointed at least a finger of suspicion at the internal combustion engine. It was, however, an attitude that began to undergo important change early in the 1950s, though it would be a decade before this resulted in significant controls on motor vehicles. The causes for change, for a new open-mindedness about vehicle emissions, were several. The most fascinating had to do with a 1949 college football game in the Northern California community of Berkeley. Air pollution in the area of the stadium during the game resembled conditions that had appeared almost ten years earlier in Los Angeles—conditions never before experienced elsewhere. The only way to account for their sudden occurrence in Berkeley was the concentration of automobiles brought on by thousands of fans driving to the game. This, at least, was the opinion of several legislators. They concluded that the stadium area had been a microcosm of Los Angeles, where traffic congestion on vast scales was an everyday affair. Vehicle exhausts, they thought, must be an important factor in the problem. Weak science, perhaps, but a nice hunch, and other evidence supported it: the studies in the 1940s that suggested the importance of internal combustion processes, the failure of stationary source controls to accomplish air quality improvements.

Soon there was evidence of a more scientific nature. Research by Professor A. J. Haagen-Smit, a biochemist who was to occupy the center of pollution control efforts in California for the next twenty-five years, suggested in 1950 that a photochemical reaction converted certain pollutants—primarily from refineries and motor vehicles—into Los Angeles smog. The mystery seemed to be solved, but for practical purposes it was not. Haagen-Smit's discovery was greeted with wide skepticism, especially by those with something at stake; the result was debate and delay. The driving public, famous in Southern California for its love of the automobile, saw a conspiracy by the "interests" to blame the smog problem on innocent citizens. One of the "interests," the oil refining industry, saw

Haagen-Smit's research as a threat to its livelihood, and worked to debunk his findings. The Stanford Research Institute, employed by one of the industry's trade associations, quickly claimed to have found fundamental flaws in Haagen-Smit's methods and conclusions. In 1953 the auto industry entered the arena, beginning its own research program on the ground that the situation was too obscure to assign blame. In the same year the Air Pollution Foundation was chartered in California as an independent organization with the mission of resolving the debate surrounding research on the air pollution question.

The Foundation played an important role throughout the decade and dissolved at the end of it. Thanks largely to it, Haagen-Smit's work was basically confirmed in 1954, and a year later the relative roles of refineries and motor vehicle exhausts were quite clearly established: that vehicle emissions in particular were an important part of the problem seemed beyond question. By 1956 the Foundation went further, concluding that vehicle exhausts were the principal contributor to Los Angeles smog, and by 1959 there was almost unanimous agreement on this point by experts. Almost, but not quite. The auto industry was unconvinced. Between 1954 and 1959, it issued a series of findings the pattern of which suggested a strategy to delay control measures. Within the time provided by that strategy, the industry undoubtedly worked on developing controls against the day when their installation would be required. (Developing them, perhaps, but keeping them under wraps as well. In 1954 the industry entered into a joint venture with the announced purpose of speeding development of controls. Some years later this agreement gave rise to antitrust charges of a conspiracy to delay introduction of pollution controls.) Others, including the Air Pollution Foundation, were also at work on control technology—though surely with more selfless motives.

The early search for technological solutions to air pollution took some bizarre forms, the focus being not so much on techniques to control emissions at the source as on means to remove pollution from the air once it had gotten there—fans to blow it away, holes in the mountains to let it escape, and so forth. More constructive efforts began about the same time, with the auto industry, the Air Pollution Foundation, the Stanford Research Institute, and other private and public agencies all taking part with varying degrees of enthusiasm. By the end of the decade the Foundation concluded that feasible vehicle control technology was little more than a year away.

Meanwhile, pollution conditions continued with little or no improvement. If anything was clear after the mid-1950s, it was that the problem was spreading throughout California and around the country. The general situation produced pressures for more state and federal involvement; periodic episodes gave them added force. In 1955 both the California legislature and the federal Congress passed laws providing for further studies on the causes, effects, and control of air pollution. The federal legislation was virtually, though not literally, the first attention directed to the problem by the national government. It had shown little interest until 1948, when a severe episode in Donora, Pennsylvania, stimulated federal investigations and plans for a national air pollution conference. A few bills were even introduced in the Congress shortly after Donora, but none of them passed. They had called only for further federal studies in any event; control was considered a state and local concern. The federal legislation finally enacted in 1955 reflected this same view: it left regulatory efforts to the states. And California's 1955 legislation mimicked the attitude, regarding control as a local responsibility—despite the troubles with the 1947 act. Moves for more significant state and federal involvement continued throughout the 1950s, especially with regard to motor vehicles. In 1959 California legislation instructed the State Department of Public Health to determine the air quality and motor vehicle emission standards needed to protect health. In the same year, the House of Representatives approved a bill calling for studies of motor vehicle exhausts by the surgeon general.

The story of the 1960s and 1970s is largely one of legislation and its implementation. In 1960 the bill for motor vehicle exhaust studies that had passed the House of Representatives a year earlier made its way through the Senate and became federal law (the Schenck Act). In the same year, California enacted far more substantial legislation. The Motor Vehicle Pollution Control Act of 1960 brought the state into the active role of control for the first time and represented the first motor vehicle emission control legislation. The act created a Motor Vehicle Pollution Control Board to certify control technology. Certified technology would be required as a condition to registration of new vehicles, and to registration or transfer of some used vehicles. Unfortunately, administration of the 1960 law was slow and clumsy. Installation of the first controls on new vehicles was not required until 1964, three years after the industry had begun voluntary installations. Used-vehicle controls were marked by an especially troubled history, and never really got on the road even by the end of the decade.

In 1962 the Congress enacted the first of a series of measures that moved the federal government, over the next eight years, out of its position of passive supporter and into one of aggressive and dominating initiator.

The transition started modestly enough: the 1962 legislation merely added a few years of life to the 1955 federal law and made the studies called for by the Schenck Act a permanent task of the surgeon general. But more was in the air, for the Congress now recognized explicitly that the motor vehicle pollution problem was of nationwide significance. The next federal step—the Clean Air Act of 1963—went only a tiny bit further in the direction of control, but it laid the groundwork and provided the namesake for the major federal incursions soon to come. The act provided a limited federal power to abate air pollution endangering health and welfare; instructed the Department of Health, Education, and Welfare to develop criteria on the effects of air pollution and its control; and stepped up efforts on motor vehicle research. Federal attention to motor vehicles increased thereafter, and the Motor Vehicle Pollution Control Act of 1965 brought the federal government into the business of motor vehicle controls. Now HEW was to set emission standards for new vehicles, and the manufacturers were to meet them. The first federal standards were established in 1966, to become applicable in 1968. It was unclear whether the states could set higher standards (whether, that is, the federal government had preempted the field).

The years 1967-1968 saw further developments of great importance on both federal and state levels. The Air Quality Act of 1967 reflected congressional dissatisfaction with the Clean Air Act, which was thought to have left too much to state initiative. The Air Quality Act required states to establish air quality standards consistent with federal criteria, and it gave HEW limited authority to step in if the states balked. As to motor vehicles, federal research was expanded, federal registration of fuel additives was required, and grants were provided for state vehicle inspection programs. Finally, the preemption issue was clarified. No state could adopt new-vehicle standards more stringent than the federal ones; no state, that is, but California, which by 1967 had undertaken bold new initiatives against motor vehicles. Problems with the California Motor Vehicle Pollution Control Board and with the fragmented approach of the 1960 law that had created it (that law had left control of stationary sources to local authorities) led to 1967 legislation abolishing the old board and creating a new Air Resources Board with ultimate jurisdiction over both mobile and stationary sources (though local authorities retained a good deal of primary responsibility as to the latter). Moreover, experience with the old board suggested that tighter standards for new vehicles were necessary, standards written explicitly into legislation. Federal law left California free to enact such standards, and it did so in the Pure Air Act of 1968. Throughout the balance of the decade, the state's program of controls on new vehicles proceeded quite smoothly, though the history of used-vehicle controls remained an unhappy one.

The federal story was also marked by gloom. The Air Quality Act of 1967 was under attack; states were slow in implementing its provisions, and the act provided the federal government insufficient power to prod them along. There was little evidence that air pollution was being reduced on a broad national scale, and experience now suggested that improvements depended heavily on drastic new reductions in vehicle emissions. These problems called for action, especially in 1969, the year of the "environmental crisis."

The decade of the 1970s marks a sharp break with the past. Before, the emphasis had been on state and local authority (with gradual incursions by the federal government) and, less noticeably, on distinctly technological solutions to the air pollution problem. Now, there would be dramatic new federal intervention, and with a mind open (at least a little) to nontechnological controls. The Clean Air Amendments of 1970 began the era of present air pollution policy. The amendments provided for uniform national air quality standards set for the states by the federal government; for uniform national emission standards for some stationary sources; for uniform and very stringent restrictions on emissions from new vehicles (restrictions to be set, at least in principle, without regard to the constraints of technical feasibility that had governed before); for "transportation controls" to reduce the miles driven by all vehicles, new and old alike. The time since 1970 has been one of struggles in California and Washington to cope with these bold new breaks, struggles that put—are still putting—the federal system to a fine test.

The foregoing sketch hardly captures the vitality of the thirty-five years of events discussed in Parts II, III, and IV, but it should suggest their drift. The two chapters (14 and 15) of Part V conclude the study; drawing and expanding on the facts and concepts introduced in chapter 1, they attempt to make a meaningful appraisal of the historical narrative. Chapter 14's primary aim is to understand, in terms of general themes of policy, what happened and why in the years 1940-1975—to identify and explain some major strains that cut across the period in question. In this chapter especially we must fall back on our earlier comment that this book is in essence an essay—tentative, speculative, impressionistic. For the themes with which we deal in chapter 14 are quite large ones, both in the sense of importance and in the sense of being difficult to capture and contain. That being so, a brief summary of the chapter's highpoints might once again make a useful preface to later discussion.

The first theme that will concern us has to do with the general character of the process of making pollution policy. Until recently, that process has been one of least steps along the path of least resistance. The history is far more one of reaction than initiative: events, not foresight, ushered in each stage of intervention. Intervention tended to consist in curative rather than preventive measures, and was designed to preserve so far as possible the prevailing social patterns—whether of business practice, citizen behavior, or the distribution of authority among local, state, and federal governments.

The reasons behind the general character of the policymaking process relate to a second theme, the practice of allocating the burdens of inertia and uncertainty to those who sought change in the status quo. This theme, like the first, has only recently shown signs of weakening. No doubt the fading of both, since about 1970, is as related as was their mutual dominance before that time. By giving to those who sought further intervention the burden of demonstrating that intervention was justified, the policy process aimed to minimize unnecessary expense and disruption and maximize the acceptability or "political feasibility" of measures actually undertaken. This ensured, however, that intervention would tend to come, if at all, in small, grudging increments. Given the great uncertainty that surrounds pollution problems (uncertainty to be discussed in some detail in chapter 1), anything more simply could not be justified by those bearing the burden to show good cause for change.

Of course, intervention did occur, even if only by small steps. The question arises how these steps were shown to be necessary. Part of the answer has to do with a third theme that recurs throughout the years and up to the present, that of crisis. Policy-by-least-steps, coupled with the very social patterns it aimed to preserve so far as possible, meant that pollution would only continue (indeed, it would increase) with growth and concentration of population, industry, automobiles, and so forth. Periodic pollution episodes would and did occur. They fomented "crises" that in turn stimulated change. Crises have been both bane and boon in the making of pollution policy.

Change came in response to crises, and in forms that reveal a fourth theme—the fixations of pollution policy on technological solutions imposed through the direct, quick, but also generally crude means of regulation (as opposed, for example, to other methods of intervention, like subsidies and emission fees, that will be described in chapter 1). The technological and regulatory fixations were tightly tied to the demands of crisis and the aims of policy-by-least-steps; both fixations persist today, but not so markedly as before. Technological solutions were an attractive response to crises because they promised to be fast and predictable. They were most compatible with policy-by-least-steps because they appeared to be cheap (little of technology's potential had been tapped) and, most importantly, because they promised minimal social disruption. They could be used to alter the operations of machines without interfering too much with those of men. The regulatory approach was similarly attractive, precisely because it too was quick, direct, and predictable; it could go forth with little information; it could aim with satisfaction at wrongdoers and threaten punishment lest they changed their ways. It had no pretensions of subtle controls on such established patterns of human behavior as when and where people drove their automobiles. Regulation was the obvious way to impose technological controls in times of crisis. If technology was the solution, then mandate it, and woe betide those who did not follow orders.

Two facts in particular suggest the tight relationship of the technological and regulatory fixations to the concern with avoiding social disruption. First, early efforts to find technological solutions focused on ways to control nature itself rather than human behavior, to make pollution "go away" once it had formed rather than to make people use controls that would keep it from forming. Second, regulatory measures often came only after the failure of programs of voluntary control.

The move from voluntary to regulatory controls, encouraged by the demonstrated failure of the former, is one illustration of a fifth theme having to do with policy and the production of knowledge. Throughout, of course, policy aimed to produce knowledge by supporting and undertaking studies. Study subsidized or conducted by government was an essential step in reducing uncertainty; it was also a nice least step by which government could do something, but in not too disruptive a fashion. Yet, arguably, policy produced much more valuable information by a much less systematic process, one we call "exfoliation." Policy-by-least-steps let happen the very events that revealed the need and suggested the direction for the next least step of intervention. As measures were tried, and as they failed in whole or part, layer upon layer of obscurity about the pollution problem was stripped away. The failure of each "solution" produced valuable information about where to look and what to do (or at least where not to look and what not to do) next. This process had much to do with discovering, and especially with showing conclusively, that motor vehicles are an important contributor to modern urban air pollution; it also largely explains the movement of control authority from local to state to federal government. And it, together with the influence of crises, has been an important instrument of change over the years. This continues to be the case, but with one difference: since the massive doses of federal intervention that began in 1970, government has been learning from the failures more of large steps than small ones.

Reference to the dramatic shift of policy in 1970 introduces the final theme to be considered—that of so-called lag, or unwarranted delay in government resolution of a social problem. Enthusiasts see present pollution policy as the realization, or at least the promise, of an end to the period of "lag." We have, to be sure, reached a point of very substantial intervention, and by an inherently gradual and halting process. But one question that remains is whether the lag period is over, whether we have in fact resolved the pollution problem. Another is whether the delay in reaching present policy has been entirely senseless.

The themes just outlined make up the primary concern of chapter 14. A secondary concern, yet still an important one, is to assess the merit of particular policies over the years—but only up to 1970. Policy since that year, present policy, is saved for chapter 15, which contains a critical analysis of some of the major features of the federal air pollution program as it stood at the end of 1975, when the research reported here largely drew to a close. (There are a few excursions beyond that cutoff date—especially in chapter 15—in order at least to mention some important subsequent developments.) Our analysis of the federal program indicates that its reliance on uniform standards set without regard to costs is inefficient and inequitable (and that proposed amendments almost enacted by Congress in 1976, while they recognized these problems, would not have resolved but only delayed them to some degree). This leads us to suggestions for a more constructive and responsible approach—one that would take due account of conditions and circumstances that vary across the nation. We hardly claim, however, that these suggestions amount to a program. The purpose of the chapter is to pinpoint some central issues. The process of making pollution policy has to date been fundamentally flawed by a failure to ask the right questions. We try to raise a few of them.

Readers primarily interested in present policy—what it is and what to do about it—but also mildly curious about how and why that policy developed, might satisfy themselves by consuming a good deal less than all of this book. For them, reading Parts I, IV, and V should result in a saving of time without a loss of coherence.

1

A FRAMEWORK FOR THE PROBLEM

Like most persistent social ills, environmental problems are exceedingly complex: characterized by a multitude of causes and effects; plagued with uncertainties and full of surprises; thoroughly studied only in terms of a wide range of disciplines; remedied by no simple tonic. This, indeed, is why they are "problems." For all of this, still we must come to grips with complexity in order to describe and assess what policymakers have been and should be doing about environmental problems. Our task in this regard is slightly easier than might otherwise appear because our explicit concern is a relatively narrow one—the problem of motor vehicle air pollution. The purpose of this chapter is to reduce the complexities of that problem to manageable (and, it is hoped, not oversimplified) form, or at least to make clear just what the complexities are, by sketching a framework that describes the problem in terms of some important technical, behavioral, and institutional components. The framework introduces facts and concepts that figure in subsequent discussion; it should also help put that discussion into perspective. While the focus is on motor vehicle air pollution, we must of necessity look occasionally at air pollution generally. Moreover, despite the close focus, many of the remarks apply to a broad array of environmental problems.

SOME TECHNICAL COMPONENTS

Air Pollution: Its Sources and Effects[1]

Air pollution simply means a concentration of pollutants in the ambient air high enough to cause adverse effects of any kind, whether to human health, factors of production, aesthetics, wildlife, or whatever. Air pollutants themselves can be divided into two general categories: primary pollutants, emitted directly from sources; secondary pollutants, formed in the air by interactions among the primary pollutants and normal atmospheric constituents. Carbon monoxide and hydrocarbons are essentially primary pollutants. Particulate matter and various sulfur and nitrogen oxides exist in both primary and secondary forms. An additional secondary pollutant, one of particular importance to us, is photochemical oxidant, the irritating condition typical of Los Angeles "smog" and hardly unknown in many other major urban areas. Photochemical oxidant forms as the result of reactions among hydrocarbons, oxides of nitrogen, and sunlight. It is generally, though not always, accompanied by a brownish haze brought on by interactions among a number of the secondary particulates and other pollutants. The same processes that encourage formation of photochemical oxidant tend to encourage production of the haze.

The major emission sources for the primary pollutants are transportation, industrial and domestic fuel combustion, industrial process losses, and refuse and solid waste disposal. Tables 1 and 2 indicate the emission contributions of each source category nationwide and in Los Angeles County.

TABLE 1

NATIONAL EMISSION ESTIMATES, 1971

(in millions of tons per year)

	Sulfur oxides	Particulate matter	Nitrogen oxides	Hydro-carbons	Carbon monoxide	Total
Fuel combustion in stationary sources	26.3	6.5	10.2	.3	1.0	44.3
Transportation	1.0	1.0	11.2	14.7	77.5	105.4
Solid waste disposal	.1	.7	.2	1.0	3.8	5.8
Industrial processes	5.1	13.6	.2	5.6	11.4	35.9
Miscellaneous	.1	5.2	.2	5.0	6.5	17.0
	32.6	27.0	22.0	26.6	100.2	208.4

Source: Adapted from U.S. Council on Environmental Quality, *Environmental Quality—1973* (4th Annual Report, 1973), p. 266.

TABLE 2

EMISSION ESTIMATES, LOS ANGELES COUNTY, 1973
(in thousands of tons per year)

	Sulfur oxides	Particulate matter	Nitrogen oxides	Hydro-carbons	Carbon monoxide	Total
Fuel combustion in stationary sources	63.9	12.8	87.6	0	0	164.3
Transportation	16.4	23.7	304.8	253.7	2,660.8	3,259.4
Solid waste disposal	0	0	0	0	0	0
Industrial processes	52.9	10.9	14.6	27.4	3.6	109.4
Miscellaneous	0	0	0	0	0	0
	133.2	47.4	407.0	281.1	2,664.4	3,533.1

Source: Adapted from Los Angeles County Air Pollution Control District, *1974 Profile of Air Pollution Control* (1974), p. 11.

As the tables show, transportation is the greatest contributor by weight to total emissions, especially in Los Angeles. (Not all transportation emissions are attributable to motor vehicles; the transportation category also includes aircraft, ships, and railroads. Emissions from these sources, however, are quite a small part of the total transportation contribution in most areas.) One should note, however, that the bulk of transportation emissions consists of carbon monoxide, the least troublesome of all the pollutants listed. Nevertheless, transportation still accounts for the largest shares of hydrocarbon and nitrogen oxide emissions. And as to these, the tables understate the transportation role to some degree, because emissions associated with refining and marketing fuels are placed under the industrial-processes category.

So much for the pollutants and their sources. What of their effects? An exhaustive answer is hardly necessary for our purposes. Substantial evidence suggests that at sufficient concentrations air pollution causes damage to human and animal health, vegetation, soils, materials, climate and visibility, safety, production processes, and aesthetics. Particulates, for example, reduce visibility, soil clothing and materials, and in some forms are corrosive. Sulfur oxides can contribute to a wide range of effects: damage to vegetation, respiratory and lung diseases, increased death rates among older persons, fog production and other impairments to visibility, and corrosive and irritating acid mists. Carbon monoxide reduces the capacity of blood to carry oxygen, with resulting impairment in time-interval discrimination and psychomotor and visual functions. The pollutant is most dangerous on congested roadways and in enclosed parking lots—places where exposure to high concentrations can occur. Oxides of nitrogen might be carcinogenic in some forms, they cause respiratory problems, and they contribute to the brown haze usually associated with photochemical pollution conditions. Photochemical oxidant deteriorates rubber, damages vegetation, aggravates respiratory conditions, and causes eye irritation. Finally, hydrocarbons themselves probably have few direct adverse effects (although some of the hydrocarbons might be carcinogenic); they do, however, play a central role in the reaction that produces photochemical oxidant.

The health effects of air pollution take both acute and chronic forms. High concentrations of pollutants, of course, produce the most noticeable adverse effects, but there is evidence suggesting that exposure to even very low concentrations can impair health over the long term, especially among the most susceptible portion of the population. Generally speaking, health effects are difficult to quantify in any terms, and thus are the subject of considerable controversy. There is no question, however, that the problem is of serious dimension. A number of episodes over the years, both in this country and abroad, have shown that severe pollution conditions can lead to death. More than 60 deaths were attributed to a 1930 episode in the Meuse Valley, Belgium; 4,000 to a 1952 episode in London; 168 to a 1966 episode in New York. Putting the problem in a different perspective, the United States Council on Environmental Quality has estimated a figure of $6.1 billion for air pollution damage to health in 1968; total air pollution damage costs for that year were put at $16.2 billion.[2]

Approaches to Control

One characteristic of the ambient air is that it, the air itself as opposed to the pollutants put into it, cannot usually be treated directly in order to improve quality. This is not the case with all environmental resources. A stream, for example, can be treated to remove pollutants from it, or to make the water better able to absorb pollutants without ill effects. Proposals to accomplish similar results for polluted air have been made from time to time, but for the most part they have proved to be technically impossible or outrageously expensive. Air conditioning is a limited exception.

Air conditioning also illustrates receptor control, yet another way to avoid or reduce the adverse effects of pollution. Receptor control aims to limit pollution damage by, for example, developing vegetation resistant to pollution, or moving people out of polluted areas, or encouraging them to live in air-filtered buildings and wear face masks outside. Some degree of receptor control does occur in the case of air pollution; it is not so unusual for people to move out of an area as polluted air moves in. But it is easy to see that receptor control is largely infeasible, especially since one significant adverse effect of air pollution is considered to be aesthetic costs to human beings.[3] In aesthetic terms, the cure would often be worse than the disease.

Since control of receptors and of the ambient air itself are both largely impracticable or impossible, the only alternative is to control the output of pollutants from sources. This could be accomplished by nontechnological means—by requiring, for example, a reduction in the number of motor vehicle miles driven annually. Measures like these have received considerable attention, especially since 1970. But the primary effort over the years, and this remains true today, has been to control pollution sources through technology. Technological controls cover a wide range, from quite simple devices added on to sources to complex alterations in their internal mechanical operations. There are, of course, a number of distinct controls for stationary and vehicular sources. Since our primary concern will be with the latter, let us concentrate on motor vehicle control technologies, describing them only with as much technical detail as absolutely necessary.[4]

Conventional motor vehicles emit pollutants in three ways: through the tailpipe as exhaust emissions (carbon monoxide, nitrogen oxides, most of the particulates, and most of the hydrocarbons); through evaporative losses from the carburetor and gas tank (hydrocarbons); through "blow-by" from the pistons to the crankcase and then into the air (hydrocarbons and particulates). The latter emission mechanism is relatively minor in importance and quite simply controlled by positive-crankcase-ventilation valves that recycle blow-by gas from the crankcase back into the engine, where it is burned in the cylinders. As to evaporative emissions, systems to collect and use the vapors exist, but they work poorly.

Exhaust emissions present a more serious and difficult problem. Since the carbon monoxide and hydrocarbon components of exhaust are the product of incomplete combustion, the obvious solution would appear to be improvements in combustion efficiency by adjusting the air-fuel ratio. (This was the main thrust of initial control efforts.) Unfortunately, however, the higher combustion temperatures resulting from the adjustments increase production of nitrogen oxides. Generally, the more a conventional engine is adjusted to control carbon monoxide and hydrocarbons, the more nitrogen oxides are produced; and the more nitrogen oxides are controlled, the more hydrocarbons and carbon monoxide. It is, in short, extremely difficult to achieve low emissions of all three pollutants simply through engine adjustments. More substantial modifications can achieve some incremental improvements in one pollutant without affecting others, but not to such an extent that other measures are unnecessary. Either exhaust control devices must be added to the vehicle, fuels must be altered, or some more or less dramatic change in basic engine technology itself must be made.

The first approach has been most relied upon to date. One technique involves exhaust gas recirculation (EGR) from the tailpipe to the carburetor; it decreases nitrogen oxide emissions but can cause malfunctions that increase emissions of hydrocarbons and carbon monoxide as well as result in large fuel and performance penalties. An additional technique often used in conjunction with EGR controls the hydrocarbon and carbon monoxide emissions with exhaust devices—thermal reactors or oxidizing catalysts that help to correct the initial effects of incomplete combustion. The technique reduces malfunction problems, but one variant, the oxidizing catalysts, can encourage formation of certain hazardous sulfur oxides ("sulfates" and sulfuric acid mist). Other exhaust controls, some in limited use, some not yet used at all, rely on dual (reducing and oxidizing) catalysts to control, first, oxides of nitrogen, then, hydrocarbons and carbon monoxide; and three-way catalysts that simultaneously oxidize the latter two pollutants and reduce the first.

Fuel alterations as a means of emissions control range from changes in the composition of conventional gasoline to substitution of alternative fuels, primarily natural gas and liquefied petroleum gas. Some improvements in emissions can be achieved by modifying the hydrocarbon content of gasoline; greater ones are possible through use of the alternative fuels. However, conversion costs as well as supply and marketing problems tend to limit the alternative-fuel techniques to fleet vehicles.

The last approach to control consists of basic changes in or substitutes for the standard internal combustion engine. Diesel, Wankel (rotary), and stratified-charge engines—all currently available on the market—use internal combustion, but they differ fundamentally from the conventional technology. This is not to say, however, that they are uniformly better. The diesel, in use for many years, is heavy, costly, and difficult to control in terms of nitrogen oxide emissions; emissions from the Wankel are inherently high, and significant reduction severely impairs fuel economy. The stratified-charge engine, on the other hand, is capable of low emissions with no apparent countervailing disadvantages—other than the fact that retooling for mass production would be expensive.

More radical departures from conventional technology include steam engines, turbines, and electric vehicles. The first two are excellent in terms of emissions, but they present serious engineering, production, and maintenance problems. Electric vehicles are a practical and available alternative, but only for short (and slow) trips. Policymakers would probably be unduly optimistic were they to rely heavily on the more dramatically different technologies to achieve air quality improvements. Each approach to vehicle emissions control has peculiar strengths and weaknesses, and none has proved ideal, but the bias is toward a mix of the more conventional approaches described above. Consumers tend to favor them on grounds of familiarity and performance, manufacturers on grounds of avoiding risk and costly changes in capital commitments. In any event, large-scale use of some of the unconventional technologies could involve long lead times. No doubt for reasons like these, pollution policy has put a heavy stress on conventional approaches, though it has also quite properly strived to encourage development of new technologies, both directly (through grants and subsidies) and indirectly (through very stringent emission standards). Progress to date has been slow. Later in this chapter we mention one technique that could accelerate it.

Areas of Uncertainty

Some of the foregoing remarks have touched on areas of uncertainty in air pollution control, but it is time to be more explicit about the subject. Uncertainty is far more pervasive than might thus far appear.

Uncertainty arises, first, because it is difficult to estimate the total output of pollutants or measure the actual physical quality of the air at any point in time. Methods vary, and so do opinions about their merit. Second, since air pollution control is expensive, it should be exercised with some notion of the benefits it will produce. Yet, hard as it is to measure the existing physical situation, it is probably more difficult to estimate control costs, and more difficult still to get even a rough quantitative handle on the benefits that marginal improvements in air quality would yield. This is especially the case because many of the benefits—to health and aesthetics, for example—cannot be reliably quantified, much less expressed in monetary terms. Third, and assuming away all of the uncertainties just described, one still needs to decide how much to control sources of pollution to produce the desired air quality level. Unfortunately, a given amount of reduction seldom leads to a nice linear improvement in air quality, so a model is needed to describe the gains that a set of proposed controls would in fact accomplish. Yet for the most part "no general theory exists that would enable us to predict ambient air quality . . . in terms of the emissions level of the primary contaminants."[5] Rather than one theory there are many, and again there is disagreement about which is best. Fourth, just as the relationships between emission reductions and air quality improvements are not linear, so neither are those among air quality improvements and the ill effects they avoid.[6] For example, the physical quality of the air might improve only a little, and yet, because some threshold effect is escaped, far better conditions be realized. The converse can also be the case.

Suppose policymakers nevertheless manage to arrive at a judgment about the level of air quality they wish to achieve. The source controls that might then be employed cover a vast range. Oxides of nitrogen might, for example, be controlled a little more, hydrocarbons a little less, or vice versa,[a] and in each case by several techniques. One technique might be used for stationary sources, another for motor vehicles, or—more realistically—a host of techniques might be applied to different classes of stationary and vehicular sources, large ones and small ones, old ones and new ones. Each technique has a different cost and a different effect than the others, yet (and this takes us back to where we began) the costs and effects are in all cases uncertain. It is, in short, exceedingly difficult to reach sound judgments about desirable air quality levels and the means to achieve them. That the difficulties are real ones will become apparent when we look at events over the years 1940 to 1975.

The Implications of Growth

It would be bad enough if the problems of how and how much to control pollution had to be confronted only once. Unhappily, the implications of growth mean they must be faced on a recurring basis.

Growth in population and consumption is obviously a central part of the pollution problem. To some degree the air can assimilate pollution without ill effects. But as population and consumption grow, so too do emissions, and at some point the air's capacity is overburdened. If growth tends to concentrate geographically—and it has—the point of strain can be reached in a relatively short period of time. Moreover, the path to the point is seldom a smooth one of gradually declining air quality. Population and consumption, and thus emissions, tend to grow exponentially to a threshold at which the pollution problem suddenly appears: there is a quick transition from stability to "crisis." This transition is typical of exponential growth, and thus of most environmental problems.

Because air quality is a function of total emissions, and because total emissions are in part a function of growth, any program to maintain a given level of quality must necessarily anticipate successively more strict emission controls as growth takes place. This necessity, of course, further

[a]The reaction time of the process that leads to photochemical oxidant varies, among other things, with the ratio of nitrogen oxides to hydrocarbons.

complicates an already difficult problem in several ways. Reliable growth projections might be difficult to make. The relationship between growth in population or consumption and growth in emissions might not be linear. And growth in emissions or population can complicate the choice of a control strategy. For example, the least-cost approach to reach air quality level *X* at emission level *Y* may not be the least-cost approach at emission level *Z*. Another example has to do with the relationship between population growth and growth in pollution costs. Growth in population might mean that pollution costs will go up even if air quality is held constant, because an *increasing* number of people are each bearing whatever pollution costs a given air quality level implies.[7] Thus the problems of how and how much to control must be faced time and again as growth changes the picture.

Growth has another aggravating implication, especially for technological controls. As sources and thus emissions increase, the costs of maintaining a given level of air quality must also increase. Part of the reason for this is obvious: each source must spend a given amount on control technology, and the number of sources is constantly growing. But a more sinister arithmetic also comes into play, one that suggests that the total costs of maintaining a given level of quality will tend to increase much faster than emissions. As mentioned earlier, any program to maintain air quality must apply more and more stringent standards to *each* source as the total number of sources increases. (This has been the history of air pollution control from the outset of the problem to the present.) Otherwise, air quality would deteriorate with growth. But the costs of further technological control of emissions tend to increase rapidly with higher degrees of control, generally as depicted in figure 1. (The situation is especially aggravated if improvements in air quality from any given emissions reduction *decrease* as emission control costs rise, for then the costs

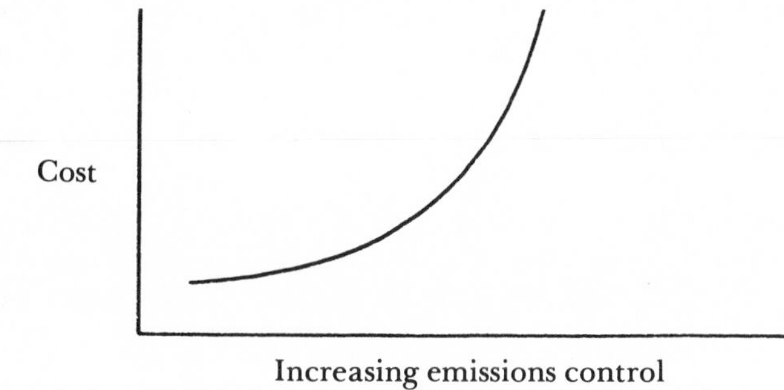

Figure 1

of air quality improvements increase even more quickly than the emission control costs.)[8]

The implications of these cost functions are, of course, serious: growth in population and consumption leads to more emissions, declining air quality, and increasing pollution costs (perhaps *vastly* increasing pollution costs). Yet the costs of further and further technical controls on emissions begin at some point to go up very rapidly.

Thus far we have been talking about the costs of applying a *given* control technology to solve the problem of growth in emissions, and we have suggested that the technological solution is constrained or limited by rapidly increasing costs. Of course, there is always the possibility that a new technology will appear that enables us to get where we wish to be at lower cost, or further at no higher cost, than the present technology. Indeed, precisely this is one meaningful way to define a "technological breakthrough": it is a development that moves the cost curve to the right, as depicted in figure 2. Can we count on technological breakthroughs,

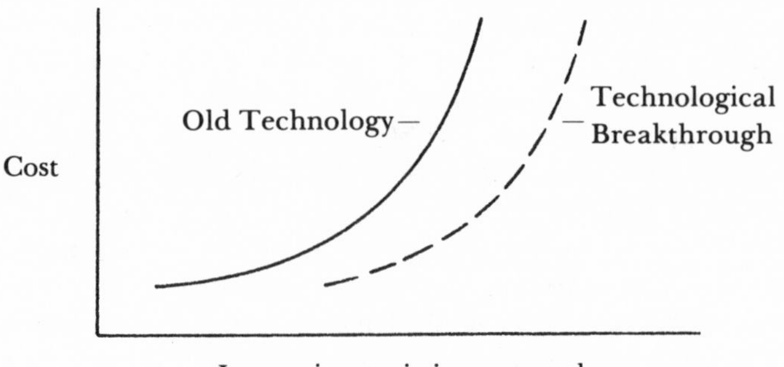

Figure 2

so that the technological solution is in fact relatively unconstrained? Several observers have answered yes, arguing that just as social problems like pollution may tend to grow exponentially, so too will the availability of technologies to respond to them.[9] These predictions of exponential growth in technology are based largely on observations of historical trends, and some writers are more cautious about the inferences that can be drawn from the past. Kenneth Arrow, for example, has noted the "limit to what can be learned even with infinitely many opportunities." In his view, "eternal exponential technological growth" is simply unreasonable. Granted, technological growth thus far conforms with observed fact. But this may simply mean "we are still in the early phases, which resemble the exponential."[10]

We can hardly resolve the issue of exponential technological growth here. We only want to underscore the real possibility of rapidly increasing costs of further technological control, with no breakthrough to set them back to a manageable point. One way to respond to these possibilities is by turning to nontechnological control techniques (such as reduced auto use) that may be cheaper than technological alternatives. The historical bias has been toward technological controls, but evidence of a modest change in attitude is developing—perhaps because satisfactory technology is simply becoming too costly, or even unavailable.

SOME BEHAVIORAL COMPONENTS

The discussion above sketched some technical components of the pollution problem; it did not consider why people behave in ways that make the problem happen. That is the inquiry now.

The Common Property Effect

As with many environmental resources, it is not feasible to define exclusive rights of private property in the air. The air is held essentially in common and each person has some rights to use it. People tend to overconsume common property, to use more of it than is in the best interests of everyone. The effect occurs for reasons that are well understood, and with implications for environmental quality that are equally clear.

When an individual uses a common property resource, he garners all the benefits but does *not* bear all the costs of his action, because the costs are spread among all other users of the common property. If, for example, 100 people own in common a forest of 1,000 trees, and one of them harvests one tree, he gains the market value of the tree but loses far less: before he held a one-hundredth interest in 1,000 trees; after his harvest he holds a one-hundredth interest in 999 trees. His net gain is ninety-nine hundredths. If one person owned the entire forest as private property, he would probably try to maximize its value by harvesting trees only when their market value exceeded their future value discounted to the present; thus the efficient (value-maximizing) harvest of trees would occur. With common ownership this seldom happens. No common owner can confidently husband the trees in order to capture future gains, for each fears other owners will take what he foregoes. And these other owners, of course, are likely to harvest out of fear that if they do not, the rest will. Assume that private ownership is impossible. In the absence of a collective agreement about how to use the trees, they will tend to be harvested even though this is not in the best interests of all common owners. Yet collective agreement is unlikely to occur because of the high "transaction costs" of organizing negotiations among all the common owners, dealing with the "holdouts" among them who will try to extract large payments in exchange for their promise to limit harvesting, and enforcing against cheating once any agreement is reached.

It should be obvious that this common property effect tends to occur whenever a person can appropriate all or most of the benefits of his actions but spread their costs among others. The costs are partially or entirely external to the actor (thus economists call them externalities) and, accordingly, are not taken into full account.[11] The problem of putting pollutants into the air is directly analagous to the problem of taking trees out of the commonly owned forest, except that private ownership in the air resource *is* impossible, and transaction costs are likely to be *very* high because so many people are involved and they are spread over such large areas. The operator of a polluting source (for example, an automobile driver) realizes the entire benefit of expenses avoided by ignoring air pollution control. The air becomes less desirable, but the costs of this are spread among everyone in the area; only a small portion at most is borne by each polluter. If a polluter does control his pollution, air quality undergoes no tangible improvement because his contribution is small and the air continues to be used by other polluters. In sum, once again the incentive of each person is to pollute, and it is too costly to reach and maintain a collective agreement that would avoid this effect. The air resource is overconsumed.[12]

The Collective Goods Effect

To some extent the collective goods effect is a corollary of the common property effect. While economists do not agree precisely on the definition of collective goods, a useful meaning for our purposes is that they are goods as to which "it is impossible to exclude nonpurchasers from consuming the good."[13] Collective goods, then, are those that can be enjoyed by people who have not contributed to their production. Return to our example of the polluting automobile driver. If he controls his pollution and cleaner air results, he bears the full control cost but any benefit of the cleaner air is enjoyed by everyone in the area. Moreover, it is difficult to see how the polluter can extract compensation from everyone to the extent of each one's benefit. In other words, the cost to the polluter is internal to him but most of the benefit is external. Common property leads to external costs; collective goods to external benefits. Both involve the market failure of externalities. Just as the common property effect creates incentives to pollute, the collective goods effect creates incentives not to control. We can be quite sure the external benefit will dampen efforts to produce clean air (or any other environmental improvement which is collective). Indeed, we can state with considerable certainty that people will tend not to take steps to produce a collective good like clean air so long as the costs to one person of doing so are greater than the benefits to that person—even though the total benefit to all individuals who enjoy the cleaner air is greater than the total cost of producing it.[14]

Collective goods yield another, subtly different but still sinister, effect. Anyone who can produce or help produce a collective good is unlikely to do so not only because of the individual cost-benefit calculus described above, but because he will reason that his contribution is hardly important, and that he cannot be denied enjoyment of the good if others produce it, whether he contributes to its production or not; he can, that is, take a "free ride." Thus, just as it is unlikely a polluter will control his activities to produce cleaner air, it is unlikely that anyone who would benefit from cleaner air will pay the polluter to produce it. Each person tends to reason this way, so none tends to help produce the collective good—even though, again, the good would be in the best interests of all. Without some form of group agreement, clean air is not likely to be produced. But again, the costs of agreement are likely to be prohibitive because of the large number of people involved and the difficulty of excluding from the benefits of the agreement the free-riders who refuse to join it.

We can see quite clearly from the discussion thus far that in the case of many environmental problems the collective goods effect is, as said earlier, to some extent a corollary of the common property effect. This is because it is often the case that the collective good of environmental quality can be produced only by reductions in the consumption (pollution) of the common resource. Thus the common property effect explains why a source has incentives to pollute, and—viewed from the other side—the collective goods effect explains why it has little incentive to cease polluting. The two effects can be seen as a unified result of externalities brought on by high transaction costs. But the reciprocal relationship does not always hold. An individual may not be a polluter, or at least not a significant one, yet still have the ability to influence the production of improved air quality by contributing payments to polluters to induce fewer emissions; or by supporting lawsuits against polluters; or by lobbying for governmental regulation of polluters. But the collective goods effect suggests that the exercise of such influence will be inhibited. This introduces a possibly substantial bias into the process of environmental resource management.

Wants

The discussion above implies that air quality will decline without controls, but this does not mean that the decline will be a "problem" in other than the technical sense. Air quality is a social problem only when people want more than they have. Health effects, for example, might be of little interest to the populace because, given its priorities and the resources to achieve them, avoiding the health effects is less worthwhile than responding to the agenda of all other social wants to the limits of all available social capital. Pollution as a technical problem is a matter of scientific absolutes, but in the context of a social problem it is a matter of relative wants.

Human wants are shaped by many factors. One of particular, but not so obvious, importance in our context is increasing affluence—rising per capita income. Enhanced environmental quality, at least beyond some minimum level required for tolerable survival, can be seen as a luxury good. As incomes go up, so too do people's tastes for such goods. Air quality that is acceptable today may be unacceptable in a richer tomorrow. The technical problem might become no worse, yet the social problem can grow.[15]

SOME INSTITUTIONAL COMPONENTS: ON THE ENDS AND MEANS OF MAKING POLLUTION POLICY

Ends

Earlier discussion described some of the costly adverse effects of air pollution, mentioned that measures designed to reduce or avoid these effects can also be expensive, and introduced the notion that policy should be reasonably sensitive to the benefits that would accompany a given level of control costs. The idea, which is of course to spend neither too much more nor too much less on control than is worthwhile, requires some elaboration and qualification.

The benefits of pollution control are clear enough: they equal all the costs of adverse effects that control avoids. The costs of pollution control, on the other hand, can be obscure; people often seem to assume that they consist only of outlays for technology. These, however, are but one entry on the ledger. Control costs also include such items as expenditures to support government control agencies, and—less tangibly—the burdens of inconvenience and limited choice if people must abjure favored and generally useful activities like driving automobiles. Put most broadly, the costs of control equal the value of the opportunities foregone by directing resources to achieving improved air quality rather than some other end or ends. These costs are as much associated with the pollution problem as are the costs of ill effects, and they, together with the costs of ill effects, should be taken into account in deciding what to do about the problem. Since pollution and avoiding it are both costly, a rational end of policy is to attempt to minimize the sum of all pollution costs and pollution avoidance costs.[b]

This end aims at efficient use of the air resource: the air should be polluted or cleaned only to the point where further pollution or greater cleanliness would not be worthwhile. Given all the uncertainties involved in making such a judgment, matters would be difficult enough even if the process could stop here, but it cannot. Distributional consequences as well as other considerations generally labeled ones of fairness or justice must also be taken into account. An efficient policy might mean clean air here and less clean air there, and if some disadvantaged or minority group lives "there," the efficient result might justifiably be considered unfair. An efficient policy might require expensive controls on old cars commonly

[b]This is simply another way of saying that the costs of a measure to control pollution should not be higher than the benefits of lower pollution levels the measure would achieve. The cost-minimization formulation, in other words, is simply a restatement of the essence of cost-benefit analysis. See G. Calabresi, *The Costs of Accidents—A Legal and Economic Analysis* (New Haven: Yale University Press, 1970), p. 26. As the formulation suggests, knowing how a change in control would affect costs and benefits at the margin is usually more relevant to policymaking than knowing the overall costs and benefits of air pollution and its control. See A. Kneese, "How Much Is Air Pollution Costing Us in the United States?" in *Proceedings of Third National Conference on Air Pollution* (Washington, D.C., 12-14 Dec. 1966), pp. 529, 531. The fact remains, however, that changes in costs and benefits at the margin cannot be assessed without at least some notion of the total situation. See H. Wolozin, "The Economics of Air Pollution: Central Problems," *Law and Contemporary Problems* 33(1968): 227, 228-231.

owned by low-income persons, or diminished production and employment that could affect low-income persons first and most. Again, policymakers might justifiably conclude that this would result in a maldistribution of the burdens of efficiency. Efficiency, in short, is hardly an exclusive end; considerations of justice or fairness must also be taken into account—whether they are seen as an ultimate constraint, or as merely an end to be traded off with that of efficiency. This increases the complexity of making policy, but it is essential to the integrity of the process.

Means

The effects of common property and collective goods, together with the high transaction costs so typical of environmental problems, suggest that any sort of policy calling only for voluntary control will fail in most instances. Unless government intervenes to limit and direct incentives to pollute, the problem will continue.[c]

[c]In a world with a *perfect* market, intervention might arguably be unnecessary; the market, characterized by people voluntarily offering and accepting payments to guide behavior, could be relied upon as the exclusive institution to resolve pollution problems, and market forces would produce an efficient amount of pollution. The reasoning behind this observation is well understood by specialists in law, economics, and some other policy disciplines. It is not worth much bother for anyone else simply because the observation has little practical importance in the case of real-world pollution problems. Suffice it to say this: Earlier discussion explained that polluting activities are likely to be too intensely pursued because of incentives created by the common property and collective goods effects, and that these incentives persist because of high transaction costs. In a perfect market there are, *by definition,* no transaction costs. Hence the incentives would not persist and the activities would not be too intensely pursued. They would stop at an efficient point.

Note, first, that nothing says this efficient point would be a fair one, and if it were thought unfair government intervention would be necessary to change it. Recall, second, that in the real and imperfect world of pollution problems, transaction costs are typically very high. Once transaction costs are introduced, some degree of intervention in the now-imperfect market becomes useful. For example, legal rules that define property rights, enforce contracts, and prohibit theft and violence facilitate voluntary transactions and thus help the market to work more smoothly. In short, real-world markets depend on some degree of intervention, so suggestions to "rely on the market" beg the question of what that intervention should be (as well as ignore issues of fairness). If transaction costs are very high, considerable government intervention to control activities beyond simply facilitating voluntary transactions is usually necessary. Even quite zealous market enthusiasts appear to agree that here pollution is a case in point. See, for example, R.

The job of making good pollution policy requires the government to intervene with controls that maximize the chances of achieving the ends of fairness and efficiency in particular cases. There are many means by which government might pursue the task. All of them involve the use of legal institutions—primarily courts, legislatures, and administrative agencies—to make and implement the necessary rules. We will be giving only a little attention to the courts. Our concern is to describe and assess state and national efforts to control vehicular air pollution, and the fact is that the courts, other than in overseeing and helping to enforce legislative and administrative actions, have not played a central part in those efforts.

This is hardly surprising; the courts are in a poor position to deal with pollution problems. Partly this is because courts cannot get involved until lawsuits are filed, and the structure of the pollution problem suggests that relatively few of them will be. Even though aggregate pollution damage to all affected persons may be very large, the damage to any one individual is generally small. As a result, the expected value of a lawsuit (equal to the amount that would be realized from a successful suit multiplied by the probability of success) to a single person is likely to be lower than the costs of suit; this inhibits individual lawsuits. Pooling of resources among all affected persons is difficult because of high organization costs, and this inhibits group lawsuits. As suggested earlier, organization costs are made high by the generally large number of affected persons (often spread over a broad area) and by the free-riders who realize that successful suits will tend to produce the collective good of cleaner air that they can enjoy without contributing to the costs of suit. These considerations suggest there will be relatively little litigation even when aggregate pollution damage is very large, and thus limited impact of even the best designed judicial rules—rules, that is, that maximize to the extent that judicial rules can the chances of achieving fairness and efficiency.

There are, of course, techniques like class actions to increase the scope and intensity of litigation, but the courts recently have been stingy in permitting the use of these, and even if fully employed they probably could not overcome sufficiently the structural barriers to litigation discussed above.[16] Moreover, these techniques leave untouched or can actually

---

Posner, *Economic Analysis of Law* (Boston: Little, Brown and Co., 1972), pp. 24-26, 159-161; H. Demsetz, "Toward a Theory of Property Rights," *American Economic Association Papers and Proceedings* 57(1969):347, 357.

This is not to say, of course, that government controls should ignore efficiency considerations, nor that they should fail to take good advantage of market forces—points to which we return.

exacerbate other problems—for example, the limited competence of courts (in terms of legitimacy as well as ability) to gather information about and deal with the technical uncertainty and ambiguous values inherent in the pollution problem, and the courts' lack of tools and resources to devise, fund, and administer appropriate solutions.[17] This is not to say that the courts have no function, only that they cannot be the exclusive or even the primary means of intervention. Most observers, including some courts, agree that legislative and administrative bodies must bear the main burden of making and impelmenting programs for pollution control.[18]

Pollution control programs can follow a number of patterns. The government might adopt a rather passive posture, simply sitting on the sidelines and exhorting polluters to "do better." This would reflect in essence a policy of voluntary control, and we have already suggested that in most instances it would likely fail. (The suggestion is based on theory, but later we will see substantial evidence for it, as well as consider some limited circumstances under which exhortations to control voluntarily might succeed.)

A slightly less passive approach, one used throughout the years that concern us, is for government to engage in research and development efforts of its own or support those of other public and private agencies. The idea, of course, is to reach a better understanding of the pollution problem and how to control it. But understanding itself is of little value unless acted upon, so again government must either exhort or, more realistically, implement active programs of control.

It is convenient to talk of these active programs in terms of regulation, subsidization, and pricing systems. These are general categories, each of which contains a variety of particular techniques of intervention.[19] Regulation refers to proscriptions and prescriptions backed up by such measures as civil or criminal fines, abatement orders, or jail sentences. Regulatory approaches include zoning controls, prohibitions of certain fuels, and ceilings on emissions from pollution sources. Concerning the last, emission ceilings might be expressed as performance standards or specification standards. The first technique sets an emission limitation that polluters must meet, but leaves to polluters the matter of how to meet it; the second goes further and specifies how the required degree of control is to be accomplished.

Subsidization refers to any form of positive (usually fiscal) payoff to polluters to induce them to undertake controls or to reduce the burden of mandatory regulations. Various sorts of tax write-offs for the installation of pollution control technology are the common example. In principle, another technique is available: cash award payments, geared to the degree of emissions reductions, could be made. This particular approach has never been applied to pollution problems in the United States (or elsewhere, so far as we know), though it has been used in other contexts—most notably, controls on agricultural production.

Pricing systems to control air pollution have also gone unused in the United States, though here and abroad they have been applied in a limited way to water quality problems. We will see, however, that they have received attention in the air pollution context, especially of late. Pricing is the conceptual opposite of subsidization. It refers to extraction from polluters of payments geared to the amount they pollute, the purpose being (unlike regulation) to leave to polluters the choice to pollute or pay. Emission fees are the familiar example, but an alternative technique—one variant of which we will later consider in some detail—would establish a limited number of rights to pollute (or to use polluting fuels like gasoline) and then set up a system to allocate these limited rights in a fair and efficient manner.[d]

The various pricing systems, as well as the regulatory technique of performance standards, are useful illustrations of a rather obscure suggestion made earlier—the suggestion that it is one thing to say pollution policy should not "rely on the market" for a solution, and another to say it should not take good advantage of market forces.[e] A classic market force is the self-interest by which working markets channel constructive behavior. People buy and sell in accord with their self-interest, and, if conditions are right, they buy and sell to the extent that accords with the self-interest of society in the aggregate. Never mind what all the conditions might be; we already know that some of them are not met in our context. Most notably, the air cannot be bought and sold in nice discrete packages by separate, identified people; and it is essentially this that argues for government control. But this hardly means government control should ignore the powerful force of self-interest, which can at times be harnessed and steered to constructive ends.

The approach of performance standards tries to do just this by leaving to polluters the decision *how* to achieve the degree of control required by law. Presumably, the polluter's self-interest is to develop and use the

[d]As we mention in chapter 14 (see footnote jj, p. 304), this alternative technique would avoid the need to make adjustments in response to growth in the number of pollution sources: holding the total number of pollution rights constant would hold the total amount of pollution constant as well, despite an increase in sources.

[e]See the end of footnote c, pp. 32–33.

cheapest controls that meet requirements. The bureaucrats who decide which controls are to be used under the specification approach, on the other hand, may have interests quite different from saving polluters money. They might be more concerned with avoiding hassle and risk in their agency careers, even if this means mandating expensive but tried-and-true controls when cheaper but less familiar ones exist. Since one end of any control program should be to minimize costs, the performance approach might often have an important advantage in these terms, and precisely because it harnesses the market force of self-interest.

Pricing systems aim to employ market forces to even greater extent and advantage. Notice that performance standards give polluters little incentive to develop measures, technological or otherwise, that could achieve considerably more than the required emission reductions, but at slightly higher cost. If anything, performance standards inhibit such behavior: if polluters did develop the measures, they would be expected to use them, and that would cost the polluters more. An additional problem with performance standards (or virtually any other regulatory technique) is that they typically apply the same requirements to all sources, or at least to all sources in a particular class, even though some sources can control more—and more cheaply—than others. They do so because it is too expensive to figure out the capabilities of each source; savings in control costs would often be more than consumed by administrative expenses. Pricing systems have the promise of responding to both these shortcomings. First, they could create continuing incentives to develop and employ technological and nontechnological measures to reduce pollution beyond the levels associated with existing practices. (A fee on vehicle emissions, for example, could be expected to stimulate greater interest in the low-pollution, unconventional engines that have been largely ignored to date; it could also be expected to reduce the total number of vehicle-miles driven.) Second, pricing systems could encourage controls to be used in the most economical fashion, the cheapest controllers controlling more, the others less. Elaboration of these points is best saved for later, when we consider in some detail why pricing has not been used, and whether it should be. Part of the answer has to do with the fact that pricing systems require a good deal of information and experience to implement successfully, and hence could be both unpredictable and slow in the results they would achieve.

Perhaps that is the note on which to conclude this chapter. Each general method and particular technique of control has strengths and shortcomings. Methods and techniques can be combined in many ways at many levels of government (federal, state, regional, local), and again each combination has peculiar advantages and disadvantages. But this is not the place to pursue them further. Eventually, we will confront the alternatives that have been or could be used in particular settings and that is a better context in which to continue the discussion. The purpose here was only to introduce the general subject of air pollution and its control so as to have a common frame of reference for more specific treatment later. At that time we can return to our observations in order to expand and clarify them, and to consider their significance in a less abstract way. It should at least be clear by now that the task of making pollution policy is terribly complex, given all the options and uncertainties. Obviously, a good amount of guesswork is involved, and perfection is not to be expected. All one can hope for is a government that makes its guesses on the basis of the best information worth developing, in the context of a sound decision process, and in response to the right questions. On this last point especially, policy to date has suffered its greatest shortcomings.

Part II

DISCOVERY, DEBATE, DELAY: THE YEARS UP TO 1960

2

A BRIEF PREHISTORY

The story that concerns us begins with the 1940s, but a glimpse of the situation before that time provides a useful backdrop for considering subsequent developments. This chapter outlines and characterizes the prehistory, giving particular attention to the physical environment, the institutional environment, and the prevailing attitudes and practices that set the stage for events beginning in 1940.

THE PHYSICAL ENVIRONMENT

It was in Southern California, more especially the Los Angeles area, that the problem of automotive air pollution first emerged. The physical characteristics of that region are of some interest, because they play important roles in our story. For some time Southern California's physical environment was considered unique insofar as air pollution is concerned; later it became clear that it is simply an extreme case of conditions more or less common to many major metropolitan areas currently beset with pollution problems.

Topography[1]

The Los Angeles area lies at the 1,600-square-mile bottom of a huge natural box. Three sides of the box are formed by mountains. The San Gabriels, thirty miles to the north and northwest from the shores of the Pacific, have a height at their base of about 2,000 feet and reach an elevation of about 10,000 feet at the peak. To the west are the Santa Monica Mountains, and to the east the San Jacinto and Santa Ana ranges. The fourth side of the box is provided by cool air currents moving gently but steadily in from the ocean.

Meteorology[2]

A semipermanent high pressure cell near the Hawaiian Islands swirls gigantic air masses in a clockwise direction in paths extending from beyond the Islands to the West Coast of the United States. The air moving from the cell toward California descends, compresses, and heats as it does so. As a consequence, when the air mass reaches the California coast, it is several degrees warmer than that at the land surface. Normally, air temperature decreases directly as altitude increases. Over the California coast and the Los Angeles region, the inverse is the case—the area is covered by a warm "inversion" layer, "a transparent sheet of air extending over the entire Los Angeles area, at a level usually varying between 1,000 and 3,000 feet. Sometimes such inversion layers are so sharp and definite a stratification that a balloon, ascending slowly, will rebound momentarily from their under surfaces."[3]

The inversion layer extends to the mountains and puts a lid on Los Angeles's topographical box. A common and long-held conception, one that influenced many of the bizarre technological solutions to pollution advanced in the mid-1950s, was that the lid was a very tight-fitting one. The inversion was seen as "a canopy over the entire Los Angeles basin, preventing contaminated air from escaping vertically, and by resting against the mountains, preventing it also from escaping toward the east"; the inversion "clamps an invisible hand down tightly over the entire metropolitan area."[4] Actually, the lid is not so tight as this, and the mountains clearly do not prevent polluted air from "escaping toward the east." For every ton of air entering the basin from the west, a ton of air leaves it through mountain passes or over mountain ridges. Otherwise, polluted air would accumulate over the area in an ever-deepening layer, and this is not observed. What happens typically is that the air, by the time it encounters the mountains, has finally been heated enough by sunlight to destroy the inversion. This allows the polluted air to move upward (vertical mixing) and escape over the mountains or through passes, reaching the desert beyond in a relatively diluted form. Seen from an airplane, the effect creates the deceptive appearance that the mountains are a barrier over which no pollution moves. In reality, the mountains mark the location where polluted air leaving the basin abruptly undergoes a marked dilution, providing the deserts with air that is quite clean in contrast with that over the basin.[5]

This is not to say, of course, that inversions are of no consequence. They do form atmospheric ceilings that limit the total volume of air available to contain the pollution in the area at any particular time. The lower the ceiling—or inversion layer—the more concentrated the pollution. Various reports claim that inversions are present over the Los Angeles region from 250 to 340 days per year. Strong inversions, where the ceiling is lowered to within 1,500 feet of the land surface, are said to occur about 120 days per year.

Air movements and sunlight complete the meteorological sketch. Weak winds are common to the Southern California area. Average wind speed in St. Louis is ten miles per hour, nine in Chicago, eight in New York. In Los Angeles the average is only six, and during the June-September period this approaches three. It is during this same period that the inversion ceilings are usually lowest, so in summer and early fall pollution conditions are typically most aggravated. Breezes in the area change direction during the day, but they do not escape the topographical box; rather, they follow a circular movement within it. Some vertical mixing occurs as air at ground level is heated by sunlight during the morning hours, moves upward, and is replaced by cooler air from above. Any contaminants that have accumulated at the base of the inversion are thus transported down to street level. As mentioned earlier, sunlight plays another role. It acts on emissions, especially those from motor vehicles, to create photochemical smog.

Population

Population figures represent an important variable throughout our story; for present purposes we need only see how matters had developed up to the time when California entered the decade of the 1940s. Table 3 presents some comparative data for the period 1920-1940.

TABLE 3

TOTAL POPULATION: UNITED STATES, CALIFORNIA, AND LOS ANGELES COUNTY

Year	Number			Percent gain since 1920		
	U.S.	California	L.A. Co.	U.S.	Calif.	L.A. Co.
1920	106,466,420	3,554,346	936,455	—	—	—
1930	123,076,741	5,708,856	2,208,492	15.6	60.6	135.8
1940	131,669,275	6,907,387	2,785,643	23.7	94.3	197.5

Source: California Department of Public Health, *Clean Air for California* (Initial Report, 1955), p. 56.

As the table reveals, California during the two decades prior to World War II experienced phenomenal growth in population, adding to its numbers far more rapidly than the heady pace set by the rest of the United States. The difference in trend over this period was even more marked for the Los Angeles area: its rate of growth was over twice that for the state as a whole. Looking back on the period ten years later, a group of California legislators regarded its population increases as "all but incredible."[6]

Not only people but their activities as well, in particular their penchant for automobiles, will be of recurring interest. To fill out the sketch of the physical environment, table 4 sets forth comparative figures dealing with motor vehicle registrations.

TABLE 4

MOTOR VEHICLE REGISTRATIONS 1930–1940
UNITED STATES, CALIFORNIA, AND LOS ANGELES COUNTY

Year	Number			Percent gain since 1930		
	U.S.	California	L.A. Co.	U.S.	Calif.	L.A. Co.
1930	26,718,900	2,136,630	871,773	—	—	—
1940	32,452,861	2,990,262	1,229,194	21.5	39.9	41.0

Source: California Department of Public Health, *Clean Air for California* (Initial Report, 1955), p. 57.

These figures were to increase dramatically in the years following the Second World War.

THE INSTITUTIONAL ENVIRONMENT

Air pollution itself was not a phenomenon solely of the 1940s, although the mysterious type of pollution that appeared in Los Angeles during the early part of that decade seemed a new variety. The problem had been familiar to California and other parts of the nation for years, and legal institutions had been called upon to deal with it over that time. This was the case in California, in other states, and to very limited degree at the federal level.

California

California had known some degree of air pollution for centuries. Accounts of the problem in the state refer, almost uniformly, to the October day in 1542 when Juan Rodriguez Cabrillo, a Spanish explorer, "discovered San Pedro Bay and noticed that although mountain peaks were visible in the distance, their bases were obscured. The smoke from Indian fires rose perpendicularly into the calm air for a few hundred feet and then spread out over the valley." San Pedro Bay is in the Los Angeles area. Cabrillo named it "La Bahia de Los Fumos" (The Bay of Smokes).[7] The name probably owed as much to natural atmospheric conditions as to a modest man-made pollution problem. Reports of explorers and early settlers "clearly indicate that there is a natural tropical haze separate and apart from and distinguished from the aggregation of air pollutants known as smog. Thus, Los Angeles would have many days of haze even if there were no people living in the area."[8]

Whatever the case in distant times, there remained by the beginning of the twentieth century little question that the Los Angeles area was susceptible to air pollution. Indeed, occasional episodes occurred even in the mid-1800s. An 1868 newspaper article spoke of six days of air "so filled with smoke as to confine the vision within a small circumference." But the event was hardly typical; the report characterized it as "an unusual atmospherical phenomenon, which, from its peculiarity, has given occasion to manifold surmises, conjectures, speculations and rumors."[9] By the turn of the century, however, such occurrences were more common and less subject to conjecture. An editorial in the *Los Angeles Herald* in early 1903 revealed the growing concern:

> During the week darkness spread over streets, lights were turned on from Third to Fifth Streets, and some thought it an eclipse of the sun. Soon all knew the truth, that it was smoke fumes that obscured the sun and drove out daylight. . . . It was like meeting a railroad train in a tunnel. The smoke nuisance is growing in volume and density . . . and councilmen are in favor of any kind of legislation that will lead to its abatement.[10]

Legislative action regarding the pollution problem in the Los Angeles area began as early as 1905, with a city ordinance aimed at emissions of dense smoke from flues, chimneys, and smokestacks. This was replaced in 1907 with a measure drafted and supported by the Public Health Committee of the Los Angeles County Medical Association. The new ordinance, adopted after a six-month fight, established the city's first position of smoke inspector and regulated discharges of smoke from industrial structures and appliances. Smoke and fumes were the subject of further local legislation in 1908, 1911, 1912, 1930, 1937, and 1945; objectionable odors, in 1930; and oil burning orchard heaters, in 1931 and 1937.[11] Other areas of the state, to the extent they acted at all, followed this pattern of local, particularized ordinances. An exception was San Bernardino County, which established an air pollution control district in 1936—over ten years before similar action by Los Angeles County.[12]

The courts, too, played a role in the years prior to 1940 as citizens, annoyed with the growing "smoke nuisance," sought redress through civil suits. A complete picture of air pollution litigation in this period is not available, but a sketch can be constructed from appellate reports. In the 1880s, there were but two civil appeals involving air pollution, and this was true of the next decade as well. By 1900–1910, the number had doubled to four such appeals, then dropped in the next decade once more to two. Appellate litigation increased again in the 1920s to four cases and reached an apparent peak of five cases in the 1930s. Measured by appellate reports, civil air pollution litigation declined markedly after this point.[13] At first glance this apparent trend seems unusual, especially because it was *after* 1940 that the pollution problem began its most intense rate of increase. As one Californian later observed, 1940 marked "the last of our smog-free decades."[14] There is one plausible explanation for the trend, but it is best considered when we turn shortly to the prevailing attitudes and practices revealed by the institutional environment as it had evolved up to 1940.

Other States

Like California, other states—especially in the Midwest and Northeast—experienced substantial air pollution problems prior to the 1940s. And like California, the legislative response in these areas occurred almost entirely on the local level and was aimed at such visible emissions as smoke. Reportedly, the first air pollution control law in the United States was an ordinance prohibiting emissions of dense smoke, passed by the Chicago

City Council in 1881.[15] Within a few years, Cincinnati, Pittsburgh, Cleveland, St. Louis, and St. Paul had passed similar laws—all of them vaguely declaring such emissions to be public nuisances.[16] The city of Boston took a more scientific approach as early as 1910, employing in its ordinance opacity standards based on the Ringelmann Smoke Chart.[17] By 1912, of the twenty-eight cities in the United States with populations over 200,000, twenty-three had smoke abatement programs.[18] But this was for some time the extent of legislative response, federal measures aside. "There were no public air pollution controls in the United States except smoke controls in the large cities prior to the photochemical smogs of Los Angeles in the 1940s."[19]

The courts appeared to play the same role with respect to civil air pollution litigation in other states as they did in California. We lack data on trends in frequency of such litigation in these other states, but can say with some confidence that nationwide the courts occupied a position of declining importance relative to the increasing intensity of the pollution problem and the legislative response.

The Federal Government

The air pollution problem had not stimulated response from state legislatures in the years prior to 1940, and largely the same situation held with respect to the federal government. What little federal activity occurred in those years was concerned with research and, as with the other levels of government, was preoccupied with such visible emissions as smoke. In 1912 the Bureau of Mines, located within the Department of the Interior, began studies concerned with smoke control; the work produced bulletins on the causes and cures of smoke emissions from coal-burning equipment. (At this time smoke was considered to be the only significant air pollutant.)[20] For a time the bureau was the chief federal agency dealing with air pollution. It established an Office of Air Pollution but later abolished it. The Public Health Service studied carbon monoxide in automotive exhaust about 1925, and became involved a few years later in claims arising from vegetation damage in the state of Washington allegedly caused by smelters in Canada.[21] But these were limited, episodic activities. Further significant federal involvement did not occur until 1948.

The almost empty history of federal air pollution control developments prior to mid-twentieth century stands in contrast to federal actions with respect to water pollution. The Refuse Act of 1899,[22] superseding a similar measure enacted in 1894, touched on the water pollution problem; it prohibited discharges of debris and refuse into navigable waters, "but it was designed to prevent impediments to navigation, not clean up the water."[23] Later activities reflected a more focused concern with water pollution as such. The federal government began stream pollution investigations in 1910; in 1912 the United States Public Health Service was authorized to carry out similar investigations as to navigable waters; the Oil Pollution Act of 1924 prohibited discharges from oceangoing vessels, though it was largely ineffective; and during the New Deal, vast sums from the federal purse were spent, as part of Roosevelt's public works programs, on the construction of waste treatment works.[24] Federal water quality measures not only preceded but to a large degree influenced the design of later legislation dealing with air quality. "Innovations in water pollution legislation [at the federal level] have generally come first, with parallel laws on air pollution following a few years later."[25] Concern over the dangers of frequent incidents of typhoid in the United States stimulated some of the early activity in the realm of water pollution control—in particular that of 1910 and 1912.[26] This is an early example of the role that crises, real or imagined, have played in the making of pollution policy.

PREVAILING ATTITUDES AND PRACTICES

The sketch of the institutional environment up to 1940 suggests a good deal about attitudes and practices that prevailed at that time. It is worthwhile to identify these and attempt to draw them into a plausible pattern, for they formed the background for response to air pollution problems that began to emerge in Los Angeles and other cities in the 1940s.

One study of the air pollution problem in Southern California, assessing the years just canvassed, observed that "in the absence of an accurate view of the air's composition and its effects on the community, control measures were instituted in a piecemeal fashion in response to particular crises and citizens' complaints. Comprehensive and effective planning for later problems did not occur."[27] This observation—with its emphasis on limited knowledge, ad hoc reaction to incident, and a general narrowness of view—is a fair characterization of the situation as it prevailed in California and across the nation as well, though one could specify some of its elements with greater particularity. There was for a time implicit reliance on the courts as the chief allocator of the air resource; there was a pronounced bias in the direction of localism, marked both by the absence of state action and an almost nonexistent federal role; there was a concern with the obvious and close-at-hand. The evidence on all the above items seems quite clear.

The limited knowledge of the period, for example, revealed itself in the ambiguous, unspecific nuisance approach of early ordinances, in their prohibitions of "unnecessary and unreasonable smoke,"[28] and indeed in the general preoccupation with visible smoke emissions. Smoke was the target "of the first attempts at control because smoke is visible and, therefore, people were aware of its existence."[29] The creation of new, particularized ordinances—aimed at specific, newly discovered sources (such as orchard heaters)—suggests narrow reaction based on limited knowledge, rather than careful, broad-gauged planning based on study and understanding. The locus and reach of these ordinances (virtually all were municipal), coupled with state and federal inactivity, illustrate the localism that marked air pollution control. That these ordinances did not appear until the end of the nineteenth century shows virtually sole reliance before that time on the courts to control use of the air resource; we know smoke emissions were not uncommon for a substantial period before appearance of the first ordinances.

The pattern of attitudes and practices is fairly obvious; its elements were consistent and, given the circumstances of the time, understandable, if not the most sensible or wise. In the preceding centuries of his occupation of the United States, man had been surrounded by an atmosphere generous enough to absorb without noticeable effect virtually all the air-contaminating output of relatively meager enterprises. There was no signal to cause a general concern about, or a general inquiry into, relationships between man's activities and the air around him. At times problems did arise—an especially obnoxious outflow of heavy smoke from one or a few sources, for example—but these were few enough in number, and offended few enough people in a widely scattered population, that it was sensible to rely on the existing system of courts for resolution, rather than to devote valuable social capital to the creation and support of new overhead in the form of legislative-administrative control. Moreover, the tradition of judicially administered nuisance law appeared well suited to a problem as simple and occasional as noxious, visible gases. So for some time the courts could efficiently serve the allocational role with respect to the air resource.

Institutional response changed as the situation did so. With the increasing growth and concentration of activities that accompanied urbanization, smoke nuisances occurred more frequently and affected larger numbers of people. This generated higher pollution costs and broader citizen concern; it generated, too, knowledge of the existence of a problem of some seriousness and persistence. Judging from our rough data, civil litigation appears to have reached its peak by the end of the period canvassed here. It dropped off after 1940, just when the air pollution problem began to increase significantly. There is a plausible hypothesis for this trend. One could expect civil suits against polluters to increase with the intensity of the problem, for a time. But as the discussion in chapter 1 suggested, at some point the courts' competence and attractiveness as a forum decline. As sources multiply, as emissions increase, as their effects become more far-reaching and complex, judicial intervention shrinks in terms of relative effectiveness. From the standpoint of both courts and potential litigants, not to mention that of the legislators, legislative action begins to look more promising.

Increased concern and increased knowledge encouraged and warranted legislative intervention, and this occurred. But there was little basis in knowledge or circumstance to believe that the problem reached beyond smoke, or the smoke beyond the boundaries of scattered cities surrounded by sparsely settled rural areas. There was, in short, little apparent justification for calling in the heavy machinery of state legislative intervention prior to 1940, whether to control or, less forcefully, even to study the problem. As new sources developed, so ordinances developed to regulate them. Matters were in hand.

There was even less basis for federal intervention. The national government had displayed a willingness to intervene in the allocation of environmental resources when problems touched then-conventional national interests—such as the condition of the Union's navigable waters, or aroused nationwide concern—as in the case of the typhoid scare. But air pollution problems had not reached such scale. There was, early in the second decade of the twentieth century, some narrow federal study of smoke emissions by the Bureau of Mines. One can speculate that this was motivated at least in part by a desire to support and assist the coal industry, which undoubtedly felt concern with local attacks on coal-burning sources of air pollution. The nation's air pollution problem was to grow far more serious in both breadth and intensity before further federal activity of any sort.

From today's vantage point, one might have wished for more foresight on the part of law and government in dealing with the air pollution problem, for initiative rather than reaction. But put in perspective, such foresight would have approached prescience. Policy dealt with matters as they occurred, and the matters themselves multiplied gradually enough so as not to signal cause for great concern. Other items, having to do with the settling and ordering of a growing populace, occupied higher places on the social agenda. To some extent, there was opinion that man was the agent of only a part of the pollution problem; nature also bore responsibility. We mentioned earlier the view that "Los Angeles would have many days of haze even if there were no people living in the area."[30] Attitudes like this, justified or not, no doubt worked to still concern and action. And to the extent man was an agent, the pollution his activities caused was a worthwhile price for progress. But there was another price, to be paid later. The attitudes and practices in the years before 1940 were sensible in context, but ill suited to deal with the massive growth in population, consumption, and technology that followed the Second World War, and with the air pollution problems such growth generated. Functional in their time, they settled into biases that hindered effective response to new pollution problems for years to come.

3

A NEW PROBLEM AND EARLY EFFORTS TO RESOLVE IT

Southern California had known air pollution before 1940, but in that year the problem seemed to take on a new and more persistent form. "A strange thing began to happen"—a "pall of haze" gradually obscured Catalina Island and the mountains around the Los Angeles basin.[1] The "strange thing" was photochemical air pollution, but no one understood this at the time. Just when it initially appeared in the Los Angeles area cannot be determined with precision. Some reports say the first episodes did not occur until 1941 or 1942, and even then they "were widely mistaken for Japanese gas attacks."[2] In September 1942 there was at least one day of serious pollution conditions—the downtown area of Los Angeles "was affected severely, the outlying districts were affected to a greater extent than at any previous time."[3]

CRISIS AND EARLY REACTIONS

Matters became more serious the following year and prompted considerable reaction. According to a newspaper account, Los Angeles in 1943 began to experience "days when the low-lying smoke and fume bank engulfed the city and environs and sent cursing citizens, coughing and crying, running for the sanctuary of air-conditioned buildings." Complaints began, the Chamber of Commerce became alarmed, and the "mayor formed a committee." At an August 1943 conference, he announced "that there would be an 'entire elimination' of smog within four months." But "something slipped on that prediction."[4] Three weeks later, on September 8, 1943,

> Los Angeles experienced its "daylight dimout" when dense smog settled over the area. According to one newspaper, "Thousands of eyes smarted, many wept, sneezed and coughed. Throughout the downtown area and into the foothills the fumes spread their irritation."
>
> Everywhere the smog went that day, it left behind it a group of irate citizens, each of whom demanded relief. Public complaints reverberated in the press. There was an outraged demand for action. Citizens committees were appointed. Elective officials were petitioned.[5]

Concern over the air pollution episodes of fall 1943, especially that of September 8th, stimulated response more substantial than the mayor's optimistic prediction of an elimination of smog within a matter of months.[6] It is safe to surmise that a motive of that prediction had been to ease the pressure of citizen complaints. The mayor and other government officials were, after all, in a bad position—the pollution appeared serious, citizens clamored for action, but government had very little notion what to do. And this was understandable: the heavy acrid hazes that began in the early 1940s were a mystery; their "peculiar nature" posed "a completely new problem."[7] In the face of this, government officials did two things once the initial prognosis of quick improvements proved incorrect. First, they took steps to abate emissions from sources they or the public considered to be major contributors to the new problem; second, and more or less simultaneously, they initiated scientific studies. Both kinds of action probably shared an element with the mayor's promise; both, that is, were probably intended in part to soothe an aroused public.

Abatement Activities

Shortly after the September 8th episode, "public officials, prodded by an angry citizenry, accused the synthetic rubber plant at Aliso Street, owned by the Southern California Gas Co., of being the chief offender."[8] A newspaper story about the incident is revealing:

> At one time, practically all the smog in Los Angeles County was blamed on this one offender. Citizens and "experts" alike apparently failed to realize that the condition was not traceable to any one source, but was the combined result of numerous factors.
>
> The butadiene plant was shut down for test periods and the installation of corrective equipment was ordered. . . . Now, following the expenditure of a sum reported to be $1,500,000, the plant is declared to be practically free of fumes.
>
> Yet the smog has persisted. On some days prior to the recent rain it was almost impossible to see the city of Los Angeles from any near-by vantage point.

The reporter had an insight as to why the plant became a target for action. "It was a new problem and nobody knew exactly what to do. Scapegoats were needed. One was found in the butadiene plant."[9] But warranted or not, the attack on the rubber plant produced a small piece of knowledge: "it . . . seemed to be proved rather conclusively when the plant was shut down at intervals and the smog persisted that it was not by any means the sole source."[10]

Scientific Studies

Ill-conceived legal action at times produced knowledge. But legal actors also undertook or commissioned a series of meetings, hearings, preliminary studies, and investigations—more systematic efforts to reach a better understanding of the air pollution problem. Considering the dearth of information, these were sensible steps; they were also a useful resort for government officials confronted with citizen demands for action but armed with little notion of what to do. Los Angeles County acted first. Its Board of Supervisors moved quickly to establish a Smoke and Fumes Commission made up of three scientists and two private citizens. The commission began its work in October 1943 and by March 1944 reported on some of the causes of air pollution, noting "ignorance and apathy" as primary problems.[11] The city of Los Angeles began investigations at about the same time.[12]

LOCAL ORDINANCES AND PROBLEMS OF FRAGMENTATION

The Los Angeles City investigations led in 1945 to an ordinance setting limits on smoke emissions from any single source. A Bureau of Smoke

Control, established in the city's Health Department and consisting of five inspectors (a chief and four subordinates), had the job of enforcing the ordinance.[13] The city ordinance, of course, had force only within municipal boundaries; it did not apply to the other forty-five cities within Los Angeles County, nor to unincorporated areas. County government could respond to part of this problem. Based on its studies and on recommendations from the County health officer, the Board of Supervisors in 1945 passed ordinances establishing a Division of Air Pollution Control in the County Health Department and setting emission standards applicable to the unincorporated areas.[14]

These moves, made necessary by the irrelevance of political boundaries to air pollution control, solved only some of the problems of fragmentation. The first ordinance applied only to the city of Los Angeles; the latter ones, only to unincorporated portions of the county. There remained untouched the forty-five incorporated cities within the county. Public officials tried to influence these cities to follow the lead of Los Angeles. The Board of Supervisors directed formal communications to mayors and city councils, urging the adoption of uniform ordinances like those of the city and county; the county offered to do the enforcement work under contract. But the urgings met with little success. By mid-October 1947 "only seven [cities had] responded favorably. Others resolutely continue[d] to do nothing."[15] By the end of the year, twenty-two had taken favorable action, yet problems remained because of variations among the ordinances. Moreover, the many incorporated cities that had passed no ordinances were in some instances significant sources of emissions.[16] Matters were sufficiently vexing that the Los Angeles District Attorney's Office announced plans to file thirteen public nuisance actions under provisions that, because they were found in *state* legislation, applied to incorporated cities.[17]

The nuisance suits, however, could hardly be an effective solution to the problem of fragmentation. They would only add yet another decision maker, the courts, applying yet another body of law, the common law of public nuisance. This frustrated the hope of a uniform approach to the problem. And there was another difficulty in the nuisance approach, a technical legal one. California nuisance law provided at this time that if a factory were located in an area where, under a local zoning ordinance, manufacturing uses were permitted, then the factory's pollution could not be enjoined as a nuisance unless proved to be the result of unnecessary operations.[18] This provision applied on its face to many pollution sources the district attorney hoped to abate, and apparently caused his office considerable concern. During the 1945 session of the legislature, he sought an amendment that would have excepted from the statute "an action to abate a public nuisance brought in the name of the people of the state of California." The measure passed the Assembly, then died in a Senate committee, thanks largely to the opposition of the Los Angeles Chamber of Commerce and the Merchants and Manufacturers Association—groups that had from the outset promoted voluntary self-policing by industry. They wanted "the prosecuting officials to leave industry alone to work out its own problems."[19]

SUBSEQUENT INVESTIGATIONS AND THE MOVE FOR STATE LEGISLATION

The small wave of ordinances that began in Southern California in 1945 proceeded on "the assumption that air pollution in Los Angeles was similar to that of other cities where it was due largely to sulphur and smoke."[20] Ordinances already in existence in eastern cities had been used as models.[21] Indications soon developed, however, that the ordinances were insufficient. Perhaps this was because they were not being (or could not be) effectively enforced. Perhaps it was because the problems at which they aimed were the wrong ones. It might be, people began to think, that air pollution conditions in Los Angeles were not similar to those of eastern cities at all.

One can find strains of all these notions. A reporter close to the problem, for example, spoke of "much talk . . . and little action." He also noted, however, that no simple solution existed—the problem was complicated. "There not only are many sources, but many different kinds of corrective measures must be taken. No one law can do the job. No one set of scientific remedies can do it." Moreover, the problem of fragmented local authority hindered effective use of corrective measures. "Apparently one of the great drawbacks of smoke and fumes control has been lack of a unified authority to deal with the question."[22] Beyond this, smoke and fumes control, however effective, might not be enough. "Even at this early date in the smog control program individuals both inside and outside the government began to realize that the problem was more complex than just black smoke."[23]

In any event, there was dissatisfaction with the pace of progress. Citizens were agitated. In the fall of 1946, after several especially aggravating episodes, "hundreds . . . marched on City Hall."[24] There was pressure for more effective response, followed by two reactions. The first was the predictable one—further, intensified studies. The second consisted of heightened attempts to overcome jurisdictional problems through new state legislation.

Public and Private Studies

In the face of citizen complaints, further study on the part of government must have seemed both a necessary response and an attractive out. The Los Angeles County air pollution control director was forced to agree "that enforcement activities alone are too slow." Accordingly, he had planned "an intensive investigation . . . to find the facts in all possible ways" and eliminate "these fumes to the extent that even especially susceptible people will not be affected."[25] As the public investigations began, private studies were also being sponsored. Most notably, the *Los Angeles Times* induced Professor Raymond Tucker, an air pollution expert from Washington University in St. Louis, to come to the city and study its problems. Tucker had earlier investigated the St. Louis pollution problem, and the cleaning up of that city's smoke was largely credited to his study and recommendations.[26]

The background of the *Times* study deserves some elaboration. The newspaper had become interested in the Los Angeles smog problem as early as 1945. Norman Chandler, publisher of the paper, later said the *Times* "had entered the campaign in the public interest with the avowed purpose, if possible, of finding all the sources of air pollution." An early step in the paper's efforts was "a series of articles on the smoke and fumes nuisance—the 'smog' in the Los Angeles area." The articles aimed to present facts and possible remedies in order "to lead the way toward a permanent solution of this blight which threatens the health and future of the metropolitan region." In the course of the series, the idea developed to engage Tucker as a consultant to study the problem.[27] St. Louis, Tucker's home base, had long suffered a persistent smoke problem, and "from 1925 to 1940, . . . went through a period of much talk but little action." Expensive educational campaigns (costing "hundreds of thousands of dollars") proved fruitless, but were supplemented by only the most paltry efforts toward public regulation. As late as 1933, there was only one official to "control" 150,000 smokestacks. With a change in the city administration, Tucker became head "of a committee to fight the smoke nuisance." By 1937 he was smoke commissioner, a job he held until 1942. During this "crucial period" of five years, "storms of controversy" surrounded the passage (in 1940) and implementation of "a revolutionary ordinance." The ordinance survived constitutional attack in 1942, and "since then, the city has enjoyed virtual freedom from really bad smoke, according to its citizens. . . . St. Louis has shown what can be done, against far greater odds than those faced in Los Angeles." The *Times,* in reporting these dramatic facts, noted that much of the St. Louis success story was "due in part to an active campaign by the St. Louis Post Dispatch and other newspapers."[28]

The idea to bring Tucker to Los Angeles must have been especially attractive to the *Times.* He was a recognized expert; he was familiar with the pressures of politics and could handle the heat of controversy; he had succeeded, where others had failed, in resolving a problem similar to that of Los Angeles; he had made good press, and good press had helped make him. (Tucker later became mayor of St. Louis.)[29] To be sure, the Los Angeles problem was perhaps quite different from that encountered by Tucker in St. Louis, but who was *more* qualified to combat the Los Angeles smog?

Tucker began a two-week investigation of the Los Angeles situation early in December 1946. He returned to St. Louis, studied his data, and prepared a report that appeared on the front page of the *Times* in mid-January 1947.[30] The report summarized the early history of Los Angeles pollution conditions, noted that they had grown more acute in the preceding five years, and commented in this regard on the fact that industries in the area had increased an estimated 85 percent during that period. Moreover, the industries tended to concentrate in locations such that morning winds carried "fumes, smoke, odors and dust . . . into the heart of the city from the industrial district." The responsible enterprises included "chemical industries, refineries, food product plants, soap plants, paint plants, building materials, nonferrous reduction refining and smelting plants, as well as numerous others of similar type." Their discharges of "sulphur dioxide, smoke, dust, aldehydes and other noxious gases" had been little supervised until recently, and Tucker recommended more controls.

Population was another factor; it too had grown dramatically during the years in question—from 1.5 million to almost 1.9 million in the city and from 2.8 million to 3.7 million in the county of Los Angeles. It was the "common practice" of this burgeoning populace "to burn combustible refuse from the homes (on orders from the Fire Department)" in backyard incinerators during the morning hours. Here, any individual's contribution meant nothing, "but collectively they are a recognizable source of air pollution"—"7,400 tons for the county" every day. "The practice of burning rubbish should be discontinued unless done in properly designed incinerators."

Tucker went on to express his concern with municipal and commercial dumps, oil burners (for heating), and railroads. But what of motor vehicles? Reports predating Tucker's had mentioned buses and diesel trucks as possible causes of the problem.[31] Tucker concluded that buses "contribute to the nuisance but not in the manner some would lead us to believe."[a] Diesel trucks, on the other hand, might be more substantial contributors—especially those not well-maintained. As to the automobile,[b] Tucker set forth his conclusions under a heading that read "Automobiles Absolved From Most of Blame."

> There are some who blame the present conditions on the automobile. . . . An inspection . . . shows that between 1941 and 1944 there was a decrease in the automobiles entering [a surveyed] area of Los Angeles; however, it was in this period that the lachrymatic [eye-watering] effect became most noticeable to the general public; in fact, the number of automobiles in January, 1946, over October, 1941, has only increased 11.7 percent. It is not the intention of this report to absolve anyone from his responsibility. It would appear, however, that although it is quite probable that the automobile does contribute to the nuisance, it is not in such proportion that it is the sole cause.

This was not quite absolution of the automobile; at the same time, it was hardly stern condemnation. And read as a whole, the report evinced even less concern with respect to the auto's contribution to the pollution problem. There were many theories, Tucker noted, as to the cause of Los Angeles smog, but none was substantiated. Further research was needed, but also drastic steps to curtail and eliminate known pollutants. The existing ordinances were inadequate for this purpose—not backed by sufficient enforcement powers, hampered by statutory nuisance laws, frustrated by fragmented authority, insufficiently comprehensive, undermanned. Changes were needed, and Tucker recommended a number of them. Not

[a]Tucker's concern here was with sulfur dioxide only, not with the reactive hydrocarbons and nitrogen oxides later learned to be major vehicle-related contributors to the smog problem.

[b]There had been earlier speculation about the auto as a cause, though the issue was "one of the unsolved questions." E. Ainsworth, "Second St. Tunnel Survey Discloses Poisoning Perils in Auto Fumes," *Los Angeles Times,* 17 Nov. 1946. The survey mentioned disclosed the presence of carbon monoxide and sulfur dioxide. Significantly, however, it revealed also that "the chief source of eye irritants was found to be aldehydes—those mysterious unidentified gases given off by engines due to lack of complete fuel combustion." Even more significantly, "a clue to respiratory irritation was found, too, in the large amounts of oxides of nitrogen."

one of them even mentioned the automobile (although there were recommendations pertaining to diesel trucks and buses).

The Tucker report was received with fanfare and praise for author and sponsor alike. "Both the Los Angeles County Board of Supervisors and the City Council prepared resolutions commending The Times and Professor Tucker for the report," the *Los Angeles Times* proudly announced. Meetings were organized "to map a campaign in behalf of the Tucker program."[32]

The Move for State Legislation

A central recommendation in Tucker's report was that "the necessary State legislation be enacted to create an air pollution control district, preferably county-wide. This legislation should be of such character as to enable the enforcement of regulations in all areas of the county, unincorporated and incorporated." The recommendation was spurred by the problems of fragmented authority that Tucker had observed. His report was hardly the initial notice of those problems, nor did it represent the first suggestion for statewide legislation to cope with them. As early as the spring of 1946, community leaders concerned with air pollution control had concluded that "there should be one overall district . . . coterminous with the boundaries of the County of Los Angeles." By the summer of the same year, this feeling had grown in intensity; "the entire citizenry was aroused to the point that it would not be patient with a jurisdictional fight as between county and city government." Coordination seemed necessary, and experience indicated it would not come about through voluntary adoption of uniform ordinances. There was a clear need "from an operating and legal jurisdiction standpoint for a single agency which would have the police power to work toward solutions."[33]

But there were problems posed by the notion of such an agency. A body of law known as "home rule" prevented Los Angeles County from simply forcing a countywide agency on the incorporated cities within it. Consent of the cities was necessary, but here the *tradition* (as opposed to the law) of home rule stood in the way. "Having in mind that for more than a hundred years at the level of both counties and cities home rule government in California was important, it was not easy to secure the consent of the cities to yield local jurisdiction to an overall County agency."[34]

Some other approach was needed, and the Los Angeles County counsel concluded he had found it in *state* legislation to "create a smoke abatement district with police power flowing directly from the State, to be in conformity with the California State Constitution."[35] By the end of 1946, he had begun drafting a measure to create unified districts with policy control and legal jurisdiction in the hands of county boards of supervisors. (The county counsel's idea for a unified district apparently found its inspiration in Metropolitan Boston, a district of many incorporated cities that had been subject to unified pollution control regulations beginning as early as 1910.)[36] The California League of Cities supported the approach, and two assemblymen agreed to sponsor the legislation when the next session of the California Assembly began in January 1947.[37] Introduced that month,[38] the bill immediately met open industrial opposition, especially from the railroads and the oil companies. Negotiations followed, and by and large the bill survived intact. The railroads were prepared to argue that the measure, if passed, would unlawfully burden interstate commerce, a position the bill's proponents did not wish to confront. At a meeting "attended by executives of nearly all of the railroads operating in California . . . the at first active opposition of the railroads was dispelled." As the Los Angeles County counsel modestly explained, "from the analysis presented by the Los Angeles County Counsel's Office as to the practical workings of the contemplated bill, the opposition of the railroads was withdrawn."[39]

The petroleum industry objected to permit provisions proposed in the bill; these would require any firm building or modifying a plant to secure a permit prior to construction in order to ensure that the new facilities would meet applicable air pollution standards. Advocates of the legislation considered the permit requirement essential; their strategy was to expose the opposition of the petroleum interests to the public through the press. The *Los Angeles Times* was more than ready to cooperate, and a few editorials and news articles did the trick. Oil company executives met with the Los Angeles County counsel and other backers of the bill, discussed the issues, and withdrew their opposition.[40] Oddly enough, the farmers were more effective. Agriculture interests were concerned that the bill would unduly restrict them in their operations without any resulting benefits in air quality. As finally passed, the measure contained broad exemptions for agricultural operations. It became law with Governor Earl Warren's signature in mid-1947.[41]

4
THE STATE LEGISLATION AND ITS IMPLEMENTATION

We need not dwell too long on the particulars of California's Air Pollution Control Act of 1947;[1] it is enough to describe its central features and their rationale, the act's early implementation in Los Angeles, and the criticisms that resulted.

CENTRAL FEATURES AND RATIONALE

The 1947 act permitted in every California county an air pollution control district with boundaries coterminous with county lines; no district, however, could be activated until the county board of supervisors (which would serve as the board of the district) found after public hearings that the air was polluted and could not be effectively managed through local ordinance.[2] The act itself contained two statutory prohibitions: against discharge of emissions more opaque than prescribed standards, and against discharges that would constitute a nuisance.[3] Beyond this, districts could enact additional requirements in the form of rules and regulations consistent with the purposes of the act.[4] In particular, districts could require sources to obtain permits before construction or operation.[5] On its face, this permit provision would appear applicable to all sources. A subsequent section of the act, however, specifically exempted motor vehicles.[6] The statutory prohibitions, on the other hand, did apply to vehicles, and presumably district rules other than those pertaining to permits could apply to them as well.

State constitutional constraints had much to do with the general approach of the 1947 legislation. The law of home rule limited county police power to unincorporated areas. A second constitutional constraint denied the police power to special districts and authorities, and a third prohibited "special legislation" explicitly applicable only to certain areas. If the legislature enacted a law, its provisions were to operate uniformly throughout the state. The difficulty here was that only Los Angeles had a problem, and it alone needed the help of state law.[7]

The 1947 act was designed to respond precisely to these constraints. As *state* legislation it applied, without regard to local political boundaries, to counties and the incorporated and unincorporated areas within them. The police power itself, however, was exercised by the state; the county districts simply adopted rules and regulations to carry out the broadly drawn legislative standards. And because districts were to be activated only upon county initiative, the law in essence achieved, without violating the requirements on uniform laws, the benefits of special legislation. Thus, the "single purpose of the state law (as has often been stated by counsel for the county) was to enable Los Angeles County to preempt authority in this problem from the cities and thus provide a uniform regional control."[8]

The statutory prohibitions—of specified emissions of smoke, and of emissions that would constitute a nuisance—were designed to take as full advantage as possible of the knowledge that then existed. The smoke emission proscriptions employed the Ringelmann Chart, "a simple objective measure for the density of visible smoke."[a] The nuisance provisions took account of the fact that the "Ringelmann Chart is, of course, of no value in testing fumes or gasses which may not be visible to the eye. . . . At the time the Air Pollution Control Act was written, no simple objective test was known for invisible substances. As to them the Act applies the common law nuisance test."[9] With time, however, new knowledge about these "invisible substances" might be generated, so that objective standards

[a]The idea was that "a statute based on some single objective test is highly desirable," presumably to avoid constitutional attack on grounds of vagueness and to ease proof of violations. See H. Kennedy and A. Porter, "Air Pollution: Its Control and Abatement," *Vanderbilt Law Review* 8(1955): 854, 864.

more specific than the nuisance test could be developed. The provisions permitting district adoption of appropriate rules and regulations were intended to permit these standards when the necessary information developed, "without the necessity of going back to the legislature for an amendment to the Act. Thus, the Act has within itself the means of forever keeping up to date with the latest advances of modern science in the matter of air pollution control."[10]

The rationale for legislative authorization of a permit system was to ease enforcement of the law and give it teeth. If a source did not comply with control requirements, its permit could simply be revoked or suspended. The source would then be out of operation until in compliance.[11] The permit approach could result in closing down a plant if the control equipment necessary for compliance were unavailable or very costly, but this of itself did not make the permit provisions invalid.[b] As a prelude to drafting the 1947 act, the Los Angeles County counsel had prepared a memorandum that "sought to embrace all of the case law that could be found in the law digests involving air pollution abatement in the United States."[12] That research convinced him that a "law is not invalid solely because it may put an establishment out of business or require the expenditure of large sums to comply with the terms of the law, nor because there are no appliances available to prevent the prohibited smoke or fumes."[13] His opinion proved sound; the 1947 act was later to withstand constitutional challenge.[14]

EARLY IMPLEMENTATION OF THE 1947 ACT IN LOS ANGELES COUNTY

The Los Angeles County Board of Supervisors acted quickly after passage of the 1947 act to put its provisions into operation. In October 1947 (the legislation could not be put into effect until ninety days after adjournment of the California legislature), the board held the required public hearings, declared the need for an air pollution control district in the county, and

[b]Moreover, noncompliance would not necessarily mean denial, revocation, or suspension of a permit. The act provided a variance procedure whereby sources could continue operations if, in essence, they had made good-faith efforts to comply and if their operations would not substantially aggravate pollution conditions, the theory being that under such circumstances there would be undue hardship in closing down a source. H. Kennedy, *The History, Legal and Administrative Aspects of Air Pollution Control in Los Angeles County* (Los Angeles County Board of Supervisors, 1954), pp. 56–57.

secured its first director, Dr. Louis McCabe—formerly with the United States Bureau of Mines and "one of the most experienced men in the nation in the field of air pollution control."[15] McCabe arrived in Los Angeles "via Union Pacific from Washington, D.C. . . . As [he] stepped from the train he was met with a view of the typical gray blanket of fog, smoke and fumes which envelops the city so often. 'There is your job!' he was told."[16]

And a job it was. The years of World War II had been accompanied by booming industrial growth in Southern California; by one estimate, there were at least 13,000 sources to control.[17] These were spread throughout the county, many of them in the incorporated areas that had previously resisted control efforts but were—thanks to the 1947 legislation—now under McCabe's domain.[18] The "aggregate discharges" of these sources were seen as "a serious menace," yet a hard line on control was not entirely feasible. "In some cases the installation of corrective equipment would be so costly as to put a small plant out of business. Yet it is producing smog. What should be done in such a case? Dr. McCabe's task as administrator will be to draw the line on these hardship cases."[19]

McCabe's response was to adopt a "fair but strict" policy consisting essentially of two phases: "reducing the atmospheric pollution which was present when the program began . . . , and . . . making certain that the growth of the community did not offset those gains."[20] Both phases reduced themselves to attacking the pollutants and sources known at the time to exist—especially smoke, fumes, and sulfur dioxide, from oil refineries, chemical plants, oil burning industries, and rubbish dumps. Controls on these were realized through rules and regulations enforced by the permit system, injunctive proceedings, and criminal actions.[21] "Reasonableness" was the hallmark of these control efforts, however. Constitutional law appeared to permit closing down offending sources, even if abatement of the source would be harshly expensive or, indeed, impossible. But this seldom if ever occurred—such an approach would be strict, the authorities reasoned, but not fair. "Practical needs of the industry" were taken into account. Variances from the strict requirements of the law were commonly granted where control technology did not exist, was not good enough, or was unduly expensive for a particular source.[22] The board of the Los Angeles Air Pollution Control District (APCD) knew that the constitution and the 1947 legislation granted "sufficient power to cause the closing of a plant or industry if it cannot comply with the reasonable rules of the district." But it was "delighted to report . . . that not one single business or industry has been put out of operation by virtue of the board's activities." The APCD's publicized position on pollution was "that wherever there is a way to control it, it must be controlled to the limits of present-day engineering knowledge."[23] There was a mixture of ironies in the statement. On the one hand, the 1947 legislation appeared to contemplate control *beyond* that attainable by "present-day engineering knowledge" where necessary; on the other, the APCD did not go nearly so far as it claimed. Its variance practice often had the effect of stopping far short of what "engineering" could in fact accomplish.

Early abatement activities in Los Angeles were based on the knowledge that existed at the time, but it was part of the APCD's agenda to enhance that knowledge through research activities. Detailed accounts of research have not been found. There apparently were some efforts to determine "what pollutants are in the air . . . and in what amounts."[24] The total research effort was small, however, and this was soon to draw criticism.

CRITICISMS OF THE ACT AND ITS EARLY IMPLEMENTATION

The years 1947–1949 were not quiet ones for the Los Angeles County APCD. Its officials had been in a bind from the beginning. Citizens demanded "using the full power of the District to force a strict compliance by all types of industry and particularly the petroleum industry."[25] McCabe, "feeling the tremendous pressures being exerted on the part of the public" (who complained that the air pollution program "was getting nowhere"), was "insistent" on pursuing the oil companies. They, in turn, were adamant that McCabe's policies "were inequitable, too strict, and would result in unwarranted expense to the petroleum industry, without sufficient proof that the removal of the sulphur would cure the overall smog problem."[26] In this atmosphere, one member of a Citizen's Smog Advisory Committee that served the APCD resigned "with a blast at McCabe and the program charging 'politics and Big Business.'" The Los Angeles County Grand Jury appointed a committee to look into the charges.[27] (This started a trend. The Los Angeles County counsel wrote in 1954 that "every Los Angeles County Grand Jury since the activation of the District has had an Air Pollution Committee to make sure that everything was being done by the District officials that could be done.")[28] In 1949 the legislature conducted an investigation of its own, and a serious episode in the fall of 1949 generated more expressions of dissatisfaction.[29] "This great wave of criticism from the public and from industry came at a time when Dr. McCabe was receiving attractive offers of lucrative employment elsewhere because of his wide and extensive experience in the

Bureau of Mines and in the Army and his outstanding qualifications in this field."[30] McCabe finally quit in 1949 to return to the Bureau of Mines. His chief assistant, Gordon Larson, took over the APCD.[31]

He also took the abuse directed at it. The various investigations of 1948 and 1949 resulted in further criticism of air pollution control in Los Angeles County—some of it aimed at the 1947 legislation itself, some at its implementation. There were at least four complaints about the basic design of the 1947 legislation. The first of these had to do with the fact that, under the act, activation of a district was purely voluntary; counties could choose to maintain the status quo. The legislation took this shape because in 1947 many areas of California were relatively (or even absolutely) free of pollution problems. This was, of course, a shortsighted position. It overlooked the fact that, with growth, other areas would come to be plagued with pollution. But the act's purpose was curative, not preventive—an observation underscored by the fact that the legislation, at least if taken literally, made activation of a district dependent upon air pollution in a county being excessive. "Preventive measures taken before air pollution becomes a problem would undoubtedly be more effective and less expensive for all concerned, including the taxpayer and industry," said one group of critics. "Too much emphasis has been placed upon cure, once smog has occurred, and not enough upon prevention and determination of cause."[32]

A second criticism related to the basic structure of the 1947 act may suggest a factor behind the charges of "politics" in the administration of the law by Los Angeles County. Under the legislation, the board of supervisors of any county activating an air pollution control district served as the board of the district. "An initial mistake was made in putting the air pollution control apparatus in the hands of the elected Los Angeles County Board of Supervisors, where it became subject to patronage and political rivalries."[33]

A third criticism concerned the irony of drawing air pollution control district lines coterminous with county boundaries. A prime mover of the legislation had been the fact that air pollution stubbornly ignored political boundaries, making it impossible for cities to cope effectively with their pollution problems through purely local ordinances. The 1947 legislation allayed this problem, but only in part, because pollution respected county lines little more than city ones. "The district being but county-wide in scope, its area of control may not extend far enough to permit a district to reach as far as the smog does."[34] In 1949 the legislature authorized the merger of activated air pollution control districts in two or more adjoining counties,[35] but this response was not entirely adequate to handle the problem. Integration was voluntary, would result in an unwieldy board consisting of several boards of supervisors, and was unavailable to noncontiguous counties.[36]

The three points above all found their roots in a fourth, more general observation about the 1947 act. The law was tailor-made for problems in Los Angeles County; only constitutional constraints motivated the masquerade of a general law. There had been no study of conditions or needs in other areas. The California Assembly Committee on Air and Water Pollution attributed to this history the absence of concern with preventive measures, the use of county lines to establish district lines, and the autonomy (isolation) of district boards (boards of supervisors). It contrasted the 1947 act to the state's water pollution control legislation, which it looked upon approvingly in all the above regards.[37]

The committee was particularly concerned with the permit provisions of the 1947 act, and with how they were being implemented in Los Angeles. Early state water pollution control legislation had used a permit system with unsatisfactory results. In implementing the system, control officials had resorted to specification standards, attempting to dictate the precise means to be used to achieve effluent reductions. The new water law "recognized the practical impossibility of doing this" and restricted officials to performance standards. These, the committee thought, had many advantages: they encouraged the use of new control techniques that might otherwise be rejected by officials who "inevitably insure their action with over-conservative demands"; they eliminated official bias toward particular control methods; they put officials in a position to "demand results without compromise."[38]

Nothing inherent in a permit system required specification standards, but Los Angeles APCD officials—like the officials under the old water pollution legislation—were using them nevertheless. Moreover, a "blanket" approach had been adopted, prescribing the same emission limitations for all sources. Such "across-the-board standards," the committee noted, could well "be too lenient for some and too harsh for others." The committee feared that, to avoid harsh results but maintain blanket standards, many sources were "permitted to *get by* with merely meeting standards established on the basis of what the least capable offenders can do."[39] There were grounds for that fear. Under the APCD's policy of "fair but strict," not one source had been put out of operation.

The entire approach of Los Angeles to standard setting was considered suspect. Standards were "admittedly" based not "upon the effects of the pollutants on the public health, but rather merely upon the ability of industry and others to meet the standards."[40] At first glance, this criticism seems unduly harsh. The committee knew there was a dearth of detailed information on the health effects of air pollution and conceded that "in the absence of this precise information, only arbitrary rules and standards can be used in requiring reduction and control of waste discharges to the air." Indeed, the committee acknowledged that standards in Los Angeles "*must* be based on what the industries can do, rather than upon a scientific determination of the significance of each particular waste with respect to public health."[41] In short, the Los Angeles approach seemed a necessary one; at the least, it appeared sensible. What better way to set standards, in the absence of sound information on health effects, than to require the best that industry can do?

The committee's criticism, however, was neither so harsh nor inconsistent as might appear. First, there was the evidence that Los Angeles set standards based on the best the *least* capable could do. More to the point, the APCD was making little effort to improve information. Its approach to standards might have made sense as an interim program that went hand in hand with research on the health effects of air pollution, but the APCD was conducting virtually no such research. Its concern was abatement, and what research it did was directed to sources and their control. As a result, the committee concluded, there was no attention to "what must be done to protect the public health."[42]

The APCD could put forth some defense on this point. Gordon Larson, its new director, argued that the "health aspect of the problem" had hardly been forgotten; the county, however, simply lacked the funds and expertise to do the task. In Larson's opinion, "the job of medical research is beyond the abilities of local government to finance."[43] There was at least partial support for this view. One study of air pollution control efforts in the state during the time in question (1947–1949) concluded, in a report published later, that local air pollution control agencies were in a poor position to conduct the required research. They were "preoccupied with immediate problems of abatement" and, in any event, too limited in resources.[44]

Appropriate research funded by and carried out at some other (higher) level of government appeared to be the answer. Larson himself said "we would encourage, in every way, its beginning under any procedure that could get it started, either locally or otherwise, as a scientific problem."[45] And the Los Angeles County counsel reported that "Los Angeles did not oppose the state entering the picture as far as research into the causes of air pollution or the effect of air pollution upon health generally."[46] Despite these remarks, the facts appear to be that Los Angeles County actively opposed proposals for statewide research, proposals initiated because of the APCD's own failures with respect to research.[47] The Committee on Air and Water Pollution summarized the history of opposition,[c] which began when its chairman "requested that the State Department of Public Health prepare an outline of study and research which could be undertaken in a single comprehensive project. . . ." The outline was subsequently prepared and proposed to the committee in the course of its hearings. "The many authorities . . . who testified were unanimous in their support of the recommended project, including the officials of the Los Angeles Air Pollution Control District." Accordingly, a bill calling for the research proposal was introduced into the legislature. But now, and even though "all changes proposed by the Los Angeles officials were accepted . . . , the legislative representative of the county appeared in opposition to this bill in any form whatever, and offered no further alternatives."[48]

The county's position on research was not entirely inflexible. Its counsel "assured the standing committee that an equivalent research would be undertaken and financed by Los Angeles County."[49] As a result of this information, the research bill was tabled. Los Angeles County, however, fell short of its assurances. It refused to appropriate to medical research the funds it had earlier promised.[50] Fresh hearings on the issue began shortly after this new rejection. These "terminated with unanimous agreement among all witnesses, including the Los Angeles officials, that a state-sponsored research project should be undertaken." The only reservation, expressed on behalf of Los Angeles County, was that the research should "be confined solely to study of the effects of air pollution upon the public health."[51] A new bill along these lines was subsequently introduced and passed unanimously by the Assembly. It was tabled in the Senate despite the fact only one witness from a local chamber of commerce testified against it. "Whatever other opposition there may have been," the committee dryly observed, "remains an interesting matter of conjecture."[52]

What might account for the Los Angeles APCD's apparent resistance to research—conducted by itself or the state—on health effects? This too is a matter of speculation, but the hint of an answer lies in the county's expressed concern that any state research be "confined solely" to health questions. Undoubtedly the APCD, under the pressure of serious pollution conditions, wanted tangible improvements—*any* improvements—to which it could point as evidence of the success of its program. Thus, it was against diverting a part of its own funds from abatement to research, for the diversion could slow the rate of "progress" as the APCD saw it. Similarly,

[c]Perhaps because of this opposition, the proposals were defeated. California Assembly Committee on Air and Water Pollution, *Final Summary Report* (n.d., c. 1952), p. 36.

proposals for research at the state level must have been threatening to abatement efforts, for several reasons: research could result in suggestions to hold up controls until the facts were in; worse, research might show that APCD controls were ill-conceived. Thus, when Los Angeles had earlier opposed, "in any form whatever," the proposed bill for state research, its county counsel explained the action by expressing "concern over the possible effect which the results of a research project might have upon the enforcement of the program."[53]

These speculations find support in the APCD's efforts to show material improvements in pollution conditions by the close of 1949. Los Angeles officials claimed improvements in three respects—visibility, concentrations of sulfur dioxide, and total amount of pollutants removed from the air. As to the last, "they concluded that an approximate 35 percent reduction of air pollution had been accomplished."[54] The claims found some support: a legislative study later concluded that pollution conditions existing in Los Angeles in 1947 had been substantially eased by the end of the decade. The conclusion, however, was based exclusively on APCD data and appeared in a report exceptionally laudatory of the 1947 legislation and the efforts in Los Angeles to implement it. The chairman of the committee making the report was himself the author of the 1947 act.[55] Other investigations, based on independent appraisal of the very data used by the APCD, found that growth in sources had eaten up gains in control, and that "sources which have not yet been touched constitute by far the greatest part of the problem." The facts "demonstrated either no improvement or actually an intensification of the problem."[56] The analysis behind this rebuttal of the APCD's claims revealed how ingenious (and ingenuous) it had been in marshaling data to show the three categories of material improvements.

As to visibility, a noted meteorologist from the California Institute of Technology focused not on *average* visibility (as the APCD had done), but on total number of hours of visibility reduction. His approach led "to an exactly opposite conclusion from that of the district."[57] The same expert pointed out the fallacy of using average visibility records, where one very clear day can have a greatly distorting effect. APCD data showed that November-December 1949 enjoyed the greatest improvement in visibility, as "shown by the averages." Yet it was in this very period "that the notorious 21-day period of intense smog occurred, a recall movement against the Supervisors was threatened, and the mayor of the city requested a grand jury investigation of the entire smog administration."[58]

As to sulfur dioxide (on which the APCD had most zealously concentrated its control efforts), substantial reductions were conceded. The problem here was that sulfur dioxide appeared insignificant "in relation to the entire pollution problem." The amount in Los Angeles was already only a fraction of that in cities with no smog at all. Thus, it appeared the pollutant was "one of the least important contaminants in the Los Angeles smog"—a conclusion that had to "be drawn even in the face of the great preoccupation of the officials of the district with this particular contaminant, and their concentration upon efforts to reduce it almost to the exclusion of work upon other causes."[59]

Finally, on the question of total amount of pollutants removed, the APCD's figures (showing a 35 percent reduction in air pollution) had not taken into account several categories of sources because of incomplete data. But since the time of these estimates, it had been found that the sources in question discharged over 2,000 tons per day of significant pollutants, as compared with a total of 741 tons per day accounted for in the APCD data. "In other words, although 35 percent of the 741 tons of known material had been removed . . . there remains the balance of this material, or 482 tons, plus 2,042 tons of material which had not been taken into consideration!"[60]

The amount of pollution going into the air by the end of the 1940s was, thanks in part to growth, probably larger than the amount prior to the beginning of active controls in 1947. Not only was it doubtful that material improvements had occurred; the situation seemed to be worse than it had ever been. "When . . . improvement does occur," observed the Committee on Air and Water Pollution, "it will not be necessary to demonstrate it with statistical data. It will be apparent to the man in the street as well as to the technical observer."[61] Not long before this remark, a newsman had reported the APCD's claim of substantial improvements in air quality. The report "brought a flood of protesting telegrams" from the public.[62]

5
CHANGES IN DIRECTION

Criticisms of the 1947 legislation and especially of its implementation in Los Angeles, the lack of significant improvements in air quality despite some seven years of effort, and several other factors and events not yet recounted led at the beginning of the 1950s to important changes in the direction of California air pollution control policy. Before considering these, we must have a clear picture of precisely where policy stood as of 1949.

THE STATUS QUO IN 1949

Air pollution control policy as it existed in 1949 attached virtually no importance to the role of motor vehicles. At the least, this was the position of the only operational implementers of policy, the Los Angeles APCD. The focus of its efforts from the outset was on industrial sources, fumes, smoke, and sulfur compounds. Motor vehicles were ignored. This stance cannot be attributed to a lack of legal authority to control automotive emissions. It is true that the 1947 legislation exempted automobiles from its permit provisions; it is also the case, however, that other statutory provisions in the act were applicable.[a] As one study observes, the APCD "had the power under [the 1947 act] . . . to enforce State standards on vehicular emissions. There is little mention of the exercise of this power in the literature."[1]

It seems quite clear that the APCD's policy was the product not of powerlessness so much as conviction that vehicles were not the culprits. For example, at the First National Air Pollution Symposium—held in Pasadena, California, in 1949—both Larson and McCabe presented papers. Larson, at the time director of the Los Angeles APCD, viewed the air pollution problem as one of controlling the traditional stationary sources; he devoted no attention to the automobile (although he made passing reference to diesel trucks).[2] McCabe, the APCD's former director, and thus probably the most instrumental in formulating its early policy, was more direct in his paper. In discussing "National Trends in Air Pollution," he referred to suspicions about the motor vehicle as "folklore":

> It is easy for the layman who experiences high concentrations of exhaust fumes from a bus only a few feet away to believe that all the air pollution in his city is from this source. There have been extensive studies of motor exhaust gases in vehicular tunnels and mine air. The maximum concentrations considered safe for these places are never reached in the city air except in the immediate vicinity of the vehicle's exhaust or in congested traffic lanes. I do not mean to imply that improperly operated and obsolete motor vehicles should be allowed to pollute the atmosphere—they should not be—but neither should folklore be encouraged that will place the onus of metropolitan area atmospheric pollution on the automobile, without proof.[3]

With respect to the automobile, Los Angeles officials made "often reiterated statements that this source of air pollution is relatively minor as compared with other industrial and municipal sources."[4]

The APCD's attitude is interesting in light of a September 1947 paper prepared by an engineer with the Los Angeles County Office of Air Pollution Control (the predecessor of the APCD). The paper, a literature survey entitled "Exhaust Gases—Their Relation to Atmospheric Pollution," observed that abatement efforts commonly aimed at "pollutants which can be seen by the populace." While these were important, "the invisible gases of various types of combustion which cause the most intolerable and irritating effects should receive much more consideration than is ordinarily given them." In particular, a survey of the literature revealed the presence of a number of contaminants, including nitrogen oxides and

[a]See the discussion at p. 63.

hydrocarbon compounds, "in the exhaust gases of internal combustion engines. . . . Some of these," the paper concluded, "might quite reasonably be among the most potent irritants and lachrimators" (tear-producing substances).[5]

There was other early evidence pointing at least the finger of suspicion at the automobile. Beginning in 1943, health agencies of the city and county of Los Angeles had conducted pollution investigations and by 1947 had concluded that smoke and dust were of minor importance, that various invisible "gaseous oxides" probably entered "into smog-forming reactions in the atmosphere," and that eye irritation was caused by "partially oxidized hydrocarbons discharged from many combustion processes."[6] Yet when the Los Angeles APCD began operations in 1947, it ignored these findings.[7] Moreover, when the air pollution personnel of the Los Angeles City Health Department were transferred to the APCD in early 1948,[8] the APCD "did not see fit to continue the research on automobile exhausts which had been conducted with significant results by the city and county health departments. . . ."[9]

CHANGES IN THE STATUS QUO

The APCD attitude with respect to the insignificance of automobiles as a pollution source—an attitude implicit in research and abatement operations and explicit in statements by Los Angeles officials—drew sharp fire from the Assembly Committee on Air and Water Pollution, the same committee that had been so disenchanted with other aspects of APCD operations. The committee "was especially critical of the seeming reluctance on the part of local officials to give proper consideration" to automobiles as one of the "prime suspects" along with refineries and refuse burning.[10]

There were good reasons to regard the auto as one of the "prime suspects," some of them already noted: the fact that the "critical conditions at Los Angeles which were the basis for state legislation in 1947 remain[ed] substantially unchanged" two years later; the fact that early (but largely ignored) research had already raised suspicions about the automobile as a significant pollution source. And there was a third, very intriguing reason that "strengthened these suspicions."[11]

One day in November 1949 the University of California and Washington State football teams played a game in Berkeley, located in the San Francisco Bay Area. The "many thousands of persons attending . . . experienced intense eye irritation." On that day, and during the entire week before, the Bay Area had been enveloped by pollution conditions similar to those of Los Angeles. It was only on the day of the game, however, that eye irritation—the hallmark of Los Angeles smog—occurred, and then only at Berkeley. The committee, aware of the event and curious about its significance, began an "intensive investigation." It found that Bay Area weather conditions during the week in question "were practically identical with those which bring on severe attacks of smog in Los Angeles." No unusual industrial operations had occurred. Sources of pollution in the area were closely comparable with those in Los Angeles. The only "unusual occurrence . . . was the concentration of automobiles at the football game in Berkeley, accentuated by the idling of motors, starting and stopping, which occurs in such a traffic jam." Circumstances at the football game presented a plausible microcosm of Los Angeles, where "the crowding of existing freeways leading to the downtown area results in daily traffic jams as the flood of cars enters the city in the morning, with resulting accentuation of the exhaust problem by idling, stopping, and starting." Comparison of the two situations revealed a similarity that was "very striking." Los Angeles-type eye irritation occurred only on the day and at the place of the game, and it "could only be concluded that the cause of this particular eye irritation was in some way directly related to automobile exhaust."[12]

Evidence of the importance of motor vehicles in the pollution problem had reached the point where Los Angeles could no longer afford simply to disregard them as a source; rather, the APCD responded by redirecting and accelerating its research. The status quo had been altered, if ever so slightly. More attention would now be directed to the nature of Los Angeles smog. Quite quickly it would appear to confirm suspicions about the importance of automotive sources. It would be years, however, before confirmation was broadly accepted, and years after that before it was implemented.

6

DISCOVERING AND DOCUMENTING THE ROLE OF THE AUTOMOBILE

The decade of the 1950s was filled with a number of overlapping events that figure in our story. These, of course, played into each other, and also into discrete lines of significant (and more or less simultaneous) developments. This chapter describes one line, the effort to discover and document the precise contribution of motor vehicles to Southern California's pollution conditions. Two subsequent chapters discuss responses to an increasing pollution problem—in particular, the search for technological solutions and the moves for further federal and state involvement—and a third considers the (advisory) air quality and motor vehicle emission standards set in California at the close of the decade.

Efforts to pin down the role of motor vehicles in Los Angeles smog, the concern of the present chapter, involved years of discovery and disagreement. First, scientific research produced further and more substantial evidence of the importance of automobiles as a pollution source. Second, controversy concerning the merits and significance of this research arose within the scientific and political communities. Third, the automobile companies began to play an active part in the process of making pollution policy, contributing to debate and delay. These developments occurred in conjunction with one another, but they are best described separately. Table 5 gives an overview.

TABLE 5

Discovery and Debate in the 1950s

Year	Events
1950	Haagen-Smit suggests photochemical process causes smog; refineries, motor vehicles, and refuse burning are major sources. Public unhappy with implications of these findings. Stanford Research Institute (SRI), working for petroleum industry, implicitly agrees with Haagen-Smit on role of motor vehicles and refuse burning, but claims industrial sources less than one-half the problem.
1951-1953	SRI attacks fundamentals of Haagen-Smit's research and says smog still a mystery. Haagen-Smit responds.
1953	Severe episode in Los Angeles; public blames refineries and weak enforcement by Los Angeles APCD. Auto companies say role of motor vehicles in Los Angeles smog not established; form committee to investigate problem.
1954	SRI regards Haagen-Smit's research findings as not yet established; considers motor vehicles by far the major factor. Scientists hold widely different views on causes and cures of smog. Air Pollution Foundation (APF) begins efforts to resolve debate. Auto industry committee visits Los Angeles and concludes much more research necessary; industry begins research and development program.
1955	APF confirms fundamental soundness of Haagen-Smit's research and determines relative role of refineries. Auto companies enter into cross-licensing agreement to share research and development; claim no proof autos cause smog.
1956	APF concludes motor vehicles the principal contributor to smog. Auto companies agree autos a major contributor, but reserve judgment on whether they are the principal source.
1957	APF considers it conclusively proved that motor vehicles the major factor and turns to work on controls. Virtually all experts agree. Auto companies say Los Angeles problem is peculiar—a position they maintain throughout the decade.

INITIAL SCIENTIFIC RESEARCH

There was general agreement by 1949-1950 that the Los Angeles smog problem was more complicated than had formerly been believed. Industrial sources of smoke, fumes, and sulfur compounds were contributors but not the sole cause; the automobile seemed to play *some* part. This much was clear, but little else. The basic makeup of smog was still a "mystery."[1] There were a large number of contaminants (close to fifty) in the Los Angeles atmosphere, but no one of them alone produced the characteristic and bothersome eye irritation. Apparently some sort of reaction took place among them, but the nature of any reaction—indeed, the precise nature of the contaminants themselves—was unknown.[2]

Partly as a result of this uncertainty, partly because of criticisms about the absence of research, and partly because pollutant reductions from control efforts were not resulting in proportional alleviation of smog conditions,[3] systematic investigations into the nature of the problem were finally begun. To describe the initiation of these investigations requires a small step back to the late 1940s, when A. J. Haagen-Smit was leading a relatively quite life as professor of biochemistry at the California Institute of Technology in Pasadena, a community in the Los Angeles metropolitan area. Haagen-Smit was working with flavor compounds—analyzing, for example, the constituents that give pineapples their taste. He was also a member of the scientific committee of the Los Angeles Chamber of Commerce, a happenstance that would lead him to a career of almost heroic proportions in the history of automotive pollution control. (Haagen-Smit, Holland-born and delightfully direct in manner, would later be described "as a combination of Old Dutch Cleanser and St. George. . . .") The Chamber of Commerce had been "buried in complaints from farmers about unusual damage to crops, and asked the scientific committee to look into it." The job fell upon Haagen-Smit. "Using the same equipment as he had on his pineapples," he began filtering "thousands of cubic feet of Pasadena air. . . ."[4] Early findings indicated that Los Angeles smog was not caused by known industrial gases.

Word of Haagen-Smit's experiments reached the Los Angeles APCD, and in March 1949 its board took steps to hire him as an advisor.[5] The Assembly Committee on Air and Water Pollution also heard of the experiments at public hearings in early 1950 and saw them as consistent with its own view of the importance of the automobile. While Haagen-Smit's findings did "little more" than confirm those already reached by the health agencies by 1947, the committee nevertheless regarded recent events as "progress." The fact that APCD officials had engaged Haagen-Smit suggested they had "expanded their viewpoint considerably beyond their original restricted concern with smoke, sulfur dioxide and metallic fume." Perhaps now there would be more attention to "genuinely significant sources of pollution."[6] The APCD had hardly engaged in a complete turnabout. During the same public hearings at which the committee heard of Haagen-Smit's research, the APCD director "stated again that the district engineers considered the automobile and truck exhausts to be only a minor part of the pollution problem." Nevertheless, the APCD employed Haagen-Smit on a full-time basis a few months later, and provided funds and staff to support his research.[7]

Haagen-Smit's work progressed fast enough that by November 1950 he could present significant conclusions to the Assembly committee: laboratory experiments suggested that through a photochemical process, sunlight transformed hydrocarbons, in the presence of oxides of nitrogen, into eye-irritating, rubber-cracking, plant-damaging smog. The hydrocarbons came from many sources, but motor vehicle exhausts, refineries, and refuse burning appeared the most important. All of this was a "hypothesis," but it "fit the known facts very well," and Haagen-Smit regarded it as sufficient for "an intensive abatement program" against sources of hydrocarbons.[8] The smog problem seemed on the verge of resolution. Granted, there "remained considerable work to be done on actual verification that these reactions take place in the air." Moreover, a survey of sources of hydrocarbons was necessary, "since they are known to arise from many sources other than automobile exhausts."[9] But surely these were details. By 1951–1952, even the Los Angeles APCD seemed convinced. It called for investigation into control of automobile emissions, and for an "attack" on the automobile problem.[10]

CONTROVERSIES WITHIN THE SCIENTIFIC AND POLITICAL COMMUNITIES

New ideas are neither easily understood nor casually accepted, especially when large interests are at stake. Haagen-Smit's research proved no exception; it generated "undreamed of" problems and considerable controversy.[11] One observer saw no prospect for "immediate relief for the Southland's eye-smarting millions unless somebody can develop a means of changing its atmosphere, moving its mountains, or getting rid of its 3 million automotive vehicles."[12] Haagen-Smit himself was more optimistic. "When automobiles and the petroleum industry are controlled," he said, "smog won't bother us any more." He intimated that the problem would be cleared up in a "short time."[13] "What Haagen-Smit hadn't realized was the thunderhead of protest his findings would cause." Some members of the public denounced his findings "as a plot by 'the interests' to put the blame on the 'little man's' automobile and incinerator." Haagen-Smit's findings also implicated the petroleum industry and it, "particularly, tried to shoot them down. One large well-financed research institute used an industry grant to conclude that Haagen-Smit was all wet."[14]

The institute referred to was the Stanford Research Institute (SRI), an independent not-for-profit research corporation with no direct connections to Stanford University. SRI had been involved with Los Angeles smog, and with the petroleum industry, for some years. It was hired in June 1947 by the Commitee on Smoke and Fumes of the Western Oil and Gas Association to carry out pollution research thought by the industry to be beyond its capabilities.[a] SRI carried on its investigations "intensively" from the outset, describing them as "probably the most extensive research program so far undertaken anywhere to learn the sources and nature of air pollution." As early as 1949, SRI found "that no single material produces the eye irritation; it results from a number of contaminants working together." Research was underway "to determine the principal processes and activities that are producing contaminating materials."[15] By 1950 the Institute concluded that the "processes and activities" primarily responsible were "automobile operation, refuse disposal and even ordinary heating processes, with only some 40 percent of the problem chargeable to industrial wastes."[16]

SRI found itself in fundamental disagreement with Haagen-Smit. It claimed to have replicated his experiments, but the "smog" resulting from them "was no more irritating than fresh outside air."[17] Thus Haagen-Smit's hypothesis, based on his experiments, seemed totally unfounded. That hypothesis, simply put, was that "smog is formed in the atmosphere from waste gasoline vapors." This was a threatening conclusion to the petroleum industry, and SRI's rebuttal must have been a comfort. For SRI considered that pollutants from "thousands" of sources "somehow meet in the air and undergo some mystic sort of chemical reaction to produce smog." Matters, in short, were still a mystery, and "from so many potential sources, no single class of violators could be condemned."[18]

These words were soothing to the petroleum industry, but vexing to Haagen-Smit. Though there were points of agreement between him and SRI (for example, the presence of a chemical reaction), the merit of his research—and thus his standing as a scientist—had been called into question, and this he did not take lightly. He worked for six months "with the dogs at his heels" to demonstrate that SRI had not replicated his experimental method at all—it had ignored several important procedures crucial to his laboratory findings. "But resistance continued."[19]

[a]Members of the Smoke and Fumes Committee were employees of Union Oil Company of California, Standard Oil Company of California, Richfield Oil Corporation, Tidewater Associated Oil Company, and Independent Refiners Association of California, Inc. Stanford Research Institute, *The Smog Problem in Los Angeles County* (Menlo Park, Calif., Third Interim Report, 1950), p. 2. See also ibid. (Second Interim Report, 1949), p. 2.

The differences between Haagen-Smit and SRI were deeply held. The press called SRI's work an "attack" on Haagen-Smit's theory, and spoke of "frequently bitter scientific disputes" between the two.[20] The differences were also important to pollution control policy. Haagen-Smit's theory suggested that automobiles and petroleum refineries were about equally responsible for Los Angeles smog, but according to SRI, "automobile exhaust alone is responsible for nearly 95 percent of the reactive hydrocarbons, while backyard incinerators, oil refineries and service stations contribute substantially less."[21] In light of conclusions like these from SRI, it was virtually impossible to overlook the strong possibilities of bias behind them. The Institute itself seemed not at all shy about occasionally adopting the posture of a petroleum industry spokesman. One SRI representative told a reporter "that the refining industry is concerned lest final judgements be taken too soon about smog's causes which could result in the industry being forced to take measures which would not eliminate smog and would be extremely costly." The press was quick to note that the Institute had been "employed" by a committee "all of whose members represent gasoline refineries."[22] It reported complaints that SRI had "spent $1,250,000 given by the oil industry and still professes to know nothing about" smog.[23]

Despite skepticism on the part of the press, SRI's resistance was effective—at least in forestalling statewide (as opposed to local) controls. Its (and the petroleum industry's) basic position was that "much more work needs to be done before any authoritative answers can be given."[24] The aim seemed not so much to absolve refineries as to show that they had not been *proven* responsible. Such doubt-mongering served its purpose well. In 1950 a bill to control industrial sources through regional agencies to be created in every part of the state was held up for a final report "on research being conducted independently" by SRI. An advance report suggested industrial sources were responsible for only about 40 percent of the pollution problem, and it would be "unwise" to expect regional agencies to improve the air "when more than half the problem would necessarily be beyond their jurisdiction." The "community interest would be best served by withholding action of an enforcement nature pending further research and development."[25]

To SRI, Haagen-Smit's work was a "suggestion";[26] matters were hardly sufficiently settled to take action. This seemed the prevailing view in 1953–1954. Scientists "were told that 'great voids' exist in scientific knowledge about smog."[27] There was at best a "vague understanding of the problem. . . ."[28] At one 1954 meeting, "more than half a hundred scientists, engineers and representatives of county and industrial organizations held widely differing views on the approach to the smog problem." One of them summed up the situation: "the more we investigate the less we seem to know."[29] Whether this uncertainty was largely a product of SRI's research cannot be determined, but it is clear that the Institute was advocating its findings. Haagen-Smit expressed his own views of SRI's criticism of his work: "That didn't help me, especially when this is all built up with public relations," the Dutchman recalled. "I didn't have any public relations agent, but they had. And it was very unpleasant."[30]

The period 1953-1954 was one of crisis as well as (perhaps because of) confusion. The Los Angeles APCD had by this time put into effect a number of regulations, yet the air seemed to be getting no cleaner. "Because smog abatement was not noticeable as a result of these regulations, public disappointment grew into intensive criticism." An especially serious "five-day siege of smog" in the fall of 1953 aggravated matters further. The press asked, "how many more Black Mondays must we have before those to whom we have delegated enforcement powers and who say these powers are sufficient to do the necessary enforcement are going to give us a demonstration?"[31] Much of the heat fell on the Los Angeles APCD and especially its director, Gordon Larson. Charges of maladministration and failure "to estimate accurately the magnitude of the problem" led to a temporary demotion, followed by investigation and reinstatement.[32] But Larson never really recovered; he was replaced in late 1954.[33]

Apparently the importance of automotive emissions got lost in the commotion. A point of implicit agreement in the great scientific controversy between Haagen-Smit and SRI was that the automobile was a significant contributor to the pollution problem. Haagen-Smit's hypothesis suggested that vehicles and refineries were each responsible for about half of the important emissions. To be sure, SRI challenged the very fundamentals of the hypothesis, thus calling all of its implications into question. But the Institute's research also implicated the automobile, only more so. Within the explicit disagreement there was a common ground. Yet the concern during this period seemed quite uniformly to be with refineries. There are indications that the general public turned debate in this direction. Larson had given a public presentation on the role of incinerators and automobiles in Los Angeles's pollution problem. But this would implicate each citizen and was, according to a Pasadena columnist, as publicly acceptable "as a dose of castor oil to a small child."[34] The problem, citizens charged, was that the "refineries were . . . dumping pollutants into the atmosphere under the black of night," and the APCD "was . . . not enforcing the law."[35] Probably because of such pressures, the APCD—despite its call not long before for an "attack" on the automobile—continued to center its own efforts on stationary sources, especially refineries.[36] There was, to be sure, a new focus to these efforts. Consistent with Haagen-Smit's findings, they now aimed at control of hydrocarbon vapors from the refineries, as well as sulfur-containing gases.[37] SRI had lost the day in Los Angeles, and motor vehicles continued to be largely ignored.

Two events in late 1953 appear ultimately to have put California pollution policy back on the track of the automobile. By October 1953 the Los Angeles situation was serious enough that the mayor "succeeded in bringing Governor [Goodwin] Knight to the city where he appointed a special committee to review the effectiveness of research and controls."[38] The committee, headed by Arnold Beckman, reported its findings at the Governor's Air Pollution Control Conference, held in Los Angeles on December 5, 1953. It registered implicit agreement with Haagen-Smit's findings, noting that the Los Angeles APCD research program, carried on with "the aid of consultants," had "been outstandingly successful." This research "established . . . that hydrocarbons and their oxidation products form eye irritating and haze producing substances in sunlight and under the action of nitrogen oxides." The very research that SRI had called into question was to the committee "one of the most important advances in the study of air pollution."[39] On the automobile, the committee sided (perhaps diplomatically) with both Haagen-Smit and SRI. While observing that the "major industrial problems are related to the petroleum industry," vehicle exhausts contained the same substances as those emitted by refineries. Vehicle emissions were steadily increasing, and "the development and enforcement of effective steps to control pollution from automotive exhausts appear to be among the most serious problems of the air pollution control program." Based on this conclusion, the committee urged "unremitting efforts . . . to secure aid from industry through improvements in engine design, better fuels, and development of corrective devices for existing motor vehicles." And lest such technological controls should take too long or accomplish too little, the "long-needed rapid transit system for Los Angeles should be implemented."[40]

About a month before publication of the Beckman Committee's report, there occurred a second event that helped eventually in the development of a broad consensus about the importance of automotive emissions. On November 2, 1953, a group of businessmen, civic leaders, and government officials met and agreed to organize the Southern California Air Pollution Foundation; by the end of the month, the Foundation had been incorporated as a nonprofit research organization.[b] (A year later the organization was to drop "Southern California" and change its name to the Air

[b]The Foundation later explained that it "was created because the best attempts of then existing agencies to solve the problem had failed, years of trial-and-error rules and regulations had proven ineffective, and confusion had increased with

Pollution Foundation—"not alone because the name is shorter but, inversely, because the problem itself is expanding and air pollution is receiving nation-wide attention.")[41] The chairman of the organizational meeting suggested the direction the Foundation's work would take: "This should be a program carried out over a period of several years without fear or favor. If we can come up with the answer and somebody gets hurt, that's too bad. If no one gets hurt, so much the better."[42]

The Air Pollution Foundation (APF) had a dramatic sense of mission. "We are," its spokesmen said, "tackling in a pioneering phase one of the biggest social problems man has yet encountered—one which requires all the scientific knowledge and technology man has, plus new knowledge."[43] Undoubtedly sensitive to the implications of bias that had attached to SRI's work, APF was determined to maintain independence. While many financial contributions came from industrial sources of pollution (including the automobile companies and the refineries, either directly or through the Automobile Manufacturers Association and the Western Oil and Gas Association), there was "constant insistence on objective independent action by the staff." The commitment was to "impartial fact-finding," and there "were no sacred cows." APF enlisted the help of the press in "proclaiming our objectivity. . . ."[44]

Early in its work, APF "became aware" of the "strong differences of opinion among scientists working on the [pollution] problem." Accordingly, it undertook to specify precisely "the areas of agreement and disagreement with respect to Dr. Haagen-Smit's theory of photochemical smog and the relative contribution of various sources to atmospheric smog formers."[45] It accomplished the job before the end of 1954.[46] Common ground seemed sparse indeed (experts agreed only that "a variety of emissions" caused the pollution), and APF set out to fill in gaps in knowledge. It contracted with a research foundation to extend Haagen-Smit's studies and, it hoped, resolve the criticisms they had received. As a result, Haagen-Smit's findings were "corroborated." Next, APF embarked on an investigation of the role of the refineries—"an area of considerable controversy and emotional indignation." And it began studies of "two important sources of pollutants about which more information was deemed necessary," motor vehicle exhaust and rubbish incineration.[47]

By 1955 APF had concluded that the contribution of refineries to the Los Angeles pollution problem proved to be about midway between estimates by the APCD and the Western Oil and Gas Association. A year later

---

the increase of smog itself." Air Pollution Foundation, *1959 Annual Report* (Los Angeles, Calif., 1959), p. 7.

the importance of incinerators had been confirmed.[c] And by the end of 1956, APF had concluded "that motor vehicles were the principal contributors to smog in Los Angeles."[48] Consensus on this point grew quickly. By 1957 it was "almost unanimously accepted by experts in air pollution that exhausts from motor vehicles must be controlled before we can hope to solve our smog problem."[49] In APF's view there was "final conclusive proof that auto exhaust is the major factor in Los Angeles smog." APF's future program "would be directed almost completely to a study of motor vehicle exhaust and its control."[50]

THE AUTO COMPANIES AND POLLUTION POLICY

It was one thing for APF, scientists, and even the general public to agree that the automobile was the primary pollution source; it was quite another to convince the auto companies, or at least to move them to admit they were convinced.

The companies had been "interested" in the pollution problem at least as early as 1953, when the Automobile Manufacturers Association established a vehicle emissions program to be carried out by a technical committee.[51] The committee was "to study this problem very carefully and to make recommendations regarding industry action"; "to investigate thoroughly all available information on technical aspects of the air pollution problem as it relates to motor vehicles and, on the basis of this work, develop an industry program for dealing with the problem."[52] The first step was a visit to Los Angeles in early 1954 to review "a peculiar new form of air pollution" that the municipality was experiencing. "The point had been raised . . . that the automobile might have something to do with it."[53] Motivations behind the meeting are ambiguous. One industry representative said the committee was "a new group with no concrete plan of action—only a strong curiosity about some recent theories indicating that incomplete combustion in car and truck engines might be playing a prominent role in Los Angeles air pollution." Another individual close to the scene claimed the committee traveled to Los Angeles at the insistence of the Los Angeles APCD.[54]

In any event, "ten leading automotive engineers" arrived in Los Angeles from Detroit in early 1954.[55] They came, they said, "to look and listen and learn"; they had "no immediate solution up their sleeves to any part

[c]Backyard incinerators were banned in Los Angeles County effective October 1, 1957. See G. Hagevik, *Decision-Making in Air Pollution Control* (New York: Praeger Publishers, 1970), pp. 84-87.

of the smog problem." But the auto industry was "vitally interested in the problem" and had, "indirectly at least, been conducting research on it for many years."[56] Within a week, the committee summed up its impressions in a report committee members "had spent all night preparing": the problem required "further investigation" to "determine the exact contribution" of all sources, to study the effects of proposed controls, to develop better measurement instruments, and to study health effects. "Impressive air pollution research had been accomplished in Los Angeles," the committee noted, but "a tremendous amount of work remains to be done." The challenge, however, would be met. The industry would "do whatever we possibly can to assist in the solution of the automobile exhaust fumes' part in air pollution. We are dead serious. We mean business. We didn't come out here to fool around."[57]

Following the Los Angeles visit, the automobile industry initiated "a one-million-dollar-a-year industry research program."[58] Some observers found this sum insignificant as compared with expenditures on executive salaries or model changes,[d] but the industry was proud of it. "Though we've grown used to glib talk of Government billions, a million dollars a year—spent on one problem by one industry—is still a substantial outlay." The money was used for work on control devices, combustion studies, and the effect of vehicle maintenance on emissions.[59]

The auto industry did something else following the committee's visit to Los Angeles—it entered into a cooperative venture under which the automobile manufacturers would share information and research efforts. The venture began on an informal basis in 1954; the following year it was formalized in a cross-licensing agreement signed by the large and small companies alike.[60] The announced purpose of the agreement was entirely laudable. "Experience has taught that, when the best interest of the public clearly demands it, such cooperative effort in scientific research brings the greatest good to the greatest number of people."[61] Sharing of effort would mean faster progress. True, it would also mean that the "intensely competitive automobile manufacturers" were giving up any edge that would otherwise result from treating breakthroughs in the ordinary proprietary fashion, but this was an "emergency."[62] The industry "felt that this program was so much in the public interest that no company should

[d]One source noted that industry had spent $9 million on air pollution research as of 1963, compared with $9.5 million for one year's (1963) salaries of twenty-two top industry executives, or with $1 billion for model changes in 1963. S. Griswold, "Reflections and Projections on Controlling the Motor Vehicle" (Paper presented at Annual Meeting of the Air Pollution Control Association, Houston, Texas, 21-25 June 1964), p. 6.

seek to reap competitive advantage from the situation." The only concern was with "licking the smog. We do not intend to make money out of this thing."[63]

Perhaps not. Yet it seemed the companies did not intend to make progress either. Speaking ten years later, an official of the Los Angeles APCD noted how little the auto companies had achieved since 1954. He alleged that the cross-licensing agreement was nothing more than a means to ensure that no one auto manufacturer would develop and install expensive control technology. The competitive edge had been dulled, but not to serve the public interest.[64] By one report, it was this statement that led to antitrust conspiracy charges filed by the United States against the auto companies in January 1969.[65] The suit, alleging that the companies and their trade association had conspired to delay the development and use of automotive air pollution control devices, was settled—over storms of protest—by consent decree in late 1969. The government charges did, however, spawn a number of private lawsuits.[e]

We need not (and probably could not) unravel the merits of the antitrust suits against the auto companies, but one point seems quite clear. Whatever its motivations, the industry was remarkably slow in conceding what had become obvious to everyone else—the importance of automotive emissions. One of the participants in the auto companies' 1954 visit to Los Angeles later said, "until then, except for giving off smoke, which could be eliminated by good auto maintenance, we thought we were clean." But after the committee finished its interviewing, "almost everyone on it was convinced it was the automobile."[66] This may have been the committee's position, but it was hardly the industry's. Both before and well after the Los Angeles visit, the industry maintained an air pollution strategy described by one critic as "minimal feasible retreat."[67] The central features

[e]For the controversial history of the government suit, see J. Esposito, *Vanishing Air* (New York: Grossman Publishers, 1970), pp. 42-47; *Congressional Record* 117 (18 May 1971): H4063-4074. For general discussion of some of the legal and other issues involved, see G. Lamb, "Antitrust Factors Involved in Sanctioning Joint or Cooperative Activities Among Private Parties," in *Air and Water Pollution: Roles of Industrial Property, Innovation and Competition* (Washington, D.C., George Washington University, PTC Institute, 1970), p. 97; R. Lanzillotti and R. Blair, "Automobiles, Reciprocal Externalities, and Antitrust Policy" (Paper presented at Meeting of Southern Economic Association, Vancouver, B.C., Aug. 1971); idem, "Automobile Pollution, Externalities, and Public Policy," *Antitrust Bulletin* 13(1973):431. With respect to private litigation, see, for example, In re Multi-district Vehicle Air Pollution, M.D.L. No. 31, 481 F.2d 122 (9th Cir. 1973); In re Multi-district Vehicle Air Pollution, M.D.L. No. 31, *Commerce Clearing House 1973 Trade Cases,* Vol. 2, Para. 74,819 (C.D. Calif., Nov. 21, 1973).

of the strategy were to insist that the automobile's role be clearly proved, and to construe any proof as narrowly as possible. In early 1953, for example, Ford Motor Company wrote to a member of the Los Angeles County Board of Supervisors its view that exhaust vapors "are dissipated in the atmosphere quickly and do not present an air-pollution problem." Ford was doing no research on eliminating the vapors because "the need for a device which will more effectively reduce exhaust vapors has not been established."[68] A month later, General Motors wrote that while Los Angeles studies indicated exhaust gases "may be a contributing factor to the smog," other cities seemed not to be suffering pollution conditions like those in Southern California. Thus "some other factors," peculiar to Los Angeles, "may be contributing to this problem."[69]

By 1954 the auto industry "recognized the automobile as the largest single source of hydrocarbons" in the Los Angeles atmosphere. There was competent scientific evidence mounting "to show that automobile exhaust gases . . . are capable of forming ozone and may be considered a definite source of smog." But further "confirmatory work" was necessary "before there is substantial agreement by all concerned on the cause-and-effect relationship between these various pollutants and the formation of smog."[70] A year later the industry agreed "that auto exhaust was probably [Los Angeles's] major source of *air pollution,* [but] said the evidence did not prove that it produced *smog,* that is, eye irritation, plant damage, and so on."[71] And in 1956 the companies "accepted the conclusion that motor vehicle exhaust did in fact produce smog and agreed that it was a 'major' contributor to smog. They reserved judgment on whether it was the principal source of smog until additional data became available."[72]

Additional data did, of course, become available; we saw that by 1957 APF and virtually everyone else considered it conclusively proved that the automobile was the primary source of photochemical air pollution. Confronted with this hard reality, the auto industry abandoned attempts to refute the irrefutable. Rather, it reverted to its earlier attitude that the Los Angeles problem was peculiar.[73] It was still holding this position as late as 1960. In that year a representative of the Automobile Manufacturers Association testified before a congressional subcommittee: "The popular term 'Los Angeles smog' . . . has a sound origin, because photochemical smog is . . . not likely to occur anywhere else on earth with the frequency and intensity found in this area."[74] The statement was surprising. For several years it had been quite clear that photochemical air pollution was not confined to Los Angeles. Smog had appeared elsewhere in California, and around the United States as well.

7
RESPONSES TO AN INCREASING PROBLEM: THE SEARCH FOR TECHNOLOGICAL SOLUTIONS

THE GROWING POLLUTION PROBLEM

The automobile companies could attest in 1960 that photochemical smog was peculiar to the Los Angeles area, but the evidence was quite to the contrary. It is true that as late as 1958 observers presumably more neutral than the automobile companies noted, with respect to the Los Angeles-type smog, that "except for a few isolated instances which are somewhat doubtful, no cities in other parts of the country have been subjected to this type of pollution." The same report pointed out, however, the "real possibility" that the problem could appear elsewhere in the nation."[1]

Inquiry suggests that it already had. A 1953 account of studies by a California scientist reported that New York City, on days of low winds and inversions, experienced "the same type of attack" as the Los Angeles area.[2] A year later, an SRI study on *The Smog Problem in Los Angeles County* found that "other large cities are beginning to have similar problems. Such widely separated cities as San Francisco and New York have experienced smog-produced eye-irritation, low visibility, and crop damage, although the frequency of such attacks has been very low."[3] In 1957 "Los Angeles-type smog" appeared, apparently for the first time, in Philadelphia.[4] More generally, by 1958 data showed that air in American cities was getting dirtier, at least as measured by suspended particulate matter. Analysis of the data revealed that the major sources were the everyday activities of the public, not industrial operations.[5] The situation in California was much the same as that around the nation, except that in California the air pollution problem was growing at a more intense rate. As early as 1950, smog damage was noted in San Francisco, Riverside, and the San Fernando Valley. By 1953 damage ranged along the coast from San Diego in the south to Ventura in the north. In 1955 areas such as Bakersfield reported problems, and by 1958 Fresno had joined the list.[6] In the span of a few years, air pollution in California had changed from a highly concentrated local problem to one that occurred in most urban and some rural areas throughout the state.[7]

The major factor accounting for the spread of Los Angeles-type smog throughout the state and nation was undoubtedly growth—in population, industry, and automobile usage. It was understood by the mid-1950s that Los Angeles's "peculiar" inversion condition was not peculiar at all—it was common to much of the state and the rest of the country.[8] With sufficient emission output, many areas could suffer the same pollution problems as Los Angeles. The situation was especially severe in California because it experienced the most marked inversions of all the states and, perhaps most important, because of its "unprecedented growth of population and industry."[9] Thus, a 1960 report noted that "the urgency of the problem has been growing along with the spectacular growth of the population, the rapid expansion of industry, and the enormous increase of automobiles."[10] Statewide, the latter figure was particularly striking. In 1945 there had been only about three million motor vehicles in California; by 1950 the figure had reached almost five million; six years later it had grown to over seven million.[11] The situation in Los Angeles was in some respects worse than in the rest of the state. The rates of population and vehicle growth in Los Angeles about matched those of the state as a whole, but industry there was expanding "at the highest rate in the nation. Total new construction in Los Angeles in 1953 exceeded the 1953 combined construction totals of Chicago, Philadelphia, Detroit, Houston, Dallas, Denver, New Orleans, Baltimore, and Boston."[12] Tables 6 and 7 set forth the relevant growth figures for the United States, California, and Los Angeles County. They show that motor vehicle registrations were growing faster than the population; it should be added that gasoline consumption was increasing even more rapidly than motor vehicle registrations.[13]

TABLE 6

TOTAL POPULATION: UNITED STATES, CALIFORNIA, AND LOS ANGELES COUNTY

Year	Number			Percent gain since 1940		
	U.S.	Calif.	L.A. Co.	U.S.	Calif.	L.A. Co.
1940	131,669,275	6,907,387	2,785,643	—	—	—
1945	133,434,000	9,344,000	NA	1.34	35.28	NA
1950	151,868,000	10,643,000	4,151,687	15.34	54.08	49.04
1956	168,088,000	13,581,000	5,394,000	27.66	96.62	93.64

NA: Data not available

TABLE 7

MOTOR VEHICLE REGISTRATIONS 1940–1956
UNITED STATES, CALIFORNIA, AND LOS ANGELES COUNTY

Year	Number			Percent gain since 1940		
	U.S.	Calif.	L.A. Co.	U.S.	Calif.	L.A. Co.
1940	32,452,861	2,990,262	1,229,194	—	—	—
1945	31,035,420	3,118,840	1,271,000	-4.38	4.30	3.40
1950	49,161,691	4,976,296	2,008,000	51.49	66.42	63.36
1956	65,153,810	7,065,699	2,889,000	100.76	136.29	135.03

Sources for Tables 6 and 7: State of California, *California Statistical Abstract* (1970); Automobile Manufacturers Association, *1971 Automobile Facts and Figures* (1971); U.S. Bureau of Census, *Historical Statistics of the U.S., Colonial Times to 1957* (1960); Los Angeles Area Chamber of Commerce, *The Researcher* (1971).

Relocations during the war years placed on California a heavy burden of industrial and demographic growth, and the postwar boom meant rapid increases in population, industry, vehicles, and air pollution for the entire nation and especially for California. In Los Angeles, growth led to deteriorating air quality despite active efforts on the part of the APCD.[14] Much of the blame was laid on "the increase in completely uncontrolled sources, such as motor vehicles."[15] Emissions of oxides of nitrogen and of hydrocarbons and other organic gases from motor vehicles showed sharp increases in the 1950s.[16]

THE SEARCH FOR TECHNOLOGICAL SOLUTIONS

The 1950s were marked by efforts to develop some technological means to control the growing pollution problem. These efforts, and some of the comments surrounding them, revealed a distinct technological fixation: if the pollution problem were to be solved, it would be through technology. This was the faith of the times—a faith that was not to weaken until the end of the next decade, and one practiced in its early years with nothing less than religious zeal.

Bizarre Technological "Solutions"

A raft of bizarre technological solutions to pollution was offered during the mid-1950s. Though serious in purpose, they lacked the scientific foundation of more substantial research and development efforts undertaken during the same period. They are nevertheless worth discussing to illustrate the extremes the technological fixation could reach in the minds of reasonably well-informed, well-meaning men, and to suggest the costs that can attend what would appear "harmless" ideas.

Atmospheric inversions were known to play a prime part in photochemical pollution,[a] and some researchers considered it sensible to do something about them. Thus there were suggestions to eliminate inversions entirely, or at least to "punch a hole" in them to permit escape of pollutants, in both cases by thermal means.[17] But these ideas were hardly practical: in each case the energy required would be at least equivalent to that produced by burning (with 100 percent efficiency) all the crude oil processed by all the refineries in the Los Angeles basin in twelve days (and of course the task would be an ongoing one)! Related proposals suggested blowing the inversion away vertically with ground-based fans from below or helicopters hovering above. These too were shown to be impractical, requiring ten twelve-ton helicopters per square kilometer of air, or as

[a]Contrary to this common understanding, a few saw inversions as playing the part of solution, not cause. "Two members of a private research firm theorized . . . that a gaseous brown cloud, which originates at sea, is a major contributor to the Los Angeles smog problem." The men, both trained scientists, presented their discovery as the "missing link" among earlier smog findings. They reasoned that the cloud, which could reach an area of 1,000 square miles, hovered over Southern California except when dispersed by wind and rain. It descended from the upper atmosphere and was carried to industrial and populated areas. The only thing *protecting* these areas was the temperature inversion! "[The] temperature inversion, long considered the lid which holds down air pollutants, actually protects Southern California from the brown mystery cloud. When the inversion is high, . . . little of the cloud reaches the earth. When the inversion ceiling is low, the cloud descends closer to the earth and air turbulence mixes the brown gas with the surface atmosphere." "Brown Cloud Claimed Cause of Smog Blanket," *Riverside* [*Calif.*] *Daily Press,* 1954. A year later, the "mystery cloud" theory had "not been confirmed." California Department of Public Health, *Clean Air for California* (Initial Report, 1955), p. 11.

many fans on the ground. Needless to say, there would be considerable expense, danger, and noise.[18] Removing inversions, then, seemed out of the question. But since sunlight was another important factor in photochemical smog, perhaps something could be done about it. Along these lines, there was at least one proposal for a "smoke umbrella"—"a gigantic parasol of white smoke laid by airplanes high over a city. . . ."[19] The idea, of course, was that the clouds would eliminate sunlight and thus smog. Whether theoretically correct or not,[b] the idea was apparently never pursued—perhaps because sunny Southern California would then be cloudy every day.[20]

A direct attack was still possible—remove the smog itself. Again there were several sorts of proposals. The least bizarre involved seeding clouds to make rain to cleanse the air. The suggestion did not appear promising. Most meteorologists considered that rain did not cleanse the air; accompanying winds did the job. In any event, the clouds seldom contained enough moisture to produce rain. A similar approach called for seeding the air with some sort of agent that would "neutralize" the smog. But a neutralizer was unknown. "Every substance that has been proposed to date is just as objectionable as the material we're trying to dispose of. In fact, most of the substances that have been suggested are those that we are trying to get rid of in the atmosphere."[21] Other proposals actually to remove the smog all involved, in some manner or other, pushing it away—with huge fans, for example. Energy requirements again posed an insuperable obstacle. To do the job in Los Angeles "would require at least 12 Hoover Dams, more than twice the electricity produced in California, or at least one-sixth of all the electricity produced in the United States."[22] Another estimate claimed the task "would require the total power produced by Boulder Dam from the time it was built in 1937, until the present—to clear our Basin of smog for just one single day."[23] Perhaps, however, removal would be feasible if tunnels were first built in the mountains around the Los Angeles basin, and the smog then blown through these. There were proposals for such measures, but study revealed that as many as fifty tunnels would likely produce no noticeable effect; even to begin doing the job right would require, by one estimate, "14,000 ventilation tunnels, each approximately 40 feet in diameter."[24]

Still other ideas came forth, but they were similar in approach and

[b]Prof. Haagen-Smit is reported to have said in 1950 that if Los Angeles were hidden continuously by heavy clouds, "we never should have heard of smog." W. Barton, "Puzzle of Smog Production Solved by Caltech Scientist," *Los Angeles Times,* 20 Nov. 1950.

substance to those discussed above.[25] All these bizarre proposals seemed harmless enough—objects of ridicule but little else. Governor Knight of California had them in mind when he referred to inventors "with clanking tools, crackpots, retired Navy chief petty officers from submarines, engineers who have been fired from oil companies, a whole army of Rube Goldbergs."[26] Some people, however, had to take the proposals more seriously. No matter how obviously ridiculous they might be, they garnered publicity, and APCD officials in particular were pressured to explain why such bright ideas were ignored. Explanation took time and money, and when the APCD suggested that proponents should demonstrate the feasibility of their ideas, the innovators in turn charged that the "vested interests" had "negative attitudes." Thus, costs in resources and ill will attached to the proposals.[27]

Research and Development on Auto Pollution Control Devices

Bizarre suggestions for technological "solutions" to air pollution reflected the implicit belief that in direct applications of technology lay the key to relief. More explicit evidence for this state of mind is found in the many statements, by persons directly involved in the problem, that pollution was a technical or engineering problem, not a social or legal one. Perhaps the attitude developed from the way pollution control objectives had been perceived at the outset. The goal was to "minimize [pollution] within the limits of scientific knowledge . . . ," to control "to the limits of present-day engineering knowledge."[28] It followed that the "solution of California's air pollution problem will depend on the successful application of engineering principles."[29] Thus, Governor Knight could say in 1954, "smog is a scientific and engineering problem and not a political or legal one."[30] Or a member of the Los Angeles County Board of Supervisors could write to an auto company in 1953: "I realize that this problem has to be solved at the engineering level by the automobile industry, rather than by legislation."[31]

Owing to these points of view, there developed a marked "emphasis on securing 'top flight scientists and engineers from government, industrial and educational institutions.' "[32] The consensus appeared to be that an effective air pollution control program especially needed "scientists, chemists, and engineers . . . for research and development of control techniques."[33] To be sure, an occasional voice called for a more open approach to the problem—one that contemplated some useful engineering of human conduct as well as of machines.[c] There was some talk, but little action, to encourage car pools, create a rapid transit system, and otherwise reduce the use[d] of motor vehicles.[34] More strikingly, there was late in the decade a prescient proposal for a "smog tax" on motor vehicle pollution to encourage drivers "to take any action to reduce pollutant emission." The authors of the proposal noted that an individual could reduce his tax bill by a number of measures—tune-ups, purchase of a cleaner car, driving less, or installing control devices. They pointed out the "important conclusion . . . that effective smog-control need not necessarily await the perfection of special anti-smog devices." Their proposal, compared with those for bizarre technological solutions, was thorough, well reasoned, and based on data. It discussed administrative considerations and made comparisons between the smog tax and other possible approaches.[35] But, like the few other suggestions for essentially nontechnological responses to the pollution problem, it was virtually ignored.[e] The faith of the times

[c]For example, a 1957 report noted: "It is not likely that an easy, quick solution to the smog problem is forthcoming. While attempts to control auto exhausts must be continued and even accelerated, it may be time to consider some fundamental measures to correct the encroachment of smog in our urban areas. . . . In a word: Social aspects broader than air pollution control, even new patterns of use and development of metropolitan centers, are involved in the search for clean air for California." California Department of Public Health, *A Progress Report of California's Fight Against Air Pollution* (Third Report, 1957), p. 32.

[d]There was also the recurring suggestion to cut down on auto emissions not by reducing the use of automobiles as such but by reducing the amount of pollution produced by a given amount of driving. This was to be accomplished by building *more* freeways! Dr. John Middleton, a California pollution expert later associated with state and federal pollution control agencies, pointed out in 1954 "that an automobile traveling at constant speed emitted only a slight amount of pollution. . . . Freeways, Middleton pointed out, reduce smog by permitting autos to proceed at constant speeds." Thus, "more freeways" were a way "that a community can reduce air pollution." "Zoning and Freeways Urged as Smog Cure," *San Diego Union,* 4 March 1954. See also, "Panel Defines Smog Problem in Terms of Its Total Effect on Riverside County," *Riverside* [*Calif.*] *Daily Press,* 26 Oct. 1954; A. Haagen-Smit, "The Control of Air Pollution in Los Angeles," *Engineering and Science,* Dec. 1954, pp. 11, 15 ("The construction of freeways is . . . an excellent means towards the reduction of exhaust fumes"). These suggestions overlooked the fact that more freeways would likely result in large increases in the total vehicle-miles driven.

[e]Two related proposals that received similar treatment suggested a "penalty tax" on uncontrolled automobiles, and a tax on driving to serve as an incentive for car pools. See M. Landsberg, "Measure Urged to Force Car Exhaust Curbs," *Riverside* [*Calif.*] *Daily Press,* 10 Oct. 1959; "Professor Points Out Smog Blame," *Riverside* [*Calif.*] *Daily Press,* 21 Nov. 1960. The rough conceptual opposite of the various

was in mechanical, not human, engineering. The search for antismog devices was the preoccupying concern.

That concern had formed at least as early as 1954, about the time when most were beginning to agree that automotive emissions played *some* important part in photochemical air pollution. In August of that year, APF arranged a conference dealing with, among other things, "automotive engineering design and exhaust control devices."[36] A year later, however, it regretted to report that "reduction of the hydrocarbon emissions from the exhaust of internal combustion engines awaits an effective method," though a few approaches showed some promise. Moreover, it was difficult to envision an "effective reduction device . . . for oxides of nitrogen."[37] To be sure, there was optimism in some quarters. People "heard" in 1955 that the auto industry "hope[d] to have a successful device ready for Los Angeles authorities to test and approve or disapprove by early 1957," and perhaps "this could be advanced one year to 1956."[38] But the hopes were not realized. By the end of 1956 there appeared to be no way available to control motor vehicle exhaust.[39] A year later there had at least been accomplished an acceleration in research activity—by research institutes like APF, by industry, and by the Los Angeles APCD.[40] But no solution was in sight, and "no definite time can be set when . . . a device can be expected. A minimum of several years appears to be the best that can be hoped for."[41] There was "little probability of major alleviation within the immediate future." The precise role of engine and fuel variables and specific exhaust components in smog formation, as well as "development of scientific principles upon which effective exhaust control devices may be based," had "thus far defied solution."[42] By one measure, 1958 appeared to hold only more frustration. There was "no doubt about the magnitude of the automobile as a primary source of air pollution." Indeed, the problem was increasing in complexity as time passed. It was clear by now, for example, that "in addition to what comes from the exhaust line, an automobile loses substantial quantities of gasoline vapor from fuel tank and carburetor vents, especially during hot weather." Studies had "raised more questions about possible control techniques than have been answered."[43]

By another measure, however, the situation was much brighter—at least in the eyes of some. While the problem had not yet been solved, it might

---

sorts of "smog taxes"—a subsidy (in the form of rapid amortization for tax purposes) to encourage use of pollution control devices—was instituted for individual and corporate taxpayers in California as early as 1955. California Revenue and Taxation Code §§ 17225, 24372 (1970).

soon be. In its 1958 report, APF announced that "a breakthrough can at last be envisioned. . . . At the rate we are going, two more years can complete the basic research needed for a final solution." APF expected "to have all the needed facts about auto exhaust fully in hand by the end of 1960."[44] The expectation was substantially realized. While problems remained in abundance, the years 1958–1959 saw significant advances in research and development on afterburners, catalytic converters, improved design and maintenance, and fuel composition. As the decade drew to a close, APF reported that "completion of the scientific work necessary for an economic end to our smog problem is now within one year's reach." And by the end of 1960, "many companies were known to have reached a point in design and development where only a knowledge of final criteria and standard test procedure precluded submission of devices for test."[45]

Factors Affecting Research and Development

A number of factors affected the research and development efforts that, by 1960, held promise of an effective auto pollution control device for the near future. Included among these were activities of the Los Angeles County APCD. In 1953 Kenneth Hahn, a member of the Los Angeles County Board of Supervisors, asked the county counsel for a legal opinion on the board's power to require installation of control devices on autos sold in the county. The response was that devices could be required "on every new motor vehicle sold for use in Los Angeles County if [a] satisfactory device is perfected and available on the market."[46] Hahn called the opinion to the attention of the auto companies.[47]

The difficulty was that in 1953 no control device existed, and this—in the opinion of the county counsel—left the board powerless: "Until such a device is perfected and on the market, any rule requiring the use of the device is arbitrary, capricious and void." A rule of mandatory installation could only be "adopted when the required device is available, but not until then."[48] This attitude stood in marked contrast to the counsel's earlier opinion, regarding control of stationary sources, that a "law is not invalid solely . . . because there are no appliances available to prevent the prohibited smoke or fumes." But as to this point, the counsel claimed that "were the situation sufficiently serious," the supervisors "would probably be sustained (in court) in prohibiting the use of motor vehicles entirely, at least in the downtown area, until such time as a control device is available." He added, however, that "obviously the situation is not critical enough now to justify such drastic action."[49]

In short, the board could require a device if one existed, but one did not; it could prohibit auto use if the pollution problem warranted, but it did not. Matters were, for the time, at a standstill. About the best the authorities could do was devote renewed effort to research and development on control devices, and this the Los Angeles APCD did. A marked "period of growth, particularly in its research activities," began in 1954. By the end of 1955 it had undertaken a number of inquiries related to automotive pollution control, and work along these lines was to continue throughout the decade.[50] The APCD's objective was not to invent a device itself, but to encourage and help others to do so, and to test any promising technologies.[51] In 1956, for example, it "called a meeting . . . of leading firms and figures in the chemical and automotive accessory fields, with the aim of stimulating their interest in the development of catalytic or other types of exhaust controls."[52] Proposed approaches were tested in an Automotive Combustion Laboratory set up by the APCD in 1955.[53]

As Los Angeles County's interest in automotive pollution control devices advanced, so too did its interest in securing legislation to require such devices. Hahn had said in 1953 that the auto pollution problem was not one to be solved by legislation. By 1956 he had changed his mind and was considering introducing an ordinance to require county-approved devices on all new vehicles sold in Los Angeles.[54] Hahn knew that such a requirement could not be "legally" enacted "until the District can demonstrate that workable control devices are, in fact, available to the public."[55] Apparently, however, he grew impatient with such technicalities. In 1958 he threatened "establishing a deadline date by which the automobile industry must come up with a satisfactory device in order to sell new cars" in Los Angeles.[56]

Los Angeles County's activities had a positive influence on research and development, but their importance must not be overstated. The county engaged in no independent efforts to develop control technology, and threats of legislation were more sound than action. According to one report, "as late as October 1958 the Los Angeles Air Pollution Control Officer never *requested* legislative action in an appearance before a legislative committee, despite his emphasis on the importance of the vehicle pollution problem."[f] Notwithstanding county contributions, most credit

[f]A. Carlin and G. Kocher, *Environmental Problems: Their Causes, Cures, and Evolution, Using Southern California Smog as an Example* (Rand Corp., Santa Monica, Calif., Report Number R-640-CC/RC, 1971), p. 64, footnote 1 (emphasis added). There is some evidence, however, that as early as 1956 the Los Angeles County Board of Supervisors had asked Governor Knight for support in achieving

for early developments in auto pollution control technology must go elsewhere. The activities of a number of private and public research organizations resulted in much of the advance. Most notable of these, perhaps, were APF and SRI, but many similar (although usually proprietary) institutions played substantial roles.[57] On the public side was the work of the University of California. Both the governor and the legislature requested that the university devote substantial attention to the pollution problem—the governor in particular suggesting giving "priority to testing and developing a 'practical and inexpensive' anti-smog motor vehicle exhaust device."[58] The pace of the university's research efforts was accelerated, and substantial contributions were and continue to be made.[59]

And there was the auto industry. Opinion on its efforts varies sharply,[60] but clearly the industry engaged in *some* research and development in the 1950s. The issue is how earnest and open its efforts were. The answer in each case is probably "not very." Perhaps even that puts it mildly, but there is some evidence on the side of the companies. By 1955 they were at work on "a device which may curtail the loss of hydrocarbons from exhaust as much as 30 to 50 percent." (In the view of the Los Angeles APCD, this was not enough; it wanted a 90 percent reduction in hydrocarbon emissions from automobiles.)[61] Industry work on various approaches to hydrocarbon control continued in 1956 and 1957, and there were even early indications that prototypes would be ready for testing by the end of 1956 and for production during the 1958 model year. While some prototype testing was indeed accomplished by 1957,[62] the results were not encouraging. In 1958 a representative of the Ford Motor Company wrote to Hahn that only a 25 percent reduction in hydrocarbons could be expected, and the APCD apparently regarded this as too little to justify the costs of a mandatory control program. Ford agreed.[63]

Industry efforts continued, spurred on perhaps by the threats from Los Angeles County. By early 1959, the industry apparently concluded that it had something concrete to report. It arranged meetings between a committee of industry representatives on the one hand and the California

---

state regulations requiring that buses and trucks be equipped with control devices. The request was denied. T. Roberts, "Motor Vehicular Air Pollution Control in California: A Case Study in Political Unresponsiveness" (Honors thesis, Harvard College, 1969), pp. 27-28. Moreover, the county counsel claimed that in "1954 we urged the Congress to adopt legislation in the field of motor vehicle pollution control and argued that the federal government should compel automobile manufacturers to install exhaust control devices as part of the necessary equipment on new automobiles." H. Kennedy and M. Weekes, "Control of Automobile Emissions—California Experience and the Federal Legislation," *Law and Contemporary Problems* 33(1968): 297, 309.

legislature and other interested officials on the other "to explain the industry's progress in five years of smog research." The companies recommended automobile inspections, tune-ups, "and eventual use of a recently-developed smog control muffler."[64] The "muffler" was probably a catalytic afterburner, which had been troubling the industry for some time.[65] Apparently it continued to do so; industry efforts soon shifted to positive crankcase ventilation techniques (blow-by devices) to control hydrocarbon emissions. These "had been employed for over 30 years for other purposes on some vehicles."[66]

8

RESPONSES TO AN INCREASING PROBLEM: MOVES FOR FEDERAL AND STATE INVOLVEMENT

The public and private research and development efforts described in the last chapter were pressed forward by a growing problem, by government pleas and cajolery, and by heavy-handed threats. They were supported to some degree with funds provided by government agencies (and a few private institutions, such as APF) and dispensed through contracts and grants.[1] The most significant pressure and support, however, came from moves for federal and state legislation that developed and grew in the 1950s. These moves, like the research and development efforts with which they coincided, were tightly related to the increasing pollution problem. Table 8 sketches federal and state developments through 1959; they are discussed in turn.

MOVES TO INVOLVE THE FEDERAL GOVERNMENT[2]

As sketched in chapter 2, federal involvement in the air pollution problem began early, but on a very limited, episodic basis. No significant activity

TABLE 8

MOVES FOR FEDERAL AND STATE LEGISLATION
1950-1959

Year	Federal government	California
1950	U.S. Technical Conference on Air Pollution recommends research role for federal government. Bills for federal research and support fail to pass (as do other such proposals in 1951-1952).	
1953	Further proposals for federal research and support; none passes.	Proposals for regional pollution control districts fail to pass.
1954	Bills proposing federal research and tax subsidies and loans for abatement equipment fail to pass. Interdepartmental committee urges federal program of research and technical assistance.	Governor's meeting on air pollution recommends that legislature appropriate funds for research. Governor makes emergency grants to University of California and Public Health Department to accelerate research. Proposals for uniform minimum state air quality standards and for state air pollution control board fail to pass.
1955	Legislation enacted providing for five-year program of research and training.	Public Health Department recommends more active state role in research; legislation enacted providing for research and assistance program in department.
1956	House subcommittee on traffic safety hears testimony on auto exhaust and air pollution.	Public Health Department report calls for more research.
1957	Advisory committee on air pollution endorses federal support role. Congressman Schenck proposes bill prohibiting use in interstate commerce of vehicles discharging amounts of unburned hydrocarbons dangerous to health.	Public Health Department report notes growing pollution problem and increasing importance of motor vehicles. Bill for state board to coordinate local controls fails to pass.
1958	Schenck bill dies in committee. National Advisory Health Council emphasizes need for further research. Surgeon general calls First National Conference on Air Pollution.	Los Angeles APCD calls for state legislation to control motor vehicle emissions.
1959	1955 legislation extended for four years. Proposal by Schenck for federal study of motor vehicle exhaust passes House. Secretary Flemming proposes legislation granting federal authority to hold public hearings and make recommendations on interstate air pollution; proposal fails after objections from the Public Health Service and Bureau of the Budget.	Los Angeles APCD secures introduction of state vehicle control legislation, but measure fails to pass. Legislation enacted requiring Public Health Department to set air quality and motor vehicle emission standards.

occurred until 1948, when a six-day siege of smog in Donora, Pennsylvania, resulted in twenty deaths, six thousand cases of illness, nationwide publicity (the *New York Times* devoted several front-page stories to the episode), and federal reaction. (The impact of Donora was no doubt heightened by a London episode following closely on its heels and reportedly responsible for close to eight hundred deaths.) The United States Public Health Service undertook an extensive study of the Donora episode and considered it the first conclusive evidence of the acute health damage that air pollution could inflict.[3]

The pace of activity on the federal level increased subsequent to the Donora episode, both in the executive and the legislative branches. Much of this may have been in response to the Public Health Service's 1949 report on Donora, which emphasized the need for study on the nature and effects of air pollution.[4] In December of that year, President Truman requested the Department of the Interior to organize an interdepartmental committee for the purpose of planning the first United States Technical Conference on Air Pollution. It was a portent of things to come that Truman ordained the conference reluctantly, believing air pollution to be largely a local problem. He did "not contemplate . . . programs which will commit the Federal Government to material expenditures from an already heavily burdened treasury, since the responsibilities for corrective action and the benefits are primarily local in character."[a] The year 1949

[a]Quoted in R. Dyck, "Evolution of Federal Air Pollution Control Policy" (Ph.D. diss., University of Pittsburgh, 1971), p. 22. See generally ibid., pp. 21-22, 183. The conference was held in 1950. See *Air Pollution: Proceedings of the U.S. Technical Conference on Air Pollution,* ed. L. McCabe (New York: McGraw Hill Book Co., Inc., 1952). Dyck, "Evolution of Federal Policy," p. 183, reports: "Although the conference [results were] subsequently published, its impact on federal policy-making was not immediate, perhaps a reflection of its sole emphasis upon technical matters." Nevertheless, implicit in the conference "was a recognition that air pollution was a problem of many facets that demanded the expertise of a number of Governmental agencies and professional areas." This recognition was typical of a new awareness "that air pollution was a complex phenomenon that encompassed exhausts other than smoke." R. Ripley, "Congress and Clean Air: The Issue of Enforcement, 1963," in *Congress and Urban Problems,* ed. F. Cleaveland et al. (Washington, D.C.: The Brookings Institution, 1969), pp. 224, 228. Other signs of the awareness can be found in those instances, between 1949-1952, in which technical and manufacturers' associations (e.g., the American Society for Testing Materials, the Manufacturing Chemists' Association) established air pollution committees, or changed the names of old organizations like the "Smoke Prevention Association of America" to the "Air Pollution and Smoke Prevention Association of America," then to the "Air Pollution Control Association." L. McCabe, "Technical Aspects—the 1950 Assessment," in *Proceedings of National Conference on Air Pollution* (Washington, D.C., 18-20 Nov. 1958), pp. 25, 28.

also saw the Public Health Service become involved in another United States-Canadian air pollution conflict (recall that there had been one in the late 1920s), while in the legislative branch, four bills were introduced. These ranged from a resolution calling for a "national Air Pollution Abatement Week" to a measure authorizing the study of airport smog; the former made reference to the Donora episode. Two other bills would have provided for Public Health Service study of the health effects of air pollution. None of the measures got out of committee, but the proposals for federal study of health effects were prophetic.[5]

The 1950 United States Technical Conference on Air Pollution, called by Truman and attended by over "750 persons from all levels of government and industry," yielded a consensus "that the federal government help identify air pollution problems and develop the technology to combat them."[6] The emphasis here was on "identify . . . and develop." As a participant in the conference later said, one of the "trends . . . in legal opinion by 1950 . . . was the feeling that the Federal Government should not enter into policing the State problems of pollution of the air. In contrast there was the feeling that the Federal agencies, independently or in collaboration with the State and local bodies, should undertake investigations and research on problems of air pollution." The participant was aware that air pollution did "have certain aspects that transcend the purely local interest"—the earlier disputes with Canada, for example, had demonstrated this.[7] But these were not seen as sufficient cause for any intervention at the federal level beyond support of research and development. Indeed, it was even intimated that federal policing action would confront constitutional problems, although they were not considered to pose "an insurmountable barrier," at least not "if the pollution of air increases as a menace to the public welfare, and if the States are unable to or fail to control it themselves."[8]

For the time being, however, opinion weighed against a heavy federal role, and the legislation introduced in 1950 and for some time thereafter reflected this. Between 1950 and 1954, a number of measures calling for increased federal support of research on air pollution were introduced in Congress. There were at least three such measures in 1950, none of which found its way out of committee. A similar cluster of measures introduced the next year suffered the same fate, despite repeated incantations of Donora. In 1952 a London smog was blamed for thousands of deaths, and congressional interest in air pollution revived. A resolution calling for "intensified research" by federal agencies passed the House but was killed in the Senate on the ground that no one knew what it would cost.[9] But this instance aside, events in London appear to have had little impact on American policy.[b] The episode occurred at the dawn of 1953 and received considerable coverage in the United States, yet that year witnessed the introduction of only two measures. Again they were proposals for federal research, and again they failed to pass.

The years 1954-1955 were the first of real significance with regard to development of a federal interest in air pollution. Perhaps this was in part due to the fact that 1953 had closed with yet another severe pollution incident—a late-November episode in New York City.[c] An additional factor may have been increased efforts by state and local governments to obtain federal support for their own efforts. The Los Angeles APCD in particular had begun a campaign along these lines. "Smog," it argued, "is a problem of . . . national import. There is no good reason why the taxpayers of Los Angeles County should bear the entire cost of research efforts which are of concern, and will be of benefit, to the entire . . . Nation."[10] Once again, bills were introduced calling for intensified federal research efforts. In addition, there were several proposing amendment of the income tax laws to provide for accelerated amortization of pollution control equipment. (Only one such measure had been introduced prior to 1954—in 1950.)[11] Most important, however, were the efforts of Senators Thomas Kuchel (California) and Homer Capehart (Indiana) to combine research support and accelerated amortization, as well as special federal loan provisions for abatement equipment, in one grand amendment to the housing bill of that year.[12] All of the measures failed; as to the latter in particular, this provoked bitter response from the Los Angeles press. Failure of the special loan provisions was "not too great a loss as far as Southern California is concerned." It was designed for small industries; in the Southland the problem was "big industry and our motor vehicles. . . . What hurt was the elimination of a provision allotting $5,000,000

[b]Dyck, "Evolution of Federal Policy," pp. 27-28. It has been suggested, however, that the episode led directly to a British study committee (the "Beaver Committee") and thereafter to the British Clean Air Act of 1956. Ibid., pp. 27-29; H. Beaver, "The Growth of Public Opinion," in *Problems and Control of Air Pollution,* ed. F. Mollette (New York: Reinhold Publishing Co., 1955), pp. 1-11; W. Wise, *Killer Smog* (New York: Ballantine Books, 1968).

One other measure introduced in Congress in 1952 called for a congressional select committee to investigate the causes, effects, and control of air pollution. Its quite unconventional premise was that federal control of air pollution was required in the absence of local control and in the case of interstate pollution. Federal control was offered as the only "sure solution" to another Donora. Dyck, "Evolution of Federal Policy," p. 26.

[c]The episode received extensive media coverage. Nevertheless, there is some opinion that it had little impact on public policy developments. See Dyck, "Evolution of Federal Policy," p. 28.

for air pollution research. Research on smog is something that is needed, needed now and needed badly."[13]

Kuchel and Capehart were not about to give up. When it became clear that their measure would not pass in 1954, they wrote President Eisenhower suggesting an interdepartmental committee to study the question of federal activities in the air pollution field. The White House found the proposal an "excellent one." The committee was established and in fall 1954 issued a report urging legislation to authorize a broad federal program of research and technical assistance. (The report helped shape subsequent Public Health Service positions favoring research and grants-in-aid and opposing federal enforcement.) Eisenhower responded early in 1955 with two messages recommending increased attention to pollution research, but proposed no administration bill. Kuchel did, however, submit a measure with administration support. His bill provided for a federal program of research, training, and demonstrations, backed up by an authorization of $3 million annually for five years. The sum was later increased to $5 million and the bill passed, with only minor controversy, in this form.[d] It was signed by the president in mid-July 1955. A number of other measures introduced the same year died in committee.[14]

The 1955 Legislation

At least from the time of Truman's 1949 request for a national pollution conference, the consensus had been that *controlling* air pollution[e] was not federal business. The view was not unanimous, but it was very nearly so.

[d]In its original form the bill contained a provision for an advisory committee composed of representatives from the federal government and from science, industry, and the general public. Federal representatives opposed the provision, preferring instead that an advisory board be set up administratively, or informally as needed. Their opposition was successful and the provision was deleted. See Dyck, "Evolution of Federal Policy," pp. 184–185; J. Fromson, "A History of Federal Air Pollution Control," *Ohio State Law Journal* 30(1969):516, 520.

[e]*Not* pollution generally. In the same year that the federal government limited itself to research and support in the air pollution field, it established federal enforcement powers to control interstate water pollution. J. Sundquist, *Politics and Policy: The Eisenhower, Kennedy, and Johnson Years* (Washington, D.C.: The Brookings Institution, 1968), p. 332. But in the view of Congress, no such authority was necessary with regard to air pollution. It was told in 1955 that "instances of troublesome interstate air pollution are few in number." The administration view was that "unlike water pollution, air pollution .... is essentially a local problem." Quoted in J. Davies, *The Politics of Pollution* (New York: Pegasus, 1970), p. 51. Federal control of air pollution was to come later, however. The path from inaction

Discussion surrounding passage of the 1955 law—indeed, the law itself—reflected as much. Kuchel had assured the Senate that his proposal would not bring the federal government into control activities. Control would remain "where it ought to remain"—with state and local government.[15] Three years later he repeated that conviction. For the senator the strengths of the federal government were in its access to data, its "unmatched position to coordinate investigation and research," and its ability to mobilize experts in science and engineering.[16] A lone voice rose to suggest that federal control was at least necessary for interstate problems, but the argument got nowhere. The chairman of the Senate subcommittee hearing the Kuchel bill made it clear he had no sympathy for federal enforcement powers. It was simple to still his concern; the bill proposed no such authority. Indeed, one draftsman of the measure never even so much as entertained the notion that it might be a first step toward federal control.[17]

The particulars of the legislation enacted in 1955[18] were in every way consistent with the philosophy of federal restraint behind it, "the philosophy that the primary responsibility for the regulatory control of air pollution rests with the States and local governments, and that the Federal role should be a supporting one of research, technical assistance to public and private organizations, and training of technical personnel."[19] The legislation did "not propose any exercise of police power by the Federal Government and no provision in it invades the sovereignty of States, counties, or cities."[20] Rather, it was limited to authorizing the Secretary of Health, Education, and Welfare to conduct, recommend, and support research and investigations; to collect and disseminate information; and to encourage cooperative activities by state and local governments.

The Public Health Service took over administration of the 1955 air pollution legislation; obviously, the service had "no law enforcement authority and . . . [could not] engage in air pollution control activities."[21] Its primary efforts took two directions—an "engineering air pollution program" and a "medical air pollution program." The first involved "limited technical assistance and research" on the nature, sources, and control of air pollution; the second was concerned with evaluation of the health effects of air pollution. The general purposes of both programs

---

to support to control characterizes federal involvement in the water pollution field just as it does that in the air pollution field. It is simply that federal activity with respect to water quality began earlier; the steps of its development have been largely mimicked by the later activities dealing with air pollution. See Davies, *The Politics of Pollution,* pp. 37-49; Ripley, "Congress and Clean Air," p. 235, footnote 1.

were to improve knowledge, provide technical assistance, and "stimulate . . . increased attention and greater resources to the prevention and control of air pollution." At least initially, the bulk of the work under the programs was handled through contracts and grants—for research, training, demonstration projects, and grants-in-aid; the service did, however, conduct its own research program as well, and also provided technical assistance, conducted surveys, made loans of personnel, and ran training programs.[22]

California, and Los Angeles in particular, received a good deal of the new federal largesse. Federal personnel in engineering, physical chemistry, meteorology, medicine, instrument science, and statistics were assigned on loan to the state in 1955–1956; in turn, several of these were reassigned to the Los Angeles APCD. In addition, federal research grants were made to the University of California, the University of Southern California, and other agencies and individuals in the state. Despite this assistance, it was observed that early impact of the federal program "on the Los Angeles problem has been small. One suspects, in fact, that the existence of a federal program, and its early form, are strongly influenced by guidelines and observations made in California."[23]

1956–1959

The last four years of the 1950s saw little in the way of new federal initiatives, but there were significant developments in terms of what was to come. Particularly with respect to control (as opposed to study) of air pollution, and with respect to automotive emissions, moves for increased federal involvement begun in these years were to culminate in the next decade.

In some respects the pattern established in the first half of the 1950s continued into the second. The familiar story of pollution episodes and of proposals for supportive legislation (tax relief, federal information gathering) was replayed. In 1957 the surgeon general created, by administrative action, the advisory committee on air pollution initially proposed (but subsequently abandoned) in the air pollution legislation passed in 1955. The committee, consisting of representatives from industry, control agencies, and health research centers, endorsed the tradition of federal-support-but-not-control. Continuing the same theme, another advisory committee—the National Advisory Health Council of the Public Health Service—in 1958 emphasized the need for further research in environmental health. In partial response, the surgeon general scheduled the First

National Conference on Air Pollution, which was held the same year.[24] The strategy of the conference was largely to show that the pollution problem consisted of more than simply emissions of visible smoke (which had been largely abated by this time). The emphasis was on demonstrating the health effects of invisible, gaseous pollutants. The strategy succeeded; the conference has been credited with passage, in 1959, of a four-year extension of the 1955 legislation.[f]

This much of the story of the last half of the 1950s is little more than that of the first half retold. But the period saw more significant developments as well; two in particular deserve some mention.

In 1956 a new House subcommittee on traffic safety held extensive hearings on highway traffic accidents. In the course of testimony, one witness dealt briefly with air pollution caused by automotive exhaust. The testimony stimulated the interest of Congressman Paul Schenck, a Republican from Ohio, in the vehicle exhaust problem. The next year, Schenck introduced a bill prohibiting the use in interstate commerce of any motor vehicle discharging unburned hydrocarbons in an amount found by the surgeon general to be dangerous to human health. The measure was the first aimed at vehicular emissions, and one of the first to propose federal control. It was the subject of hearings in 1958, but died in committee. The Department of Health, Education, and Welfare had opposed the bill because suitable criteria for "dangerous" were not yet developed. When Schenck pushed the same measure a year later, HEW again voiced opposition. It repeated the need for more information, and objected to federal enforcement (citing the prevailing federal philosophy of local responsibility for pollution problems). Schenck's bill was watered down to provide merely for studies of vehicle exhaust by the surgeon general; it passed the House in this form in 1959, and was to become law in 1960.[25]

In the course of Congressman Schenck's unsuccessful efforts to achieve federal control of automotive emissions, Arthur Flemming—the new Secretary of HEW—exhibited an interest in some federal control of air pollution generally. Flemming's experience in the administration of the Water Pollution Control Act had convinced him by 1958 that the federal

[f]Dyck, "Evolution of Federal Policy," p. 188. The author of the extension measure, Senator Kuchel, employed the conference as a platform to promote the legislation. See T. Kuchel, "Public Interest Demands Clean Air," in *Proceedings of National Conference on Air Pollution* (Washington, D.C., 18-20 Nov. 1958), p. 15. Passage of the bill did not come without controversy. Kuchel wanted a permanent extension and an authorization of all necessary appropriations. The Senate would go only so far as a four-year extension at $5 million a year. The conference committee concurred and the bill became law in this form. Dyck, "Evolution of Federal Policy," pp. 38-40; Ripley, "Congress and Clean Air," pp. 233-234.

government needed enforcement powers to abate interstate air pollution. (The government had some power under the water quality legislation to abate interstate water pollution.) The Public Health Service (which, under Flemming's proposal, would have the new enforcement powers) objected, and Flemming retreated—a bit. In a December 1958 news conference, he suggested that the federal government should be empowered to hold hearings on interstate air pollution on its own initiative and make findings and recommendations. But the Public Health Service opposed even this limited role. It "construed the power to hold public hearings as a kind of enforcement power and it was 'conservative' in terms of seeking enforcement powers for itself." The service saw itself as a professional, apolitical research organization. It enjoyed good relations with state and local government and feared that enforcement powers would disrupt these. As the surgeon general put it, "the advantages of such independent action would be likely to be outweighed by resentment and opposition engendered by what might be construed as an encroachment on state responsibilities."[26] Moreover, in the service's view air pollution was not yet a serious interstate problem, and in any event the agency considered that scientific standards were not sufficiently established that they could serve as a basis for enforcement.

These were general reasons for the Public Health Service's stand, but there was a specific one as well. The service was anxious to have the 1955 legislation extended (as it subsequently was), and feared that a negative reaction by Congress to Flemming's proposal would jeopardize the chances of extension. Despite this opposition from within, the secretary ordered preparation of draft legislation that incorporated his views. But while he could overrule Public Health Service objections, he could not deal similarly with opposition from the Bureau of the Budget, which forced a caveat from Flemming during presentation of his proposal to the House. The bureau, Flemming was made to say, believed "the need for and desirability of" his legislation required "further study." Thereafter, Flemming tried repeatedly to achieve passage of the public hearing provision—without success.[27] The 1950s closed with but the most limited intervention at the federal level.

MOVES TO INVOLVE STATE GOVERNMENT

Moves in the 1950s to bring state government into more active confrontation with California's pollution problems followed the same lines as efforts at the federal level—one directed toward research and development activities, the other toward state control. The difference was that attempts to realize state control began somewhat earlier and accomplished slightly more.

Research and Development

It was clear at least as early as 1952 that too little was known about California's air pollution generally and its medical aspects in particular. As a result, there was no "clear-cut blueprint" to employ in measuring air pollution, gauging its health effects, directing research, or establishing reasonably specific standards to serve as the basis for control measures. One medical expert testified to a legislative committee, "the function of this committee is primarily to find out what effect this whole smog problem has on public health, and frankly we don't know."[28] There was, to be sure, *some* evidence of adverse health effects: authorities could say with confidence in 1952 "that sufficient exposure to the right kind of air contaminants can adversely influence health."[29] In the same year, one study "found a significant relation between the death rate from cardiorespiratory diseases and the intensity and duration of smog."[30] The problems related to what was *not* known, and it was resolution of these unknowns that was thought essential to reasonably effective standard setting.

Considerable public interest in the general subject of health effects had developed by now, and it tended to be heightened during times of critical air pollution episodes.[31] In spite of this public interest, and notwithstanding the paucity of information on the subject, there was at first no forceful move to involve state government in research activity. Indeed, the Beckman Committee appointed by Governor Knight to study and make recommendations regarding the air pollution problem found it "advisable," in late 1953, "to use all funds and manpower available to the district in an allout effort to get rid of smog and not to divert them into channels which, however interesting, will not assist in reducing air pollution." These "channels" apparently included medical research, the committee's view being that "smog abatement is the all-important goal; with the elimination of air pollution, associated health hazards automatically vanish."[g]

[g]A. Beckman et al., *Report of Special Committee on Air Pollution to Governor Goodwin Knight's Air Pollution Control Conference* (Los Angeles, Calif., 1953), p. 15. The report itself is somewhat ambiguous regarding whether the committee meant to suggest virtually ignoring health research. A contemporary account of the conference at which the report was delivered, however, appears to make the committee's views on the matter clear: "Money should not be diverted from the

This sanguine view, however, was soon to be changed by events. Crisis moved the state to act, just as it had the federal government. An especially severe, two-week smog episode in Los Angeles in early October 1954 "spurred renewed demands there for immediate action to halt the smog menace."[32] Governor Knight responded by calling a "hastily arranged meeting" with doctors, scientists, and government officials. By the time of the meeting, however, "the smog appeared to abate," and the governor was assured that there was no need to declare a state of emergency—action he had been considering. He was also assured that there was "no danger to health or human life at the time." The need for further study was nevertheless recognized: the meeting resulted in recommendations that the legislature be asked to appropriate funds "for further research work in the cause and effect of air pollution"; that the University of California's facilities be employed in the research; and that long-range (fifteen- to eighteen-year) studies be undertaken.[33]

Little more than a week later, the governor announced an "all-out smog war . . . , a four-step program to throw all the state's technical resources" into the battle. The university was granted emergency funds and asked to give priority to testing and developing a "practical and inexpensive" automotive pollution control device; other university smog control projects were to be stepped up; telegrams were sent to the heads of the major auto companies "emphasizing their responsibility . . . and urging their full cooperation" with the university; President Eisenhower and Interior Secretary Douglas McKay also were sent telegrams requesting that "all the experience of the Bureau of Mines smog and fumes division . . . be made available to the University." Governor Knight took one additional step that was to prove more significant for the near term—he made an emergency grant of $100,000 to the California Department of Public Health in order that the department might undertake "a broad attack on health problems arising from smog conditions."[34] In the view of one active participant in California's pollution story, it was this step, spurred by the coincidence of a severe episode in an election year, that led directly to the first assumption of responsibility by the state with regard to the pollution problem.[35]

There is some basis for this view. The October episode did stimulate official reaction, including the governor's commissioning of the Public

---

L.A. smog control district for medical research on the effects of smog because, when smog is eliminated, the health hazards will 'automatically vanish.'" B. Barger, "Drastic Refinery Controls Urged by Governor's Group," *Riverside* [*Calif.*] *Daily Press,* 7 Dec. 1953.

Health Department study. That study, in turn, produced a report that surveyed California's air pollution problem and recommended, among other things, that the state government become more actively concerned in the problem's study and resolution. The department was unwilling to go so far as to recommend actual state *control*; "local responsibility for regulation and control in most fields of interest" was the "well-established pattern," and there seemed no "sound reason for departing from this principle in air pollution." The department did recommend, however, legislation by which the state would recognize air pollution as a matter of state concern and designate state agencies to undertake integrated programs of surveillance, research, development of control methods, and local assistance.[36]

Control

There were others who wanted to go further. As early as 1953, a strong view had formed that the jurisdictional problems created by the constraints of local boundary lines called for some sort of state or regional control. In September of that year, for example, the Los Angeles City Council, "spurred by public outcry over . . . recent waves of smog, . . . asked for state and federal help in fighting the menace." The council noted that "we can't put a fence around Los Angeles and expect to keep other counties' smog out of here." Because "smog does not respect city boundaries," the council concluded that "the state should bear all or a major portion of the cost of cleaning up the atmosphere."[37] Proposals for regional (multicounty) pollution control districts made during 1953 reflected the same sort of frustration. At least one bill calling for regional control found its way into (but never out of) the legislature in 1953, and by the end of the year a California Assembly committee had been established to investigate regional control.[38]

By 1954 the move for state intervention appears to have accelerated. Now some quarters wanted direct control at the state level. The city of Riverside, in particular, asked the state "to establish uniform, minimum state air contamination standards," its reason being that "smog has no boundaries."[39] An unsuccessful resolution in the Assembly called for a state air pollution control board and nine regional "sub-boards."[40] Several bills were submitted to establish "elaborate schemes for state and regional control boards."[41] None of these measures succeeded, perhaps because the issues were posed in the most extreme terms. There seemed to be no mention of fruitful state-local cooperation, but rather only of co-optation.

The press saw the matter as one of "either-or"—"whether state or local authorities should handle smog problems"; "whether the state should take over air pollution control or enact drastic legislation which would force local officials to take stronger action."[42] This posture put powerful interests in a dilemma. Some sort of state authority seemed justified, but local autonomy ("home rule") also appeared to be at stake. The pull and haul of these competing considerations produced ambivalence and vacillation. For example, the League of California Cities was expected to support the request by the city of Riverside, mentioned above, for minimum state air standards. But the League refused, deciding "instead that a special committee be formed to determine whether the organization should call for state legislation on smog." The League's executive director was "somewhat uncertain as to whether Riverside actually wants to surrender its local autonomy concerning air pollution to a state agency."[43] Another example is found in the attitudes expressed by Governor Knight and his advisors. In September 1953 Knight announced the view that "smog is no longer merely a Los Angeles problem but is within the realm of state agencies. . . . I'm inclined to believe that much stronger action must be taken immediately." Less than a month later, Knight "expressed the belief that smog should be locally controlled," then two days later "said he might put the smog problem before the State Legislature. . . . There was no explanation given for the Governor's seeming change in mind." And Knight's mind may subsequently have changed once again. In December 1953 his advisors "went on record in opposition to state action against the smog problem, expressing belief that the problem is essentially local."[44]

The situation, in short, was one of marked confusion about the direction pollution policy ought to take. The net result was a compromise—legislation that reflected some assumption of responsibility by the state, but in a manner that left local authority largely intact. The legislation, passed in 1955, required the California Department of Public Health to maintain an "air sanitation" program consisting of research on the causes and effects of air pollution, monitoring, emergency controls, and assistance to local agencies.[45] The views of the Department of Public Health had prevailed: state government took a more active concern, but local government was left with "the prime responsibility to regulate and control air pollution."[h]

[h]California Department of Public Health, *Clean Air for California* (Initial Report, 1955), p. 49. One might conclude that those calling for more substantial intervention by the state—for example, through regional boards—also enjoyed some success in 1955. An additional piece of legislation enacted that year established the Bay Area Air Pollution Control District—a regional board encompassing nine counties in the San Francisco Bay Area, all of which suffered from a common

In order to implement its new responsibilities under the 1955 legislation, the Public Health Department established a Bureau of Air Sanitation with the primary task of conducting an active program of research (on the causes and health effects of air pollution), monitoring, coordination, and assistance.[46] By 1957 the program had resulted in, among other things, two reports on the air pollution situation in California. The first of these stressed the absence of fundamental information and the need for further research; the second, while repeating the same points, also mentioned the great complexity of the problem, its growth in terms both of intensity and geographical distribution, and the increasing importance of automotive emissions.[47] These observations were hardly unique; on the contrary, they were coming to be shared by most people involved in the state's pollution problem, and to many they suggested the need for a larger and more active state role.[48] Severe pollution episodes (or warnings of their coming) in 1955, 1956, and 1957[49] probably served to strengthen opinion in this regard. By the last date a measure had once again been introduced to give state government a more dominant position in air pollution control. The bill was hardly a drastic step—it called only for "a state board of smog control with a director to coordinate the work of local air pollution districts."[50] Nevertheless, it failed to pass.

But the move for significant state intervention was once again under way and gathering steam. By the fall of 1958 it was going full force, especially with regard to automotive emissions.[51] At an October legislative hearing, officials of the Los Angeles APCD testified on the need for vehicular emissions control. They had for some time been talking of such control on the local level. Now, however, they "first officially expressed

---

air pollution problem. (Six counties were in the initial district; three others could opt in by vote of their boards of supervisors.) Ch. 1797, § 1, [1955] Cal. Stats. Reg. Sess. 3317, adding to California Health and Safety Code §§ 24213, 24214, 24345-24372. But this was hardly state control in the sense it had been urged by those advocating a regional approach (under a general state board). Indeed, the enabling legislation declared that "this legislation is local." California Health and Safety Code § 24346.2 (1955). What the legislation accomplished was little more than to establish a board much like those under the 1947 legislation, but which could have control over nine counties as opposed to one. The state itself would exercise no active authority over the problem. For an excellent study of the Bay Area district, see Stanford Workshop on Air Pollution, *Air Pollution in the San Francisco Bay Area* (Stanford, Calif., 1970). See also California Assembly Committee on Air and Water Pollution, *Final Summary Report* (n.d., c. 1952), pp. 28-35; F. Fredrick and L. Lowry, "Legislative History and Analysis of the Bay Area Air Pollution Control District" (mimeo., n.d.); M. Walker, "Enforcement of Performance Requirements with Injunctive Procedure," *Arizona Law Review* 10(1968):81.

to the State of California the need for state legislation to abate vehicular contamination problems in Los Angeles County and throughout the state."[52] The APCD representatives outlined the essentials of the legislation they sought—statewide emission standards, testing and mandatory installation of devices, inspection and enforcement.

The APCD's growing inclination to favor state automotive controls over (or in addition to) local ones was strengthened when the California Department of Motor Vehicles, in January 1959, denied Los Angeles County's request to employ the department's vehicle registration system in the course of administering a contemplated requirement that vehicles in the county be equipped with approved devices. The commissioner of the department ruled that only a state control program could have access to the registration system. The response of the county was to devote more effort to the development of a statewide program; county officials drafted legislation introduced before the California Assembly and Senate in early 1959. Briefly, the bills would have required new and used motor vehicles to be equipped with approved control devices by specified dates (with provisions for suspension of the deadlines in the event no devices were approved); established a state vehicle exhaust laboratory (in the Department of Motor Vehicles) responsible for determining standards and certifying control devices; and provided for establishment of inspection stations (again under the auspices of the Department of Motor Vehicles) to aid in enforcement. Despite support from the Southern California Regional Air Pollution Coordinating Council and from the County Supervisors Association of California, the measures failed to pass. Opponents argued that statewide legislation was unnecessary because Los Angeles smog was a local phenomenon distinctly different from that "in any other city in the world"; that no device existed that would "guarantee" a reduction in photochemical air pollution; and that until there was such a device, manufacturers would be "reluctant to invest large sums to develop devices which might not be required by the state and which would then find no market."[53]

Yet 1959 did not close without some significant state intervention. Legislation providing for actual state control had to await 1960, but an important (perhaps a necessary) precursor was realized in 1959. For in the course of the movement for actual state control, there had also been introduced legislation that would require the California Department of Public Health to determine the air quality and motor vehicle emission standards necessary to protect health and avoid interference with visibility and damage to vegetation. Governor Edmund G. ("Pat") Brown supported the measures. While he "emphasized" that he was "all for continued home rule on smog control and enforcement because of wide variations in geography, weather and population," he nevertheless recognized that smog had "become a statewide program." In any event, the proposed air quality standards would not detract from local authority, but rather would simply "give local control officials and health officers a measuring stick for smog." As for the motor vehicle emission standards, to determine them hardly meant that they would also be applied. Brown "said that eventually it may be necessary to compel the use of anti-smog devices [on vehicles], but he laid down no deadline."[54] Thus the air quality and emission standards to be set under the proposed legislation would be more or less advisory: they would serve as guidelines for further action on the state or local level.[55]

The air quality-emission standards legislation, posed as a useful and not very threatening compromise between state and local interests (and perhaps as well between the auto companies and the growing numbers who sought to control vehicular emissions), found its way successfully through the legislature and became law in mid-1959. Its central provisions have already been suggested. The California Department of Public Health was directed to hold public hearings and publish state air quality standards by February 1, 1960. The standards were to reflect the relationships between the composition and intensity of air pollution on the one hand, and its effects (on human health, visibility, and vegetation) on the other; they could be revised from time to time. In addition, the department was to determine, by the same deadline and according to the same procedures as above, the maximum allowable vehicular emissions compatible with preserving public health and preventing damage to vegetation and interference with visibility.[56]

9

DEALING WITH UNCERTAINTY: SETTING POLLUTION STANDARDS

There is no need to dwell on the details of the legislation enacted in California in 1959 nor on the standards that finally resulted pursuant to it. It is important, however, to trace a few aspects of the scientific and technical background and approach that led to those standards.

Earlier discussion described the considerable uncertainty that existed in 1950–1955 regarding the health effects of air pollution. The situation in this respect appears to have improved very little by 1959, despite the Public Health Department's air pollution research program that had begun with California's 1955 legislation. The department itself, in 1955, spoke of "the paucity of information" on the subject. How much air pollution humans could tolerate was unknown, though indications were "serious enough that we ought to find out as soon as possible." The evidence that existed was rough at best and regarded accordingly by the department. Limited data on Los Angeles was available ("similar studies for the rest of the State have not been done because of the absence of adequate continuous measurements of air pollution"); it "did not reveal any measurable effect from smog." But this did "not necessarily mean that no such effect occurred. Failure to demonstrate it may mean only that the methods of measurement were too crude."[1] A year later, in 1956, it was still unknown whether smog was "definitely a hazard to public health." Practicing physicians felt that under any broad general definition of public health ("complete physical, mental, and social well-being") there was no question that air pollution was detrimental.[2] But, as was the case a year earlier, "statistically there is nothing that can be blamed on the smog at present," perhaps (again) because methods of measurement were "simply too crude."[3] By 1957 more extensive air measurement studies were "beginning to describe the quality of air in California on a far broader basis than has ever been done in the past." Further evidence of health effects had been gathered, but still it did "not meet the expectations of the scientific community as to what should be properly done in the field by the Department of Public Health." The effects of air pollution on human health remained an "unanswered question . . . , the area about which the least is known at present."[4] And this situation, at least as to chronic diseases, had changed little by the time the Public Health Department began the standard-setting process in 1959.[5]

The question of just how to respond to air pollution in the absence of hard evidence of its effects could be viewed as one of burden-of-proof, but in these terms the issue had never been satisfactorily resolved. At one extreme were those who insisted on positive proof before any control measures could be undertaken. An air pollution control official took this view in 1955 with respect to motor vehicular pollution: "Should the exhaust of vehicles be an item for control measures, and evidence is pointing in that direction, the contribution to general air pollution of a community, or to the street level nuisance effect, must be clearly demonstrated before any action can be taken."[6] This refrain of proof-before-action had been heard from the outset of modern attempts to control air pollution in California.[a] At its opposite extreme was the view of many medical experts that control should go forth despite poor information on pollution effects. This represented the official position of the Department of Public Health in 1955, when it said that "control should not await the demonstration of severe health effects. Pollution should be controlled whether or not a severe health effect has been demonstrated."[7] But not all public health experts agreed. One noted in 1956 that there were suspicions about health

[a]In 1948, for example, the oil companies complained that the newly created Los Angeles APCD was attempting to go too far in controlling sulfur compounds "without sufficient proof that the removal of the sulfur would cure the overall smog problem." H. Kennedy, *The History, Legal and Administrative Aspects of Air Pollution Control in Los Angeles County* (Los Angeles County Board of Supervisors, 1954), pp. 14–15.

effects, but "no definite proof." And he saw the requirement of just such "absolute or definite proof" as "a very reasonable thing in that, if the public health departments and divisions were to state, 'This is detrimental,' it then becomes something that the public can take as a fact. This would be bad in that it causes the public to worry about things which we do not know to be true."[8]

The issues of conclusive proof of health effects, and of what to do in the absence of such proof, had particular bearing on the matter of standard-setting. It was not simply that better information would lead to better standards. Rather it was that better information was necessary for *any* specific standards to be established. This at least had been the assumption of the past—usually implicit, but at times explicit. In part the assumption was premised on the belief that unless standards had a sound scientific base, they could not be legally enforced.[b] Perhaps in larger part, the premise was simply that it would be improper (whether technically legal or not) for government to impose the inconvenience of standards without first knowing precisely the bases that justified them. Even the Department of Public Health, in the same report in which it urged controlling pollution despite the absence of demonstrated health effects, found it "a matter of great importance that all of the public health effects of air pollution be understood and considered in the establishment of policy as to what level of damage will warrant exercising police power to protect the public."[9]

Whatever the premise, filling the information void had become an operational prerequisite to establishing standards. In 1956 the department took the position that effective pollution control was then impossible "because of a lack of fundamental information." It spoke in particular of the need for "additional knowledge so that standards for air quality can be established."[10] In the following year it pointed out that "community air quality standards have not yet been established," and added "it is unlikely that we will have enough information to set such standards for some years."[11] Yet not two years later, when the understanding of health effects was little improved at best, the department found itself charged by the 1959 legislation with the responsibility to develop standards a primary purpose of which was to protect against those very effects.[12]

[b]Thus a study in the mid-1950s concluded that "additional research is badly needed in the field of permissible toxic levels for humans, animals, and agricultural crops and to furnish a legal basis for enforcement." University of California Bureau of Public Administration, *Air Pollution Control* (prepared for Joint Subcommittee on Air Pollution, California Assembly Committees on Conservation, Planning, Public Works, and Health, 1955), p. 27. See also California Department of Public Health, *Clean Air for California* (Initial Report, 1955), p. 48.

The Department of Public Health had been handed a task it said could not yet be done; nevertheless, it could no longer assert, as it had two years earlier, that the time was not ripe for the setting of standards. As the department's director pointed out, the "Legislature enacted laws making it *mandatory* for the State Department of Public Health to develop standards for air quality and motor vehicle exhaust."[13] The department, quite obviously, was obligated to do the best it could. Still, in a way it went about fulfilling the obligation in a manner consistent with its earlier expression of untimeliness. What it did was to set standards in some instances, but not in others; to employ as fully as possible what little information did exist, while at the same time refusing to set any standards at all in cases of strong uncertainty. Whether this approach was consistent with the legislative mandate to the department (or with sound policymaking) seems questionable, but in any event it was tolerated with respect both to standards for air quality and standards for motor vehicle emissions.

Air Quality Standards

The Department of Public Health's expressed concept of the proper approach to air quality standards reflected a sound view, at least from the standpoint of efficiency. "The total cost of pollution," the department said in its report on standards, "must be weighed against the total cost of control in order to reach a rational and acceptable solution." Put this way, the statement is actually somewhat erroneous, since the most relevant factors are marginal, not *total,* costs and benefits. But error in this regard was corrected a page later in the report, where the department concluded that the "logical" approach "minimizes the sum of the costs" of pollution effects and pollution control.[c] Yet this approach could not be used, at least not explicitly. The legislature had directed the department to determine not the optimal (cost-minimizing) air quality standards, but those that

[c]California Department of Public Health, *Technical Report: California Standards for Ambient Air Quality and Motor Vehicle Exhaust* (n.d., c. 1960), p. 26 (hereafter cited as *Technical Report*). See the discussion in footnote b, p. 31. It is possible that the Public Health Department's *apparent* concept of the proper approach to standard-setting was a masquerade, that in fact (and despite saying otherwise) it saw the objective of standard-setting as protection of health at any cost. At least one participant in the standard-setting process, an employee of the department, gave this impression. The sole objective of the department, he said, was to ensure protection of public health, and the department was prepared to ignore costs to realize this end. Whatever had to be spent would be spent. Interview with L. Breslow, 2 Aug. 1971.

would protect public health. This, at least, appears to be the manner in which the department read its mandate,[d] and it acted accordingly. Standards were set for various pollutants at three levels—"adverse," "serious," and "emergency." The first of these was the level above which a particular pollutant would lead to "untoward symptoms or discomfort" (though not necessarily disease) in "the groups of persons in the population who are most sensitive to air pollution effects." Vehicle emission standards were then set such that adverse levels would be avoided in Los Angeles; since it had the most severe problem, other areas would have "a margin of safety in preventing air quality from deteriorating to the 'adverse' level."[14]

The approach taken to air quality standards, then, was not (except by chance) that of cost minimization. The standards were uniform across California, despite the fact that discrete air pollution sheds, varying in terms both of pollution costs and control costs, existed within the state. Moreover, the approach aimed at protecting the most sensitive portion of the population from the first appearance of discomfort, without apparent regard to the costs of doing so. If one needs further evidence that the approach taken did *not* reflect the announced logic of minimizing the sum of pollution costs and control costs, it is found in the fact that motor vehicle emission standards were *inferred* from the air quality standards established initially to protect health without regard to control cost.[e] Here

[d]The legislation itself was mildly ambiguous. The provision on air quality standards only directed the department to develop standards so "as to reflect the relationship between the intensity and composition of air pollution and the health, illness, including irritation to the senses, and death of human beings, as well as damage to vegetation and interference with visibility." California Health and Safety Code § 426.1 (1961). But the provision on motor vehicle emission standards directed the department to determine the emission limitations "compatible with the preservation of the public health." California Health and Safety Code § 426.5 (1961). A fair reading is that the emission limitations were to achieve the ambient standards, which in turn were to be "compatible with the preservation of the public health." This appears to be the way the department took the statute, as the discussion in the text will suggest. See also *Technical Report,* p. 30; interview with R. Brattain, 21 July 1971; interview with J. Goldsmith, 17 Aug. 1971; interview with G. Hass, 2 July 1971 (each interviewee commented that the standards were based only on consideration of health effects); J. Goldsmith, "Evolution of Air Quality Criteria and Standards," in *Development of Air Quality Standards,* ed. A. Atkisson and R. Gaines (Columbus, Ohio: Charles E. Merrill Publishing Company, 1970), pp. 1-18.

[e]"It was . . . decided that the [exhaust emission] standards would be based on calculations of the allowable emission of pollutants from an individual vehicle which, if representative of all vehicles in a community, would result in atmospheric concentrations not in excess of the 'adverse' levels expressed in the Air Quality Standards." *Technical Report,* p. 96.

again the department had been forced, by its (fair) reading of the legislature's charge, to ignore what it saw as the proper approach. In the department's explicit view, and consistent with a cost-minimization approach, the control costs accompanying the attainment of various air quality levels were factors necessarily to be considered in choosing the cost-minimizing air quality level.[15] But the legislative mandate was seen as not leaving room for that approach. Rather, air quality standards were to be determined solely with reference to health consequences; auto emission standards were then to be determined solely with reference to those air quality standards. The discussion of cost minimization was purely academic; its principles were ignored.[f]

In sum, the report of the Department of Public Health, closely read, reveals a tension between the views of the department and those of the legislature—a tension resolved, perhaps not surprisingly, in favor of the latter. The tension itself was simply one example of the recurrent friction between those who consider the sole objective of environmental control to be protection against adverse effects, and those who take a larger view and look to the cost of avoiding those adverse effects. The department seemed quite clearly to occupy the latter camp.[g] "In any environmental control program," its report stressed, "it is necessary to examine closely the costs which such measures may involve."[16] Nevertheless, the department's recommendations, presumably made in an honest attempt to comply with the legislative will, reflected the attitude that protection against adverse effects should be the sole concern.

If the department lost (or gave up) the day regarding its views on cost minimization, it appears to have succeeded in another implicit conflict with the legislature, one having to do with how to confront uncertainty. The legislature had directed the department to establish air quality standards despite the latter's expressed view that there was insufficient information to do the job. It would not be unreasonable to conclude that the

[f]Perhaps even more than suggested. Cost minimization would dictate consideration of *more* than vehicular emission controls—consideration, for example, of the costs and benefits of stationary controls, of controls aimed at limiting motor vehicle use, and of measures designed to induce avoidance of pollution effects by receptors (e.g., air conditioning). The department was well aware of these points. See *Technical Report,* p. 31. It is doubtful the legislature had them in mind; it seemed solely concerned with the vehicular emission limitations that would achieve the ambient standards. It is impossible to discern the extent to which the department, in setting vehicular emission standards, took into account reductions that could be expected through the other possible control measures mentioned. It does appear that at least expected reductions from stationary controls were considered to some extent. See, e.g., ibid., pp. 104-106.

[g]But see footnote c, p. 122.

legislature desired the best guesses possible—that standards be set for all pollutants with whatever little information existed. This the department refused to do. Rather it established standards only when they could be "based on and supported by available reliable data"; where such data did not exist, it was the intention "of the Department to adopt standards . . . in the future,"[17] beyond (it would seem) the statutory deadline.

The department was not unwilling to deal with *any* uncertainties. Indeed, uncertainties were so abundant with regard to setting air quality standards that a strict requirement of reliability would have resulted in no standards at all.[18] For example, the department noted the difficulty of determining reliable relationships between ambient levels and eye irritation, yet standards were established at levels thought (as a best guess) to avoid such irritation.[19] But where uncertainties were, in the eyes of the department, very strong, it opted for further studies rather than best guesses. "No [air quality] standards were set for ozone because of the scarcity of exposure data for human beings."[20] Yet there was *some* information on human effects, as the department itself pointed out.[21] Similarly, with respect to nitrogen dioxide there existed some evidence on adverse effects, but again no standards were established, the department noting that "more data on human exposures will be needed."[h] And this general approach to uncertainty was adopted in more than a few instances; there were "a number of air pollutants . . . that are important and . . . for which there are no standards" because of "the lack of data."[22]

The department's approach to uncertainty was one reflection of the "scientific" technique it had adopted. The propriety of the technique was hardly a foregone conclusion. Considering the many uncertainties and vagaries involved in standard-setting—for example, health effects and their costs, indirect costs of control (unemployment, inconvenience), ethical and distributional questions, and the fact that only one quality level can be provided for any particular area, despite a mix of wants and attitudes among the citizens within it—considering all these one might have thought that standard-setting should be viewed not as a scientific but as a distinctly political problem. The role of the scientist would be to give advice, not to make ultimate judgments. The latter would be the responsibility of a collective decision-making process. The department was aware of these issues. It recognized the factors that would suggest the propriety

[h]See *Technical Report,* p. 16, footnote 3; pp. 79-80. An "oxidant index" was established as an indirect surrogate for some of the effects of ozone and nitrogen dioxide, as well as of hydrocarbons, but it was not intended to impose a standard for these pollutants. See J. Maga and G. Hass, "The Development of Motor Vehicle Exhaust Emission Standards in California" (Paper presented at 53d Annual Meeting of the Air Pollution Control Association, Cincinnati, Ohio, 25 May 1960), p. 4.

of something other than "scientific" decision making; indeed, there are even hints it recognized that scientific decision making might well be inappropriate, or at least inadequate. The department was acutely aware of the many uncertainties in determining health and other effects and saw the difficulty of attaching reliable cost estimates to these effects. (" 'Costs' . . . refers to social, biological, medical, psychological, and economic burdens, no matter how difficult it is to place a dollar value on them.") Effects costs were especially elusive because they included the feeling of citizens "that living in air polluted areas is too unpleasant. The magnitude of this problem is difficult to estimate but a feeling of stability in one's home or work community is of great importance to social health." The department was equally aware of valuation problems on the side of control costs. Finally, it recognized the ethical issues involved in deciding whether controls should aim to protect average citizens, the sensitive, or the most sensitive.[23]

There are hints that, out of concerns like these, the department saw the need for a political approach to choosing standards—an approach based not so much on "scientific proof" as "public concern."[24] Yet the department explicitly refused to plumb the public mind. "Decisions," it noted, "are often reached by weighing both sides of the question, and by arriving at a point of equilibrium of pressures from interested parties. However, the control of air pollution calls for decisions based as much as possible on scientific knowledge."[25]

Why this conclusion was reached is not explained, but its implications for dealing with uncertainty seem quite clear. A "scientific" method was to be used. "It was agreed from the outset that any standards set *must* be based on sound data and *concurred in by scientists* in the air pollution and related fields." For advice the department went not to the public but to "scientists with recognized competence in their fields." The approach implicitly required *reliable* evidence and scientific agreement.[i] "Standards are based on and supported by available reliable data."[26] Where existing data was not "reliable" in the eyes of the department or the scientific community, standards were not established, whatever the state of public concern might have been. The department was required by the legislature to hold public hearings *before* "developing" standards,[27] but there is no evidence that these were employed by the agency as other than pro forma ratifying procedures; its 135-page report gives no sign of soliciting and utilizing public opinion. The State Board of Public Health adopted the department's scientific recommendations in December 1959.[28]

[i]Participants involved in the standard-setting process were insistent that the standards had "a scientific basis" and were "impeccable." Interview with J. Goldsmith; interview with L. Breslow.

Vehicle Emission Standards

The process of setting vehicle emission standards was in some ways similar, and in others dissimilar, to that of setting air quality standards. First, after *explicit* discussion, the department opted for uniform statewide emission standards. (This is in contrast to the decision on air quality standards, where an implicit judgment on uniformity was made, and without reported discussion. Arguably, the legislature mandated uniform quality standards, since effects of a *given concentration* on an individual's health are largely, if not entirely, independent of the location of the concentration.) The reasons it did so, however, are hardly clear. There is some clue that the department simply concluded that state-set standards almost necessarily meant statewide standards. State-set standards, the department's report argued, were justified by the "unique" problems of motor vehicle emissions control: vehicles moved freely across local boundaries, were taxed and licensed by the state, and were common to all areas "in contrast to the varying pattern of" stationary sources. "It is true that vehicle emissions have more serious effects in the more densely populated areas, but the difference is one of degree rather than kind." Several pages subsequent to these statements, but without further discussion, the department announced its "policy decision" that "a single standard would be adopted [for each pollutant], rather than a series of standards for the several regions of the State."[29]

A second decision was that emission standards would relate only to exhausts from vehicle tailpipes. They would "not cover losses such as from the fuel tank, carburetor and crankcase vents."[30] Just why this narrow approach was adopted is again not entirely clear. An initial explanation was that the department read the phrase "exhaust contanimants" in the statute as referring *only* to tailpipe emissions.[31] This could be accepted as an entirely reasonable rationale were it not for a subsequent statement that crankcase (blow-by) emissions were not covered because, at the time the standards were developed, such emissions were thought negligible (a matter about which opinion had changed by the time the department's report was published).[32] The latter statement suggests that "exhaust contaminants" was in fact read rather broadly, calling the first explanation into question. In any event, only tailpipe emissions were addressed.

Third (and implicit in the decision on uniformity), emission standards were to be calculated so as to achieve acceptable air quality in the most severely affected area (Los Angeles), leaving a "margin of safety" in other areas. Calculations would be based on an "averaging" method. The aim, that is, was not that each vehicle would achieve the standard, but that the average output of all vehicles would. Put another way, "it seems reasonable to require an individual vehicle to meet standards based on average performance and assume that the violations will approximately counter-balance those vehicles which do better than the standard."[33] In inferring emission controls from the air quality standards, the year 1970 was used as a target date "since a number of years would be required for implementation of controls for the exhaust emissions."[34]

Methodology from this point was straightforward in principle but complicated in practice.[35] Despite the complications (uncertainties in measurements, "rudimentary" knowledge of quantitative relationships), exhaust standards were established for two pollutants—hydrocarbons and carbon monoxide. For the former, two methods of calculation were employed. The first, the "rollback" method, assumed that total Los Angeles hydrocarbon emissions in 1940 resulted in smog effects consistent with a satisfactory air quality level for oxidant. Inventories of the pollutant from all sources were then prepared for 1940 and projected to 1970. Estimates of 1970 hydrocarbon output from nonvehicular sources were made, and the "calculation for the motor vehicle exhaust emissions, therefore, reduced to determining the degree of control required to reduce" vehicle emissions from the 1970 to the 1940 level.[36] The second method, the "indicator contaminant" technique, is somewhat more difficult to explain. Suffice it to say the method used concentrations of carbon monoxide (the "indicator contaminant") as a relevant measure. A comparison of the concentration on a severely smoggy day with that on a day when the oxidant air quality level was satisfactory was taken to define the general reduction in emissions required to improve air quality on a day of heavy smog to an acceptable level. A reasonably constant ratio between carbon monoxide emissions and hydrocarbon emissions was assumed—an assumption justified on the ground that both emissions came from the tailpipe. Together, the methods dictated an 80 percent reduction in current average exhaust hydrocarbon emissions. As to carbon monoxide, the department had air-monitoring data and was able to calculate the reduction needed to achieve the ambient standard in the data years. That figure was then extrapolated to 1970 and led to the conclusion that a 60 percent reduction was necessary.[37] Standards reflecting these reductions were adopted by the State Board of Public Health in 1959. They were to serve as a "basis for control legislation."[38] At this point, both they and the air quality standards were only advisory.

One cannot help but be struck by the difference of approach as between air quality and emission standards. Regarding the former, the standard-setters tolerated some uncertainty but refused to render judgments in the case of hard uncertainties—in instances, that is, where the data that existed was regarded as insufficient. (The criteria for gauging sufficiency were not expressed.) With respect to emission standards, on the other hand, there appeared to be considerably more tolerance of uncertainty; the standard-setters plunged in despite severely limited knowledge. That information was at best scant (especially regarding hydrocarbons) seems unquestionable: there were considerable "uncertainties . . . in the measurement of motor vehicle emissions as well as of pollutants from other sources" and there was only "rudimentary" knowledge "regarding the quantitative relationships between photochemical smog effects and the concentrations of the primary smog-producing pollutants."[39] The department's report itself noted a number of strong uncertainties. For example: "The measurement of the exhaust hydrocarbon loss is a very difficult problem. The analytical methods for hydrocarbons have never been considered to be satisfactory enough for standardization." Or, regarding the averaging approach, it was recognized that its reliability depended upon "the distribution of emissions after a control program is implemented." Yet the department recognized that "it is not possible to predict the shape of [the distributional] curve." As a final example, emission standard-setting required "a hypothesis concerning the relationship of smog effects to contaminant concentrations. . . . The evidence currently available is not conclusive and in some instances appears to be contradictory. *Nevertheless, the following equations are proposed as a basis for predicting the effects of exhaust contaminant control.*" The report went on to propose (and employ) the relationships, though it recognized they were "simplifications which do not fully comprehend the variety of photochemical reactions possible." It did so on the basis of "experimental evidence."[40]

Why were the standard-setters willing to confront strong uncertainties (both empirical and theoretical) in the case of emission standards, choosing to rely on "assumptions"[j] and experimental evidence, yet unwilling to proceed in the same manner with respect to air quality standards, insisting instead on "more data on human exposures"?[41] Put more succinctly, why were judgments made on the basis of "present knowledge"[42] in the case of emission standards, but often abjured in the case of quality standards? One can suggest several answers, but none of them is satisfactory. For example, two staff participants in the process noted the uncertainties attending the setting of emission standards, but added: "Nevertheless, the urgency of California's air pollution problem required action on the basis of what is presently known."[43] Yet the same urgency did not move the

[j]See *Technical Report,* Appendix A, p. 110 ("Smog Theory Assumptions for the Design of Vehicle Exhaust Standards").

department to adopt this approach to air quality standards, even where what was "presently known" revealed evidence of health effects.[44] Still another answer might be that the standard-setters considered uncertainties regarding emissions less intractable than those regarding air quality. This is at best a guess, of course, especially because (as mentioned earlier) criteria for judging tractability were never expressed. But it must be noted that most shortcomings in knowledge about air quality related only to empirical matters (measurement, experimental data on the effects of human exposures), while those regarding emissions included both the empirical *and* the theoretical.

Perhaps the answer is that the standard-setters had played into their own hands. They first set *some* air quality standards, and then found themselves confronted with the task of guessing about how to meet them. The legislature, after all, instructed the standard-setters "to determine . . . the maximum allowable standards of emissions of exhaust contaminants from motor vehicles which are compatible with" the air quality standards.[45] (Quite consistent with this mandate, "questions of practical attainability and cost were not considered in determining the reduction in emissions that would be required to meet the air quality standards.")[46] Thus the first process forced resolution of the uncertainties posed in the second. Indirect evidence for this suggestion is found in those instances (for example, lead) where information was considered too scant to justify an ambient standard, and then no emission standard was established for the very reason no ambient standard required one.[47] Turned around, the approach would suggest that where ambient standards were established, best guesses (based on "present knowledge") had to be made about how to meet them. And the standard-setters might have concluded that best guesses were justified in this context (though not in the context of air quality standards) because the problem confronting them was different, less value laden. It is one thing, perhaps, to reach a "scientific" conclusion about health effects in the absence of any "reliable" evidence (especially when the judgment could be costly and thus have significant implications for social policy), but quite another to make the "technical" or "engineering" judgment about how to reach an ambient goal once it is concluded that reaching that goal is important. The scientist might be understandably reluctant to judge how clean the air should be in the absence of good information about the judgment's implications, while the engineer steps in as a matter of course to venture judgment on the operational decision of how best to meet a given goal—the ambient standards initially established. He especially does so when the legislature has charged that he must.

This answer might be satisfactory were it not for the handling of oxides of nitrogen, where it appears the standard-setters acted so contrary to the procedures otherwise employed as to leave the entire approach to standard-setting a mystery. No emission standard was established for oxides of nitrogen because there was no ambient standard for this pollutant or for ozone (in the formation of which nitrogen oxides play a part).[48] Characteristically, ambient standards had not been set because of inadequate data.[49] Thus far, then, the picture is consistent with that sketched above: no ambient standard meant no emission standard. But there is more to the story. An "oxidant index" had been established, and it was recognized in the report that oxides of nitrogen were "necessary" to the reactions measured in the index. Indeed, it was acknowledged that oxides of nitrogen were a crucial factor in the general problem of photochemical smog (to which the oxidant index related), especially regarding their ratio to hydrocarbons.[50] Thus there *was* a relevant ambient goal in the index. Staff members suggested that hydrocarbons, but not nitrogen oxides, were controlled because they would have to be anyway. But so too would nitrogen oxides, since they are a central factor in the production of the unwanted effects (and especially since nitrogen oxide emissions, if uncontrolled, *increase* with controls on hydrocarbons and carbon monoxide—a point that was perhaps overlooked or poorly understood at the time, but one that later became painfully clear).[k] The same staff members suggested that control of automotive oxides of nitrogen would accomplish little because stationary source output of the pollutant was uncontrolled.[51] But future control could be foreseen,[52] and (as with the other pollutants) could have been projected to 1970. Or very large automotive reductions could have been recommended to take up the slack in stationary source controls. But, instead, no emission standard was set. The standard-setters refused to make a needed, and mandated, operational decision. Indeed, that refusal itself may have been a value judgment. Another participant in the process, Haagen-Smit, explained a few years after the standards were set: "In 1958, 1959, when the law on hydrocarbons was contemplated, we knew that at some time we had to control oxides of nitrogen, but this was set aside because at that time we didn't have any feasible way and so it was never forgotten."[53] Not forgotten, perhaps, but nevertheless put aside—and for reasons of practicability. The nitrogen oxides story appears to belie the claim that emission standards were set without regard to cost or practical attainability. It refutes the suggestion that emission standard-setting was seen as involving only operational decisions, devoid of value judgments,

[k]See the discussion at pp. 21, 192.

and mandated by a legislative charge.[1] (Accepting Haagen-Smit's explanation, one can view the handling of nitrogen oxides as clearly inconsistent with the legislature's instructions, not to mention the approach to standard-setting otherwise employed.) Finally, it empties the process of standard-setting of any consistent rhyme or reason, leaving this important phase of the California history at least a minor mystery.

[1]Another view of the nitrogen oxides story suggests more clearly that even as to operational decisions involving emission standards, the standard-setters were at times reluctant to confront very hard uncertainties. The view, held by one familiar with the standard-setting process, though not an active participant in it, is that no nitrogen oxides standard was set not only because of doubt about a control method but because there was little data, measurement was difficult, and the role of the pollutant was largely unknown. Interview with W. King, 24 Aug. 1971.

Part III

STATE AND FEDERAL CONTROL: 1960–1969

10

STATE EXPERIMENTATION AND FEDERAL INTERVENTION: 1960–1966

The decade of the 1960s was eventful in terms of legislation concerning air pollution generally and motor vehicle emissions in particular. Significant developments occurred at both the state and federal levels, with California usually taking the lead and the national government following—at times almost literally—in its footsteps. But that the federal government followed did not necessarily mean it deferred. The trend over the years was one of more and more federal encroachment upon (preemption of) authority that had previously been regarded as belonging to the states.

The ten years in question break into two quite distinct periods. The first, 1960 to 1966, was marked by state experimentation coupled with a good deal of failure, frustration, and controversy. During the same time, there took place a gradual, rather docile federal involvement in vehicle pollution control. The second period, covering the balance of the decade, witnessed major reorganizations of state efforts and adoption by the federal government of a much more dominant position regarding state authority and vehicle pollution control. This chapter and the next consider the two periods in turn; table 9 outlines the state and federal legislation on which the two chapters focus.

TABLE 9
STATE AND FEDERAL LEGISLATION
1960–1969

Year	California	Federal
1960	California Motor Vehicle Pollution Control Act creates Motor Vehicle Pollution Control Board (MVPCB) to certify control technology and require its installation.	Schenck Act directs surgeon general to study effects of motor vehicle exhausts on public health.
1962		Research and training program of 1955 extended through 1964; surgeon general directed to continue study of motor vehicle exhausts.
1963		Congress, after long efforts by HEW Secretary Flemming, enacts legislation strengthening federal role in air pollution control: Clean Air Act provides for federal air quality criteria, creates limited federal abatement authority, and encourages industry to develop solution to vehicle pollution.
1965		Motor Vehicle Air Pollution Control Act provides for national emission standards for new vehicles; legislation unclear regarding state authority to set more stringent standards.
1967	MVPCB, after seven years of controversial and frustrating administration, is replaced by an Air Resources Board (ARB) with expanded authority over mobile and stationary sources.	Congress, dissatisfied with progress under Clean Air Act, enacts Air Quality Act directing states to adopt and implement air quality standards consistent with federal criteria, expanding federal abatement authority, and clarifying preemption issue: federal emission standards for new vehicles to be binding in all states but California, which is permitted under certain circumstances to adopt more stringent standards for new vehicles.
1968	Pure Air Act prescribes, in the legislation itself, increasingly stringent standards for emissions of hydrocarbons, carbon monoxide, and oxides of nitrogen from 1970–1974 models; ARB given authority to promulgate stricter standards.	

CALIFORNIA'S 1960 LEGISLATION

Background

In 1959 Los Angeles County had secured the introduction of motor vehicle pollution control legislation that would have required installation of control devices, established a laboratory to certify them and stations to inspect them, and granted to the Department of Motor Vehicles enforcement powers against noncomplying vehicles and their owners. While the measure failed to pass (despite fairly broad support), it nevertheless had some impact. A Committee on Motor Vehicle Exhaust Control, appointed to advise the California Assembly Public Health Committee, recommended legislation that closely paralleled the Los Angeles County proposals; the committee's recommendations in turn formed the basis for a bill introduced the next year—1960.[1]

The bill that reached the California legislature in 1960 did not contain all of the provisions desired by Los Angeles County. There were, for example, no provisions for statewide inspection and enforcement. More important, the bill did not contemplate a mandatory statewide program; rather, it provided a local option by which each county could decide—through a vote of its board of supervisors—to opt out of the requirement of certified pollution control devices on new and used cars. The local option issue was a sore point, one that generated a good deal of heat in the course of considering the control legislation. A preprint version of the bill had contained no local exceptions. Opposition to this approach came from northern and rural legislators, described by one observer as "relatively conservative,"[2] and who in any event had no wish to impose costs on their constituents that would not in turn work a benefit for them. (Many still believed at this time that automotive air pollution was a distinctly Southern California phenomenon.) When proponents of the original measure offered to amend their bill to permit counties to opt out, the objection arose that local politicians should not bear the onus of voting to leave the program; rather, it was suggested that the legislation should apply nowhere until a county acted to implement its provisions (the approach of the 1947 act). It was only after considerable argument, joined with threats from the press and influential antipollution lobbyists to move for reapportionment, that the opt-out provision was saved. Even then, many antagonists refused to back the compromise; others "expressed dissatisfaction, but said they would support the bill because of their 'statewide duty.' "[3]

Of course, the opt-out provision (endorsed by Governor Pat Brown, among others) was a substantial weakening of the bill as originally conceived, and advocates of a strong measure were quick to point this out. Some of them, including Los Angeles County, countered with proposals to return to the original conception (no local option whatsoever),[4] others with the suggestion that the option should apply only to used cars—the program on new cars would be mandatory statewide.[5] Brown, backing down, accepted the latter approach and proposed it to the legislature.[6] The measure passed in this form, despite last minute efforts to reinstate the local option as to new cars.[7] One conservative senator who voted against the final bill thought the state was "rushing too fast into such a program. . . . 'I'm concerned about results and proof,' " he said, "explaining he doesn't think it can be argued at this stage that control devices should be required on all new cars, regardless of the area they're in."[8]

Details of the Legislation and Its Early Implementation

The California Motor Vehicle Pollution Control Act[9] established, within the Department of Public Health, a Motor Vehicle Pollution Control Board (MVPCB) "responsible directly to the Governor."[10] The MVPCB was to consist of four representatives from state agencies (Public Health, Agriculture, Highway Patrol, and Department of Motor Vehicles), plus nine members appointed by the governor with Senate consent. Appointees would hold four-year terms and receive no compensation other than reimbursement for expenses; they were to be "selected in such a fashion that the interests of various affected groups throughout the state, including agriculture, labor, organizations of motor vehicle users, the motor vehicle industry, science, air pollution control officials, and the general public are represented to the fullest extent possible."[11] Once in operation, the MVPCB was to establish criteria for approval of control devices, taking into account purchase and installation cost, durability, ease of ensuring continuing reliability once installed, and any other factors thought to bear on the suitability of the devices.[a] It was to test devices; in this connection it could, if it wished, designate and contract with testing laboratories operated by any qualified public or private agency (whether or not located

[a]California Health and Safety Code § 24386(3) (1960). Such "other factors" might include financial stability of the manufacturer; the manufacturer's ability to produce and distribute, as well as to maintain, adequate stocks of models to fit most cars; safety of the device; and absence of objectionable odors. P. Magill, "Cleaner Exhaust from Automobiles" (Consultant's Report, 1960), p. 7.

in California).[12] The MVPCB was to issue certificates of approval for any device it found upon testing to meet its criteria and the emission standards established by the Department of Public Health.[b] Finally, it was to report periodically to the governor and the legislature, making recommendations for any legislation or other actions needed to better implement and enforce the law.[13]

When the MVPCB certified *two* devices, the law required their installation on new and used vehicles.[14] New vehicles were not to be registered in the state after one year following certification unless equipped with a certified device; used vehicles were not to be registered *upon transfer by a registered owner* after one year following certification unless so equipped, and in any event were not to be registered after three years following certification unless an approved device had been installed. The provisions regarding used vehicles were inapplicable, however, in any county or portion of a county determined by its board of supervisors not to need them in order to achieve the air quality standards set by the Public Health Department in 1959.[15] The statute made it unlawful to drive unregistered vehicles, those registered in violation of the control device requirements, and those not equipped as required with a certified device.[16] The MVPCB could, however, exempt any classes of vehicles for which no devices were available, or which met state standards without control equipment, as well as motorcycles, implements of husbandry, and a limited group of other vehicles.[17] The law made no provision for the periodic inspections Los Angeles County had originally proposed.

All the provisions outlined above followed in quite straightforward fashion from the assumptions underlying the 1960 legislation. The view of the legislature apparently had been that automotive air pollution control technology existed and would soon be perfected; thus, the vehicular pollution problem could be "solved" simply by requiring installation of the control technology. The Department of Public Health had established

[b]California Health and Safety Code § 24386(4) (1960). The motor vehicle emission standards adopted by the Department of Public Health pursuant to the 1959 legislation were, when first established, only advisory, "applicable solely as guides for the Motor Vehicle Pollution Control Board in its consideration of control devices." 42 Opinions of the California Attorney General 47, 48 (1963) (Opinion No. 63-144). The 1960 legislation gave the standards the force of law, the legislature declaring that "the State has a responsibility to establish uniform procedures for compliance with these standards." California Health and Safety Code § 24378(c) (1960). As noted, the MVPCB was not to approve a device unless the device were found to meet the emission standards established by the Public Health Department. The attorney general's opinion suggests the MVPCB had wide (if not unlimited) discretion in determining whether a device met those standards.

vehicular emission standards that set the goal for the technology. There was, however, no standard procedure by which to test the merit of inventors' claims as to the effectiveness of their control devices, and some provision for one was necessary.[18] The stationary source problem was thought to be well in hand, thanks to the 1947 act. The districts enabled by the act, however, could not deal effectively with vehicular pollution: their jurisdiction was limited in most instances to county boundaries, and thus any efforts to control mobile sources would be seriously constrained.[19] A law requiring devices on a statewide basis (at least as to new vehicles),[c] and establishing a board to certify that such devices met the Public Health Department standards and board criteria, while at the same time leaving intact the existing system of control for stationary sources, would respond to all the assumptions behind the legislation.

The Motor Vehicle Pollution Control Act was, of course, just such a law; the assumptions underlying it help explain its rather limited approach. They do not, however, answer every question about the act. One of these is why a Motor Vehicle Pollution Control Board was even created. An official of the Department of Public Health at the time of the act's passage later noted, in strong terms, that a control body separate from the department was unnecessary;[d] the latter was more than capable of performing the functions of implementation and enforcement of the emission standards it had established. The analogy drawn was the case of milk purity, where the department set pasteurization standards, implemented them, and supervised their enforcement. What was good for the purity of milk, the argument ran, would appear to be good for the purity of air. The difference in treatment, the official believed, arose at least in part as a response to industry pressure to limit the powers of the department. The mandate of the agency was to protect public health; that goal was its "sole interest" in the air pollution field. To realize the goal, the department was said to be prepared to ignore costs: "Whatever had to be spent to achieve the goal would be spent, whether it was a million or a billion." (The department's 1959 report hardly reflected such a philosophy.) Those in the controlled industry, of course, were "fundamentally opposed" to such

[c]The compromise on new versus used vehicles was somewhat inconsistent with the assumption that local control authorities could not deal effectively with mobile sources; there is little reason why problems of local control of used vehicles would not be virtually the same as problems of local control of new vehicles.

[d]Of course, in a sense the MVPCB was not separate from the department—it existed within the department. Under the legislation, however, it was responsible directly to the governor, and thus was apparently independent of department control. California Health and Safety Code § 24383 (1960).

an approach, and accordingly preferred that the Department of Public Health be "shorn" of any power. The conclusion is that pressures to this end succeeded, at least partially, with creation of the MVPCB.[20]

Correct or not, this view of the MVPCB's creation does introduce points that were later to recur, and that are worth mention here. One is the declining importance of the Public Health Department in California air pollution control, a decline that continued subsequent to 1960 until the department's role was virtually emasculated in 1967-1968.[e] Another is the suggestion that the MVPCB could be expected to view sympathetically the interests of the automobile companies, a suggestion that would quickly grow to allegations that accompanied the agency throughout its life. The charges were probably to be expected. Under the 1960 act, appointed members of the MVPCB were to be selected so as to ensure representation of "the interests of various affected groups throughout the State, including . . . organizations of motor vehicle users [and] the motor vehicle industry." The approach was to appoint what one observer termed a "public interest" board, one that combined interests and specialties from a variety of backgrounds—the notion apparently being that if the recipe includes all the relevant ingredients, the final product will represent the public interest.[21] Another active participant in MVPCB activities suggested a more candid and perhaps more realistic view, although the substance of his remarks is the same. He argued that a board composed of completely "unvested" interests is very difficult to obtain, and that a more desirable approach is a "balanced" body, where each member's interests are "heavily vested" but balanced by the similarly vested interests of others, the ultimate product being sound compromises in decision making.[22] Whatever the name, and whatever the theory behind it, the fact remains that the MVPCB was explicitly required to represent relevant interests, and Governor Brown proceeded accordingly in making appointments. Thus, an early appointee selected to represent the motoring public was drawn from the management of the Automobile Club of Southern California, a group hardly notorious for its support of motor vehicle pollution control measures. At the time of his appointment, he was unconvinced that motor vehicles contributed to air pollution.[23]

Charges of misplaced sympathies can grow quite easily in the context of a board that includes representation of groups supposedly among those to

[e]The declining role of the department became a general trend. Over the years, it lost standard-setting and enforcement powers that it once held in many environmental fields. The development has been attributed to a combination of "industry pressure" and a lack of aggressiveness on the part of the department. Interview with L. Breslow, 2 Aug. 1971.

be regulated. But a "vested-interest" board's vulnerability to such charges can reach beyond this rather obvious instance. Charges can come from without, as they did when Governor Brown appointed to the MVPCB a woman who had been a staunch campaigner for strong pollution control; the allegation was that the appointment dampened her role as an effective critic of state pollution control efforts.[24] Or charges can come from within. The MVPCB was to include representation of "air pollution control officials." At the time, this could only mean local officials concerned with stationary source control; there were no vehicle controls prior to the 1960 act. In the event of poor air quality, it would be tempting for stationary controllers sitting on the MVPCB to blame conditions on the other members, charging that they acted out of sympathy for the concerns of the regulated. The tactic would, of course, deflect pressure from local officials. Perhaps this is part of the explanation for later allegations by the Los Angeles APCD that the MVPCB had adopted a conciliatory approach to the automotive industry.[25]

There was another noteworthy feature of the MVPCB's composition. None of its appointed members was required to have any particular expertise that related directly to vehicular pollution control, and—with at most a few exceptions—none did. The body was, in short, what one of its staff described as a "lay board." While all of the members were generally competent and knowledgeable, most of them could hardly be described as experts.[f] In instances of lay boards, the legislation creating them commonly provides for appointment of some sort of technical advisory body to assist in studies and deliberations. This technique was not employed, however, in the case of the MVPCB. It had authority to employ a staff, of course, and the statute contemplated that it would hire "technical . . . personnel."[g] But not too great reliance could be placed on staff, not only because

[f]Interview with M. Sweeney, 23 July 1971. A former executive officer of the MVPCB did say that its members at least had relative expertise—that they were better equipped in this regard than the members of the Air Resources Board created in 1967. Interview with E. Grant, 12 July 1971. At least two early MVPCB members, A.J. Haagen-Smit and J.T. Middleton, could probably be said to have had about as much expertise (other than in the most narrow technical sense) as existed at the time.

[g]California Health and Safety Code § 24386(2) (1960). The MVPCB also had power to contract with public or private laboratories for the testing of devices. California Health and Safety Code § 24398 (1960). Most of the early contracts were with private laboratories; the staff man in charge considered the government laboratories inefficient. Interview with M. Sweeney. Indeed, for a time the auto companies themselves were certified to test devices. Interview with J. Askew, 26 July 1971.

of limited manpower (the staff numbered about sixteen),[26] but more importantly because to do so would result in the staff, rather than the MVPCB, becoming the effective decision-making body.

The general impression appears to be that the staff was of high quality and did not occupy an unduly dominating position. The MVPCB was by no means a "rubber stamp"; its members took their obligations seriously, often discussed issues at great length, and at times made decisions directly contrary to staff recommendations. The facts remain, however, that the staff determined meeting agendas, prepared materials for members' consideration, and took virtually all the initiative on policy issues. Its executive officer usually did most of the "interacting with industry" necessary to define policy questions and answers, and then went to the members for approval. Usually, though not always, such approval was forthcoming.[27] The power to initiate policy questions, and to determine the matters that bear on their resolution, is virtually the power to make policy when no independent sources are submitting questions and answers of their own, and when the formal decision maker is essentially a lay body. Rubber stamp or not, the MVPCB must have been strongly influenced by staff positions.

The MVPCB did, however, take steps to capitalize on its own limited resources, perhaps in an effort to avoid staff domination. First, it did much of its work through committees comprised of MVPCB members; since the members themselves chose their committee assignments based on their own particular interests, the technique could be an effective way to focus the knowledge of each on those areas where it might be most valuable (not all the committees were technical in nature—there was, for example, a committee on public information).[h] An apparently quite typical work flow was from staff to committee (with staff recommendations usually accepted), then from committee to the whole MVPCB (with committee recommendations usually accepted).[28]

Second, the MVPCB established for itself the technical advisory committees that the legislature had failed to provide.[29] These committees served on an ad hoc basis; when issues developed that required special knowledge not available from internal resources, an advisory committee would be formed (usually from names recommended by staff). The advisory committee would work with the MVPCB and its committees, and would be abolished upon completion of the task at hand. If necessary, the

[h]A former chairman of the MVPCB said the committees were intended among other things to serve as watchdogs over staff, but that in fact the staff was really on its own. Interview with E. Plesset, 27 Aug. 1971.

advisors—usually experts from industry and pollution control agencies—would be reimbursed for any necessary expenses.[30] At times, however, reimbursement was not necessary; industry advisors were often provided at industry expense. Indeed, on occasion the MVPCB received technical advice, at no cost, from auto industry engineers who actually worked for the agency while their industrial employers paid their salaries.[31]

There were, of course, a number of obvious dangers in such arrangements, but generally speaking the ad hoc advisory committee structure probably worked to improve the balance between staff and members (although recall that the staff usually recommended the people appointed to advisory committees), and in any event can be assumed to have supplemented the knowledge of both groups. A former chairman of the MVPCB thought the ad hoc approach proved superior to that of a permanent technical advisory committee. Ad hoc committees, given the specific task of examining a difficult technical question or gathering particularly elusive information, and composed of constantly new personnel, tended as a result to be both confined in authority and open-minded in approach. Standing technical advisory committees, on the other hand, developed positions over time which they then felt an obligation to defend, rather than taking a fresh look at the evidence.[32]

The absence of any requirement of technical expertise among MVPCB members might appear especially striking in light of the 1960 act's total reliance on a technological solution to the pollution problem. The legislation's sole weapon to combat automotive air pollution was prohibition of registration, after specified dates, of new and used vehicles not equipped with approved devices. The chief job of the MVPCB was to approve devices, and whenever two or more devices had been approved, the time periods specified in the statute began to run and at their termination the prohibitions on registration became operative. Just why the act took these lines is, like the creation of the MVPCB itself, hardly apparent from the assumptions behind the legislation. It would have been possible for the statute simply to provide that after a specified date (say one year after enactment of the legislation), automobiles could not be registered unless their emissions met state standards.[i] At least as to new automobiles, this was a conceivable approach (one later adopted at both the state and federal

[i]This would have been a pure performance-standard approach. See the discussion at p. 34. The 1960 act was a mix of the performance-standard and specification-standard approaches. Any device that met the standards and criteria could be certified (performance approach), but only those certified devices would then meet the law's requirements (specification approach). California Health and Safety Code §§ 24386(4), 24390–24393, 24395 (1960); California Vehicle Code §§ 4000, 4750, 27156 (1960).

levels), especially conceivable given the assumption at the time of the act's passage that the necessary control technology existed. Instead, however, the statute's requirements were tied to the development and approval of technology: the statutory periods did not begin to run until devices had been certified.

Perhaps the reason for this approach related to the 1953 legal opinion of the Los Angeles County counsel. That opinion suggested that installation could be required only if a "satisfactory device is perfected and available on the market." Devices had to be "shown to be effective for the purpose and practicable in operation." Until such a demonstration, "any rule requiring the use of the device [would be] arbitrary, capricious and void." Thus installation could be made mandatory only "when the required device is available, but not until then."[j] It would follow that the law, to be enforceable, would necessarily need some sort of provision—like the certification procedure—that ensured that the required devices were in fact available. In theory, certification did not leave the development of a device dependent upon efforts by the automotive industry or other inventors; the MVPCB, for example, could have been charged to develop a device for itself. But the statute contained no provision for research and development by the MVPCB, and the assumption must have been that the automobile companies or independent manufacturers would produce the needed devices. Apparently no one "had carefully thought through the likely response . . . to such a law."[33]

The foregoing discussion may explain the certification procedure, but it sheds no light whatever on the question of why *two* devices had to be certified before controls became mandatory. There is a straightforward answer, but not a very sensible one. According to an opinion of the California attorney general, the two-device requirement was "to insure competition between two or more companies in the manufacture of the device, thus preventing an excessive price due to monopolistic conditions." In

[j]See the discussion at p. 98. See also H. Kennedy, "The Legal Aspects of Air Pollution Control with Particular Reference to the County of Los Angeles," *Southern California Law Review* 27(1954):373, 393-394; idem, *The History, Legal and Administrative Aspects of Air Pollution Control in Los Angeles County* (Los Angeles County Board of Supervisors, 1954), p. 49; A. Carlin and G. Kocher, *Environmental Problems: Their Causes, Cures, and Evolution, Using Southern California Smog as an Example* (Rand Corp., Santa Monica, Calif., Report Number R-640-CC/RC, 1971), pp. 66-67, footnote 2. Carlin and Kocher join us in speculating that the 1953 opinion may have been behind the approach of the 1960 legislation. There is also, of course, another plausible explanation for that approach: pressure by the auto industry. There is no specific evidence of such pressure, but there are reasons to suspect its presence. The certification prerequisite would facilitate industry delay.

addition, the requirement was supposed "to encourage research."[34] Just how the two-device requirement would encourage research was not explained; if anything, it would arguably discourage it.[k] In any event, it would surely delay implementation—an argument opponents made by employing the analogy of Salk vaccine ("if the use of the Salk vaccine had been deferred until such time as a competing vaccine were developed, many more children would have known the crippling effects of polio").[35] As to monopoly prices, other restraints were of course available, such as price controls, compulsory licensing, or making licensing of a certified device a condition precedent to its mandatory installation. These restraints did find their way into later state and federal law; there is no evidence they were even considered in 1960.

The two-device approach was used in California for about five years. The job of the MVPCB was to get on with the business of adopting criteria and certifying two devices that met those criteria and the standards set by the Department of Public Health. Vehicle exhaust standards had been established by the department in 1959; in late 1960 crankcase emission standards were set as well.[l] Soon thereafter the MVPCB adopted its own criteria for crankcase devices and designed test procedures to implement them.[36] (Shortly before, Los Angeles County had offered the services of its testing laboratory to the state in order to assist in the certification process; eventually a gift of the laboratory was made.)[37] The criteria concerned the ability of a device to control emissions to the required degree; its safety, odor, and endurance; and the financial resources of the manufacturer. There was also, apparently, concern with the cost of devices.[38]

[k]See Magill, "Cleaner Exhaust," p. 3: "Two devices must be certified because the legislators did not want to incur the possibility of legislating a monopoly into being. This produces the interesting situation that any company ready and anxious to offer their [*sic*] device for sale should be cheering for its competitor." Of course, the cheers might well go unanswered—if the competitor were an automobile manufacturer. And such a likelihood might in turn inhibit independent inventors from developing a device in the first instance. In fact, of course, monopoly profits might have been just the incentive necessary to encourage development of a control device; a primary purpose of the patent laws is to provide strong research and development incentives by permitting such profits for a limited time.

[l]T. Roberts, "Motor Vehicular Air Pollution Control in California: A Case Study in Political Unresponsiveness" (Honors thesis, Harvard College, 1969), p. 31. The 1959 legislation had directed the Department of Public Health to establish exhaust emission standards; in 1960 the department was instructed to "take into account all emissions from motor vehicles rather than exhaust emissions only" in any standards established after February 1, 1960. Ch. 37, § 1, [1960] Calif. Stats. 1st Ex. Sess. 380, amending California Health and Safety Code § 426.5.

In September 1961 the MVPCB certified the first new-car crankcase device, manufactured by a division of General Motors.[39] General Motors had had the device ready for use at least as early as 1960, and similar devices had been in use in some applications twenty years before.[40] General Motors had, in fact, already voluntarily installed the device on its 1961 models sold in California (in the view of the press, "arbitrarily").[41] Other manufacturers quickly followed suit.[42] (Subsequently the industry installed the same controls on 1963 models of its automobiles sold in the United States—again voluntarily.)[43] Despite this history, certification of the General Motors device was not without controversy. Four dissenters voted against approval, apparently convinced by the Los Angeles APCD's position that the device failed to meet Department of Public Health standards.[44] According to one account, the APCD's objections were especially distasteful to the majority of MVPCB members because the test findings on which they were based, though known to the APCD for some time, were not revealed to the MVPCB until the certification hearing. The MVPCB was put in a rather embarrassing position by the revelation of the APCD's representatives, and its chairman, "outraged at their action, publicly denounced them and ruled approval for the device." (The friction generated by the event was to grow over time to heated controversy between the MVPCB and the Los Angeles APCD.) Unfortunately, the approval was a bit hasty; the device proved defective and had to be removed from the market after a year.[45] By this time, however, a number of crankcase devices had been certified,[46] and thus were required on all new automobiles sold in California beginning with the 1964 model year—three years subsequent to the time the industry had begun installation on its own.

The certifications described above were not applicable to used cars, but by December 1962 two used-car crankcase devices had been certified.[47] The 1960 legislation, of course, provided a timetable for mandatory installation of these devices, but the MVPCB—apparently because it wished to require used-vehicle installation on a staggered basis—sought further legislation granting it that power.[48] The measure requested became law in mid-1963. Under its provisions, used vehicles transferred from one owner to another after January 1, 1964, would have to be equipped with one of the two devices; used commercial vehicles would need one of the devices during the first ten months of 1964, depending upon a schedule keyed to the last digit of the vehicle's license number, and no such vehicle without a device could be registered after December 1, 1964; and all used passenger vehicles—whether transferred or not—would need one of the devices during the first ten months of 1965 according to the same schedule, registration of an unequipped vehicle being prohibited after December 31, 1965.[m] (Among other things, the statute also amended the 1960 law to provide that motor vehicle pollution control devices included not only equipment designed for installation, but system or engine modifications as well.)[49] The used-vehicle requirements applied, of course, only to those counties which had not opted out of the program, but these included eleven counties containing almost 80 percent of all vehicles registered in the state.[50]

Two Controversies

The rest of the legislative and administrative developments in California through 1966 are best described in connection with two controversies that began in late 1962 and early 1963. While not every event in the four succeeding years related directly to one or the other of these controversies, most of the important ones had at least some tie to them.

The "pink-slip" controversy.—At the same meeting during which it voted to request legislative approval of a staggered system for requiring

[m]The used-vehicle crankcase requirements applied to 1950–1960 models only. The MVPCB adopted a policy that assumed that vehicles after 1960 were equipped with crankcase devices (the auto companies, it will be recalled, began voluntary installation with 1961 models). *California Motor Vehicle Pollution Control Board Bulletin,* Vol. 2, No. 8, August 1963; ibid., No. 9, September 1963. As to pre-1950 vehicles, the MVPCB no doubt assumed, quite sensibly, that there were so few of those models on the road as not to be worth worrying about.

The MVPCB probably wished to employ a staggered installation schedule as a means of avoiding a concentrated burst of business with which those making, selling, and installing devices would be unable to cope. Interview with J. Maga, 23 June 1971. It was for this reason that the automobile clubs had urged a time-of-transfer rather than time-of-registration requirement for the 1960 act, a matter in which, of course, they succeeded only partially. Interview with J. Havenner, 20 July 1971. (At the time of the interview, Mr. Havenner was executive vice president of the Automobile Club of Southern California.) The rush-of-business problem had been a concern for some time in the context of requiring devices on used vehicles. In the years 1955–1957 there were a number of suggestions that installation alone on all the vehicles in Los Angeles would require at least a year, and that considering production and delivery the required time would be three years. See, e.g., "Press Session," in *Proceedings of Southern California Conference on Elimination of Air Pollution* (Los Angeles, Calif., 10 Nov. 1955), pp. 45, 61; Air Pollution Foundation, *1956 President's Report* (Los Angeles, Calif., 1956), pp. 26–27; T. Patterson, "Control of Automobile Exhaust Next Step in Clearing Air of Pollutants," *Riverside [Calif.] Daily Press,* 12 April 1957.

crankcase controls on used vehicles, the MVPCB referred to a technical advisory committee a "scheme" to enforce those requirements.[51] In particular, the MVPCB wished to consider a program of mandatory vehicle inspection. The primary advocate of such an inspection program appears to have been Los Angeles County. From the outset, the county had viewed mandatory inspection as a central feature of any vehicular pollution control program. By late 1962 its position was supported by the Southern California Air Pollution Coordinating Council, the County Supervisors Association, and "virtually all Boards of Supervisors representing those counties in which air pollution control districts had been activated."[52] Perhaps because of such strong endorsement, the MVPCB approved the idea despite an unfavorable report from the technical advisory committee.[53]

Subsequently, in early 1963, an inspection bill was introduced into the California Senate. It provided for Department of Motor Vehicle licensing of installation and inspection stations; inspection of devices at least every twelve months after installation; and issuance of a certificate and sticker to vehicles passing inspection. In this form the bill met strong opposition from the automobile clubs and rural counties, and its final passage in 1963 reflected a much weakened measure. The original provision for establishment of a statewide system of inspection stations under Department of Motor Vehicles supervision was deleted; instead, implementation was left to county initiative and the inspection program would apply only in those counties that affirmatively opted for it. As a compromise (one originally offered by the MVPCB as an alternative to *any* inspection program, over the objection of Los Angeles County), a person registering a vehicle after December 1, 1963, was required to state under penalty of perjury that the vehicle was equipped with a control device as specified by law. As with the 1960 legislation, counties were permitted to opt out of this latter requirement.[54]

When the MVPCB set about administering the new law, it became clear that its powers were poorly defined. The statute, for example, did not explicitly provide for *annual* inspection, nor did it provide that a sticker issued for a satisfactorily inspected vehicle must be retained on the vehicle. The attorney general, however, found it "the obvious intent of the Legislature that properly working devices be maintained," and without "some means of enforcing the maintenance of properly working devices on the vehicles, the Motor Vehicle Pollution Control Law will be meaningless." Accordingly, he ruled it within the power of the MVPCB to require annual inspections and retention of inspection stickers.[55] Nevertheless, Los Angeles County was not convinced, and for a time declined to activate the inspection program. By the end of 1963, however, it and the twelve other major urban counties had implemented installation and inspection programs.[56]

The requirement of crankcase devices for used vehicles became effective on January 1, 1964; vehicles transferred from one owner to another after that date could not be registered unless properly equipped, and used commercial vehicles, transferred or not, were required to have an approved device on a staggered basis during 1964. Apparently this portion of the program went forward with little controversy.[n]

The requirement of installation on *all* used vehicles, beginning on a staggered basis January 1, 1965, produced a dramatically different reaction—one that quickly killed the program. The MVPCB had anticipated some problems. As early as August 1963 the developer of one approved device warned the agency of "a big psychological problem because of general resentment against anything mandatory. . . . He urged a public relations program to win public acceptance."[57] Subsequently, the MVPCB allocated $24,000 to that purpose, and by early 1964 had plans for billboards, brochures, spot films on television, radio announcements, a smog theme, and a "smog animal"—all designed to explain the crankcase requirements.[58] By mid-1964, however, the publicity campaign had taken a more defensive note, the MVPCB now trying to combat rumors and newspaper accounts that questioned the wisdom of the device requirement.[59]

Typical public worries were that the devices interfered with vehicle operation and could not be properly maintained.[60] These were legitimate concerns; maintenance was especially troublesome inasmuch as under the inspection program an improperly maintained device could result in failure of inspection and lead to additional expense or even inability to register the vehicle. The basis for the worries might, however, have been illegitimate; much of the concern was generated by claims of the automobile clubs, mechanics, and used-car dealers that the devices were nuisances. The auto clubs had long been opposed to mandatory installation. One might think the mechanics would favor the program—and the faultier the devices, the better—because it would generate business. But mechanics felt otherwise, apparently out of the fear that the presence of devices would make ordinary repairs and maintenance more difficult, resulting in unhappy customers, the need to do jobs over, and so forth. Used-car dealers were opposed because much of the burden of the program would fall on them: after the deadline, they would be unable to sell cars

[n]There were a few exceptions. A former executive director of the MVPCB recalled a series of complaints, some of them appearing in the press, registered by a California newsman unhappy to discover that, upon transferring the old family car to his son, a device was required as a condition to registration. Interview with D. Jensen, 25 Feb. 1975.

unless devices were first installed.[61] (Since used-car dealers were similarly affected by the earlier, time-of-transfer requirement, one would have expected an uproar from them at the time that requirement went into effect; oddly, however, there is no evidence of such a reaction.)

In December 1964 motorists began to receive 1965 registration materials in the mail; enclosed was a pink slip explaining the installation requirements.[62] The slip had been drafted by an official of the MVPCB, but subsequently changed in various respects by the Department of Motor Vehicles.[63] There is unanimous agreement that the slip was virtually unintelligible to motorists; the immediate result was considerable public protest.[64] One legislator later said that 100 percent of the mail he received on the matter was in opposition to the program.[65] Others claim that public disenchantment was relatively mild, that only about 2 percent of the population complained, but that the legislature was eager to pick up on the issue.[66] In any event, the Senate and Assembly Transportation Committees scheduled hearings in January 1965, and by February the legislature had passed a measure, signed immediately by the governor, providing that failure to install the required crankcase device did "not constitute a crime."[67] (The Department of Motor Vehicles had already eliminated inspections and affidavits of compliance from its 1965 registration requirements; thus, after the law, one could register an unequipped vehicle—and with no fear of punishment.)[68] The act stated that it was "an emergency measure" made necessary by the "great concern and confusion" caused by the crankcase requirements. The legislature, it said, "is currently considering legislation designed to substantially eliminate this requirement . . . , and it appears probable that" this legislation will pass.[69]

Pass it did. In June 1965 the legislature abolished the annual inspection and installation programs, required installation of crankcase devices on used vehicles (back to 1955 models) only upon transfer of ownership to an owner residing in a county with a functioning air pollution control district, limited the county option to withdraw from the used-vehicle installation program, set a maximum installed price of $65 for future certifications,[o] and established a system of roadside inspections to be conducted by the Highway Patrol. To assist the inspection program, the legislature doubled the size of the patrol (the necessary funds to come from increased registration fees) and gave it authority to train and license operators of installation-inspection stations to be set up on a statewide basis. Devices were required

[o]There had been earlier efforts, in the course of passage of the 1960 act, to put an explicit ceiling on the cost of control devices, but these had failed. See C. Greenberg, "Assembly OKs Auto Smog Law," and "Auto Smog Board Provided," both in *Los Angeles Examiner,* 30 March 1960.

to be installed at the stations or inspected and approved by a station in the event of installation elsewhere; moveover, inspection was required upon each transfer of ownership.[70]

How account for this rapid enormous setback to the used-vehicle installation program? A number of factors probably played a part. First, testimony at public hearings preceding passage of the 1965 legislation suggested problems with the approved devices: the auto clubs, mechanics, and used-car dealers voiced their objections; others testified that the crankcase controls increased emissions of nitrogen oxides; still others complained that the devices certified by the MVPCB were less effective than some that had been rejected.[71] The MVPCB tried to counter these assertions. The "only thing wrong with the devices," it argued, "was old mechanics' tales, combined with bad mechanics' installations." (Later it began checking complaints, "only to find that the devices had been installed incorrectly . . . , upside down and backward.")[72] The MVPCB also urged the legislature and the governor not to abandon the mandatory installation schedule, but to no avail.[73]

A second factor in the setback to used-car controls may have related to design and implementation of the program. Several persons involved in the incident believe the legislature and the MVPCB had not thought the matter through sufficiently, had attempted to establish the network of installation and inspection stations too quickly, had failed in the public information program, had inadequately supervised arrangements for informing owners of the law's requirements, and had taken insufficient account of the effect of devices on old, low-valued, poorly operating "clunkers."[74] But perhaps the largest reason for failure is that suggested by one participant's observation that the crankcase requirement resulted in a "confrontation with the general public."[75] As an executive of General Motors put it, "this past experience indicates the futility of getting public acceptance" of troublesome used-car controls.[76] It was one thing to require devices on new vehicles (or even on transfer of used vehicles), where the consumer buys a package that happens to include emission control devices; the marginal cost is hidden and the purchaser is saved the trouble of having the device installed. A member of the California Senate observed that "none of these problems ever appeared with respect to new cars because the devices were factory installed."[p] But when the car owner has to

[p]He added, "and the enforcement problem was easy." Interview with T. Carrell, 26 Aug. 1971. But the statement on enforcement with respect to new vehicles is rather misleading. For a long while, it was simply *assumed* that, after testing and certification, manufacturers were installing approved devices on their new vehicles. That is, there was no enforcement. Interview with D. Jensen.

take steps himself to have a device installed, both trouble and expense are painfully obvious. "Perception of control costs," in short, is very "different for used as compared to new cars."[77]

There is some dispute about whether the pink-slip controversy had much impact on automotive pollution control policy in California. A few observers think it did not. Others believe that it most surely did, that it killed any notion that one could effectively control used vehicles, and in any event killed the incentives of manufacturers of used-car control devices, who were "scared off" by the legislature's change of heart and generally "disgusted" with its behavior.[78] The legislature took some steps to make amends for its fickleness. Noting that the program of mandatory installation of crankcase devices had been abolished, that dealers had large stocks of the devices, and that "as high as 80 percent of their inventory will not be sold," it passed another law in 1965—this one instructing property tax assessors to take account of the effect of the program's abolition on the value of the devices.[79]

The "averaging" controversy.—The pink-slip controversy grew out of controls on crankcase emissions; the averaging controversy found its roots in efforts to control exhausts. The Department of Public Health had adopted exhaust emission standards in 1959. Crankcase standards were not adopted until 1960, but the MVPCB turned most of its first efforts to implementation of those later crankcase standards as opposed to earlier exhaust standards, perhaps because the crankcase problem was thought more tractable. The delay in confronting the exhaust problem, though rather slight, went neither unnoticed nor uncriticized. By the fall of 1960, prior to the time the MVPCB had adopted criteria and test procedures for exhaust devices, one company (under contract with General Motors) had developed a catalytic muffler to control exhaust emissions. The device, which had been road tested for six years, was expected to be ready for manufacture within a month. A news account suggests that other devices were about ready, but added that manufacturers were not "tipping their hands until the State Motor Vehicle Pollution Control Board sets some standards and specifications."[80] The story prompted an editorial complaining that the MVPCB was proceeding "snail-slow." A device seemed available, it observed, and the Los Angeles APCD laboratory was available for testing, but the MVPCB had not even settled on procedures; such "unnecessary delays" were intolerable.[81]

This criticism was the first entry in a long series of complaints about delay, responses from the MVPCB, followed by new complaints that it had capitulated in the name of expediency. Throughout its life, the

MVPCB was caught in a middle from which it never escaped, due partly to the feebleness of its attempts to do so, but owing also to the fragmented system of authority and responsibility in which it operated. Only when the state's air pollution control program was attacked by an outside foe (the federal government) did the MVPCB find friends among its former enemies, and only when a new agency with expanded authority was created in 1967 did there seem much hope of avoiding infighting among the various control authorities in the state (and that hope was soon to vanish).

The averaging controversy was at the heart of these developments. Perhaps in an effort to quiet its critics, the MVPCB by January 1961 had established some criteria for exhaust devices. Though additional criteria as well as testing procedures remained to be settled, the agency nevertheless sent out application forms to companies interested in submitting devices for testing, and the first returns were in within a month.[82] But critics remained unsatisfied; in June 1961 Los Angeles County made its official entry as the leading adversary of the MVPCB. There was no evidence, its board of supervisors said, of progress in certifying devices for installation at the earliest possible time.[83] Perhaps once more in response, the MVPCB managed by the following September to complete the exhaust device criteria, and a month later adopted the necessary test procedures. The criteria reflected a number of technical standards, required that devices be of reasonable cost, and—of most interest here—further required that devices comply with state emission standards for at least 12,000 miles.[84] The exhaust control program appeared to be on the road.

But not for long. By May 1962 three exhaust devices had been accepted for testing; by September the number was six; and by November, seven.[85] But new problems were appearing. Testing, for one thing, was terribly time consuming, requiring about an eight-month period. The MVPCB, perhaps anticipating further attacks on its progress, began devoting hard thought to means by which to speed up the test procedures.[86] By midyear the auto companies were proclaiming that "the extreme complexity" of the exhaust problem made it much less tractable than the "relatively simple" matter of crankcase emissions. One company found existing devices inadequate and unacceptable and was "very interested in seeing if these devices can meet the Board's requirements for the minimum 12,000 miles." It could offer nothing better of its own until completion of "a considerable engineering and materials development program [that] lies ahead. . . . In the meantime," efforts would be continued "in cooperation with the rest of the automobile industry."[87]

This rather pessimistic note was written with the then-existing criteria in mind; at virtually the same point in time, however, those criteria were being called into question. Evidence was developing that the estimates used by the Department of Public Health in determining hydrocarbon and carbon monoxide emission standards in 1959 were inaccurate, and that standards more strict than those adopted would be necessary to realize the air quality of 1940 by the target year of 1970.[q] By September 1962 even those with a heavy interest in the success of exhaust devices—the manufacturers seeking approval of their inventions—were skeptical of an entirely favorable outcome; a majority of the applicants were said to "believe it is highly doubtful that exhaust devices on used cars will ever be practical."[88] Once again the Los Angeles County Board of Supervisors was disenchanted; in October 1962 it reiterated its "disappointment over failure to achieve control of smog." As high an authority as Haagen-Smit told the board an effective device was yet to be perfected.[89] The year closed on the same low note; six days before Christmas the public was told that exhaust emissions accounted for 65 to 70 percent of total vehicle pollutants, but "so far no devices to control these emissions have been found satisfactory."[90]

As complaints about the slow pace of progress continued, the MVPCB began to counterattack. In January 1963 it directed some criticism of its own at Los Angeles, resolving that the area should give prompt consideration to rapid transit as a means to help control the pollution problem. The motivations behind the move were clear to all: "The resolution . . . was directed at the Los Angeles County Board of Supervisors and the City Council as a retort to attacks by supervisors that the auto smog board has been dragging its feet in the vehicle emission control program."[r] In midyear the MVPCB received the power, as a result of the 1963 legislation discussed earlier, to consider system or engine modifications—as well as separately installed technologies—for approval as pollution control "devices." The agency had requested such power in 1962, presumably in the hope it would speed up the certification process.[91] But as the year passed, still no approaches to exhaust control were found satisfactory. In April 1963 the MVPCB considered a suggestion to require that only the *average*

[q]*California Motor Vehicle Pollution Control Board Bulletin,* Vol. 1, No. 6, June 1962. Apparently the automobile industry provided the data for the original estimates. See A. Wiman, "A Breath of Death" (Transcript of special report on KLAC Radio, Metromedia, Inc., Los Angeles, Calif., 1967), p. 25.

[r]H. Nelson, "Rapid Transit System Termed Smog Deterrent," *Los Angeles Times,* 18 Jan. 1963. Perhaps to quiet such criticisms, Los Angeles County had, by the end of 1963, taken a few steps of its own to reduce vehicular emissions: it voted to purchase only Chrysler Corporation vehicles for its fleet, tests having indicated better emissions performance for Chrysler than other vehicles. See Roberts, "Motor Vehicular Air Pollution Control," pp. 33-34.

emissions of a vehicle tested over 12,000 miles comply with the state hydrocarbon standards, as opposed to requiring (as was then the case) that at no point within the 12,000-mile test could emissions exceed the standard. This approach could speed up certification, but it was rejected.[92] Pressure on the MVPCB continued; by the end of the year, the city of Los Angeles had joined the bandwagon. Mayor Sam Yorty called on the MVPCB "to end its 'interminable delay' in approving an auto exhaust control system to eliminate smog." Yorty noted that four years had seen the approval of no exhaust device, and expressed disappointment at the "failure to ask for authority to establish more realistic standards" for the devices. Instead of "rigidly adhering" to the original standards, Yorty suggested settling for less now, "with a hope for more perfection later."[93]

As 1964 began, so too did the buckling of the board. To be sure, it tried to absolve itself in some part by passing blame on to the auto companies. In 1962 its view had been that the manufacturers had "constantly demonstrated complete cooperation."[94] By early 1964 the message had changed considerably. The companies were criticized for their delay in developing control devices and accused of being evasive about progress; there was a suspicion among MVPCB members that the industry in fact already had control techniques at its command—a proposition the industry denied.[95] The MVPCB's statements suggested a hard line, but the occasion for making them was a more reliable sign that the agency was, in fact, backing down under pressure. The occasion was a January 1964 hearing to reconsider the averaging approach rejected a year earlier.[96]

There is little doubt about why the MVPCB had so suddenly reversed its position and opened its mind to averaging; all the evidence suggests that the move came in response to the constant complaints of footdragging. At the time, the reason was not stated so bluntly, at least not publicly; at the January hearing the MVPCB's executive director (who was a leading proponent of the approach) explained the averaging proposal as a way to encourage invention and production of control devices.[97] But subsequent statements by the executive director and other MVPCB members were more candid: averaging was a "concession" to get the exhaust program "into action"; it was needed because the state's program was not accomplishing much; it was a way to make real progress on certifications as quickly as possible; it was needed to get going. The executive director admitted he had finally reached the point where he wanted "to rush into something." He was aware that the averaging proposal would be controversial, and accordingly consulted with the governor and his cabinet before going forward.[98] Assurances that averaging would be consistent with the mandate from the legislature were also sought from the attorney general's office—and obtained. (At one point the MVPCB claimed it had received an opinion of the attorney general to that effect, but in fact the "legal advice" was received in a brief and informal conversation with a member of the attorney general's legal staff.)[99] With all this backing the averaging policy was adopted, but only by the closest of margins.[s]

The MVPCB may have had in mind more than simply quieting its critics when it adopted averaging early in 1964; an additional purpose might have been to intimidate the auto companies into action by reducing certification requirements and thus increasing the chances of approval of devices developed by independent manufacturers. With the approval of two such devices, the automobile industry would be obliged to install them. The automobile companies could not have been unaware of the strategy; the MVPCB stated publicly its belief that the industry had devices and "that some strong triggering action is needed to get them out of the lab and into production."[100] Thus the companies were put in the ironic position of having to contest the averaging proposal, though they could hardly express what were almost surely their true reasons for objection. Rather, they based their case against averaging "on the grounds it was really a

[s]Averaging was officially adopted by the MVPCB in January 1964. *California Motor Vehicle Pollution Control Board Bulletin,* Vol. 3, No. 2, February 1964; G. Siegel, "The Status of Motor Vehicle Emission Control Public Policy" (mimeo., n.d.), pp. 36–37. The averaging approach adopted by the MVPCB is to be distinguished from the concept of averaging used by the Department of Public Health in 1959 to determine the necessary emission standards. As discussed earlier, the department knew of great variations in emissions performance over the entire population of vehicles, and it reasoned that the desired ambient quality could be achieved so long as the average emissions of the total population reflected control to the necessary degree; the emissions of any single vehicle were considered unimportant. It appears that the MVPCB had employed this sort of averaging from the outset; when it tested emissions from a device under question, it considered not whether individual test vehicles exceeded the standards but rather whether the average for the whole test fleet met them over 12,000 miles. But the policy adopted in January 1964 eased this last, 12,000-mile, constraint. In order for a device to be approved under the new policy, average emissions for the test fleet could exceed the standard within 12,000 miles so long as the average of the fleet's performance as a whole over the 12,000-mile range complied with the standard—that is, so long as average emissions *below* the standard were sufficiently below so as to cancel the effect of average emissions exceeding the standard due to deterioration toward the end of the 12,000-mile test. In other words, under the original policy average emissions were not permitted to exceed the standard within 12,000 miles; under the new policy they were so long as the average of the average emissions within 12,000 miles did not. For this reason, the new policy has been described as "double averaging." Roberts, "Motor Vehicular Air Pollution Control," pp. 32, 96. See also M. Brubacher, Testimony before the California Assembly Transportation and Commerce Committee (mimeo., 15 Nov. 1967), pp. 2–4.

lowering of the standard."[101] But no one doubted their motivations: the "concern was that the State was going to certify catalytic mufflers, devices which were not being manufactured by the Detroit auto makers."[102] More broadly, the concern was that two devices of *any* sort would be approved, thus requiring the industry to control exhausts, whether with its own devices or those approved by the MVPCB.[103] In either case there would be increased purchase prices for new automobiles, and unhappy consumers who would be unlikely to see how they, *as individuals,* benefited from the exhaust control devices they were forced to purchase. (Exhaust controls produce a collective good that benefits people in addition to the consumer; chrome on a car, a fancy interior, and similar "extras" produce private benefits that the consumer can appropriate to himself. A purchaser might happily buy extras even though they mean a higher price for the automobile. Since most extras are optional, purchasers who do not want them have the choice to forego them. But the purchase of exhaust controls, of course, would be mandatory.) A probable result of the increased purchase prices would thus be some decline in demand for the auto manufacturers' product—a matter of considerable concern to the industry. As one of its spokesmen said a few years later:

> The automotive industry is extremely sensitive about anything that adds cost to the vehicle [without at the same time yielding commensurate private benefits to the consumer]. What you have to keep in mind is that each increment that's added to the price of the vehicle prices a certain number of people out of the market, and this reduces the demand for the product. So this is why we, inherently, are opsed to increasing the price of the vehicle.[104]

If the tactic was to scare the auto companies into action, there was no quick sign of success. In March 1964, two months after the adoption of averaging, the major companies announced a hard-line position. They could attain the required degree of exhaust control on 1967 models, but not earlier.[105] But the position was soon to soften. In June the MVPCB approved four exhaust devices developed by independent manufacturers (three catalytic mufflers and one direct flame afterburner). Though only one of these was certified for used vehicles, all were suitable for new.[106] Thus, under the two-device requirement discussed earlier, installation of exhaust controls would be mandatory on new vehicles beginning in 1966. The press noted the impact of the decision on the auto manufacturers: they would have to come up with exhaust controls of their own a year earlier than they had said they could, "or install one of the approved devices . . . if they want to sell cars in California."[107] The auto companies proved miraculously equal to the new challenge. Despite earlier disclaimers, they announced in August 1964 that they had developed systems (engine modifications)[t] of their own that were superior to the add-on techniques already approved and that could be ready for introduction beginning with 1966 models.[108] The companies sought and received certification of their methods;[109] "interestingly enough, none of the exhaust control devices developed by independent companies and initially certified for use on 1966 models was actually used on new vehicles by the big three automobile manufacturers."[110]

The promise of exhaust controls for new vehicles beginning in 1966 was a step forward, but much remained to be accomplished. The MVPCB said that smog would "begin to disappear visibly" with the appearance of exhaust control, but "scoffed at reports that smog could be eliminated 'in four or five years' through control devices. It may take that long just to get devices on all cars."[111] The press agreed. It applauded approval of the exhaust devices but cautioned that "those who expect the skies to be cleansed as soon as the state board issues its edicts will be disappointed. . . . Much more remains to be done."[112] In the meantime, the MVPCB was struggling to hold on to what had already been accomplished. As mentioned earlier, evidence had developed that the hydrocarbon and carbon monoxide emission standards originally adopted by the Department of Public Health were in fact insufficient, and just as the MVPCB was about to approve the exhaust devices, the department was considering adoption of more stringent requirements; moreover, it was considering standards for nitrogen oxides (none had been set in 1959). The MVPCB objected to new standards on the ground that they would discourage the industry in its attempts to achieve those already established. "It is imperative," the MVPCB argued, "that no action be taken by a state agency to discourage private industry from producing and installing devices geared to the present . . . standards before such devices are installed on California vehicles."[113]

The outcome was at least a partial victory for the MVPCB. Emission standards for oxides of nitrogen were not adopted; rather a report was to

[t]Chrysler had a "Clean Air Package" consisting primarily of carburetor adjustments and timing changes; Ford and General Motors had developed an air injection system. The companies preferred these integrated approaches to the approved add-on devices, which they viewed as less efficient. In 1968 Ford and General Motors adopted the Chrysler system because it was less expensive than their own. The air injection system they had been employing up to that time was "known and available for use" in the 1950s. Interview with J. Askew.

be prepared on the subject of their role in the pollution problem.[u] Stricter standards for hydrocarbons and carbon monoxide were established, but they would not become effective until 1970.[114] (The MVPCB had recommended that if changes were to be made, they should only become effective at a future date so that the existing rate of progress could continue.)[115] Nevertheless, the MVPCB's position on the issue was an uncomfortable one, inasmuch as it aligned the agency with the auto industry, which sent its "leading air pollution experts" to argue against stricter standards. The industry was especially concerned with the proposal for controlling oxides of nitrogen; its "experts assailed the lack of experimental evidence to support the proposition that control of oxides of nitrogen would be beneficial, yet they produced only a fragment of experimental data of their own to controvert it."[116]

Despite its success in retaining, at least for the near future, the exhaust emission standards initially established in 1959, and despite its success in certifying exhaust devices for new vehicles, the MVPCB was confronting difficulties in getting the used-vehicle exhaust control program on the road. One used-vehicle device had been certified, but approval of two such devices was necessary to activate an installation requirement, and the MVPCB was unable to find a second suitable device. The difficulty had to do with costs. In December 1964, at the height of the controversy over crankcase devices for used vehicles, the MVPCB considered an exhaust device that would cost $150 to install and $50 annually to maintain. No doubt because of the crankcase controversy, the agency proceeded with considerable caution on the exhaust device. It asked for "a reaction from the public on whether it would go along with the relatively high costs of exhaust devices for used cars." The answer was a resounding no. "Of 1,277 letters, telegrams, resolutions and petitions, only nine approved installation of the device if it would cost as much as current available models." At a public hearing on the issue (a hearing that the Los Angeles County Board of Supervisors had demanded), objections ran higher: "numerous witnesses," including the board of supervisors, "opposed requiring the exhaust device on used cars because of the high cost." The device was not approved, and another year of the battle against air pollution ended on a gloomy note.[117]

Events in 1965–1966 made it clear that the MVPCB had slowly been maneuvered (or had maneuvered itself) into a position where it could do nothing right, where a step in any direction would bring cries of protest

[u]Subsequently, in 1965, ambient and emission standards were adopted for nitrogen dioxide and oxides of nitrogen. See Siegel, "Status of Policy," p. 41.

from some strong interest. In January 1965 it heard demands from one member of the Los Angeles County Board of Supervisors that the county "withdraw" from MVPCB jurisdiction and that the legislature investigate its activities—all because the approved devices only increased pollution and imposed a tax on citizens.[118] The MVPCB, which had taken steps to speed up its program in response to complaints about being too slow, was now being charged with sloppiness in an effort to go too far too fast. The charge must have been especially nettlesome inasmuch as the agency had, just a month before, eased off on used-vehicle exhaust devices out of a new sensitivity to costs and public sentiment.

The MVPCB's position was to become even more vulnerable to conflicting demands. As recounted earlier, the legislature in mid-1965 had passed a measure that limited used-vehicle crankcase requirements, abolished the inspection program, and set a $65 ceiling on the price of devices to be certified in the future. The same legislation also required exhaust controls on 1966 and later model vehicles.[119] But now it was discovered, or at least alleged, that the measure had accomplished something else—it had inadvertently repealed the two-device requirement.[120] Soon thereafter, the manufacturer of the device approved earlier for used vehicles appeared with the claim that it could meet the new $65 ceiling and that, accordingly, the requirement for exhaust controls on used vehicles should be put into effect. The situation presented a chance to put the used-vehicle exhaust program into action, and one might have expected the MVPCB to take advantage of the opportunity. At the end of 1964, the agency itself had asked the attorney general whether it could meet the then-existing two-device requirement by certifying one device subject to the condition (to ensure competition) that its manufacturer agree to license the device to others; moreover, in April 1965 a legislative resolution calling for swift approval of a second used-vehicle exhaust device had been introduced.[121] Finally, since the one approved device could apparently meet the new $65 price ceiling, the MVPCB could probably anticipate less public objection than had been generated by proposals for a much more expensive device late in 1964. Yet the MVPCB took the position that the two-device requirement still existed, despite an opinion from the legislative counsel to the contrary.[122] No doubt the board had been burned once too often, and concluded that the safest course was legislative clarification.[v]

[v]It was hardly clear beyond all doubt that the 1965 legislation had repealed the two-device requirement, but a strong case to that effect could be made. The original legislation passed in 1960 tied mandatory installation to certification of two or more devices. The 1963 legislation repeated this approach. The 1965 legislation

Such caution, though understandable in light of the ambiguity of the "repeal" and the history of bad relations with an aroused and warring public, nevertheless had a heavy impact on the incentives of independent manufacturers. Those incentives had already been dampened by the pink-slip controversy, and by the auto industry's "unexpected triumph" in discovering its own exhaust controls shortly after approval of nonindustry devices. (According to one report, that discovery cost the independent manufacturers millions in research and development expenditures.)[123] The December 1964 objections to the cost of a second exhaust device further aggravated the independents and led to the development of generally "cynical and negative feelings."[124] One manufacturer had submitted cost data that suggested a price to the consumer considerably below the estimates discussed at the public hearing in December 1964, but apparently the MVPCB made public its belief that the data were suspect. The manufacturer noted that similar concerns over cost had not been expressed about auto industry control methods and added that it would abandon work (and an investment of "millions of dollars") on used-vehicle devices if the MVPCB could not be more cooperative; later it did curtail its program, detecting "a complete change in attitude . . . which by coincidence occurred right after the automobile manufacturers agreed to equip new cars with control devices."[125] When the MVPCB took the position that the 1965 legislation had not repealed the two-device requirement, another independent manufacturer wrote that he could not recommend that his company pursue marketing its device; citing the ineptitude revealed by the pink-slip controversy, the abandonment of yearly inspections, and the limitation of the crankcase program, he concluded that the MVPCB was too unreliable.[126]

In a way, independent manufacturers were to be discouraged further in 1966, though not as a direct result of MVPCB action. In late 1965 Governor Brown asked the legislature to clarify the status of the two-device requirement—in particular, to repeal the requirement in clear fashion. The measure failed to pass in that year, probably in part because of disclosures that the governor's son-in-law was employed by the firm whose used-vehicle device had already been approved—the firm that "might gain a monopoly" if the two-device requirement were abolished.[127] At

---

left standing a provision that the MVPCB was to notify the Department of Motor Vehicles when it had approved two or more devices, but it *repealed* an immediately subsequent provision that had appeared in the first two enactments and that tied the mandatory installation to the notification. See Ch. 23, § 1, [1960] Cal. Stats. 1st Ex. Sess. 346, 348; Ch. 999, §§ 3 and 4, [1963] Cal. Stats. Reg. Sess. 2264, 2266; Ch. 2031, § 2, [1965] Cal. Stats. Reg. Sess. 4606, 4609.

the beginning of 1966, the governor was before the legislature once again, asking for repeal of the two-device requirement as one of a number of proposals for air pollution legislation.[w] A measure to require only one device was introduced and became law in mid-1966. In order to protect "against a potential monopoly situation," it required the "manufacturer of the first successful device with a maximum cost of $65 to agree to MVPCB conditions prior to certification," including agreement "to make specifications available to others by cross-licensing to insure competition." (Alternatively, the manufacturer could be required to agree to price setting by the MVPCB.)[128] This was the very approach the MVPCB had considered undertaking on its own initiative in late 1964.

One might have thought that the legislature's repeal of the two-device requirement would have acted as an effective spur to research and development by independent manufacturers. By all appearances, however, it did not. The legislation, in addition to making clear that approval of only one device would activate the installation requirement, went on to impose performance criteria for new certifications that were more stringent than those that had previously existed. Annual maintenance costs were not to exceed $15; moreover, devices were to have an expected useful life of 50,000 miles. Apparently these new requirements were a final straw to the frustrated independents. One of them abandoned all hope of working through the certification process and announced that it would instead market its devices for voluntary installation. The MVPCB, however, cut off that approach, advising the company that such a marketing program would be illegal.[x] Almost five years were to pass before any further significant progress on exhaust controls for used vehicles.

[w]See Editorial, "Action Required on Air Pollution," *Los Angeles Times,* 8 March 1966. The MVPCB had urged the governor to take the question to the legislature once more in 1966, and the legislature itself wanted to reconsider. It had resolved late in 1965 that the governor submit the two-device question to it early in 1966; the resolution observed that the 1965 air pollution legislation had "inadvertently deleted" the two-device requirement. Ch. 17, [1966] Cal. Stats. 1965 2d Ex. Sess. 169.

[x]Roberts, "Motor Vehicular Air Pollution Control," pp. 93–94. Roberts suggests that the MVPCB's opinion was incorrect, that in fact "the law forbade only the representation as certified of those devices which were uncertified." Ibid., p. 94. But a close look at the statute appears to sustain the MVPCB. The law did state as Roberts claims, but added: "No person shall install or sell for installation upon any motor vehicle any motor vehicle pollution control device which has not been certified by the Motor Vehicle Pollution Control Board." Ch. 2031, § 9.5, [1965] Cal. Stats. Reg. Sess. 4611, amending California Health and Safety Code § 24395 (1965). It is worth noting that the company involved had developed the one used-vehicle exhaust device that had been certified in 1964. At the time of that certification,

It had become a fact of life for the MVPCB that each year ended on a downbeat, and 1966 was no exception. Not only had the used-vehicle exhaust device program stalled but in November 1966 the very technique that had been adopted in part to help get that program on the road—averaging—came under furious attack from Los Angeles. Louis Fuller, the county's air pollution control officer, revealed data indicating that even when tested at less than 2,000 miles, 37 percent of a sample of 1966-model vehicles exceeded state standards. The figure grew to 63 percent for vehicles with more than 2,000 miles and reached 85 percent for vehicles with 20,000 miles. Virtually no 1966 models, the tests indicated, would meet state standards by 50,000 miles. Overall, 52 percent of the models failed to meet state hydrocarbon standards. The thrust of the attack was that the vehicles were not meeting standards and that averaging aggravated the deficiency.[129] The Los Angeles County Board of Supervisors, in reaction to the information, suggested a grand jury investigation.[130] The MVPCB offered some feeble defenses. Its executive officer argued "that despite decreasing efficiency of the smog devices, the exhaust control systems alone are keeping more than 100,000 gallons of unburnt gasoline out of Los Angeles air every day."[131] Its chairman, implicitly acknowledging the accuracy of the data at issue, countered that "50% [of the vehicles] were better than state requirements."[132] The MVPCB also took the offensive, warning the auto companies that their vehicles would be barred from sale in California if they failed with increasing mileage to meet state standards; the bar would probably apply only to 1968 models, but there was talk of recalling vehicles that had already been sold.[y] But the situation was beyond redemption. The MVPCB's averaging policy, not to mention the agency's overall performance, had reached a terminal state. The MVPCB's troubled life was to end in 1967; averaging would be officially killed the following year.

---

the two-device requirement was in effect. By the time that requirement was abandoned, new requirements (e.g., initial cost) had become applicable, and the formerly approved device did not meet these. It must have been the thought that the 1964 certification was ineffective until another device was certified; thus the device had not yet been *effectively* certified and apparently could not be under the new, more stringent performance requirements.

[y]"State to Warn Auto Makers on Smog Devices," *Los Angeles Times,* 4 Nov. 1966; "State Smog Group Critical of Devices," *Los Angeles Times,* 17 Nov. 1966. It should be understood that in giving its warning the MVPCB was not abandoning averaging. It warned the companies that approval would be withheld unless average emissions were controlled within state standards. Its chairman insisted that if the MVPCB required every vehicle to meet the hydrocarbon standard there would be no control system at all. *California Motor Vehicle Pollution Control Board Bulletin,* Vol. 4, No. 11, November 1966.

Most observers have found in the averaging policy a go-easy attitude with respect to the automobile industry. The policy has been variously described as a way to allow the auto companies to "get around" the state standards; as "a sop to the automobile manufacturers and . . . a horrible hoax on the public."[133] According to this general view, averaging (combined with the MVPCB's requests that the Department of Public Health refrain from tightening emission standards) was part of a "conciliatory policy" by an agency "more bent on cooperation with . . . the manufacturers" than on fulfilling a legislative mandate.[134] To be sure, there is some (scant) evidence that averaging came about as a result of pressure from the automobile companies (not the independent device manufacturers), although the evidence amounts only to innuendo or assertions by persons who had long been critical of both the concept and the MVPCB.[135] It is also the case that a representative of one automobile manufacturer took the position, some years after averaging had been implemented, that "the averaging concept is a reasonable one . . . *if* you had a population of vehicles which would [achieve the] average . . . *and* the original structure was mathematically correct."[136] There is no more concrete evidence of industry pressure than these assertions and statements, and the fact remains that the industry took a public position against the concept. It could not, on that occasion, state its real motivations, which had to do with the likely impact of certification on industry sales. The legislature had tied mandatory installation to certification of control methods, and there is no reason to suppose that the industry would welcome that certification. To the contrary, certification—or any technique, such as averaging, that would hasten it—would likely be opposed out of the industry's perception that it would increase prices and reduce demand for its products. (The independents *did* have a real interest in early certification, since it would result in *creating* a demand for their primary product, control devices. But there is no evidence of pressure from the independents other than that implicit in their increasing frustration with the MVPCB as it failed to certify exhaust controls.)

In short, it is doubtful that averaging in fact came about in response to industry pressure. Charges to that effect were probably motivated simply by a larger belief, partly myth and partly not, that the MVPCB was engaged in a cooperative venture with industry.[z] "Pressure" did indeed

[z]This conclusion seems almost inescapable when one considers that a much publicized radio program that heavily insinuated the presence of industry pressure behind averaging at the same time summed up the evidence leading to a directly contrary conclusion. See Wiman, "A Breath of Death," pp. 24-25 (implication of pressure), p. 22 (auto industry objected to averaging out of the "concern . . . that the State was going to certify catalytic type muffler devices which were not being

push the agency to averaging, but its source was not the auto industry. Rather it was the demands, mostly from Los Angeles, for the MVPCB to speed up its program, to end its "interminable" delays in certifying exhaust devices even if this meant settling for less rigid standards now, with a hope for perfection later. The averaging decision followed on the heels of these demands, and was accompanied by statements that the decision was intended to put the exhaust program into action, to make progress on exhaust certifications as quickly as possible, to get the program going. Several years after the decision, MVPCB members repeated the same motivations. In 1967 the MVPCB's executive director told the Los Angeles County Board of Supervisors that averaging was "a response to demands from all of you, and this Board, back in 1961, 2, 3 and 4, that systems should be approved as soon as possible, even though not perfect."[137] Others hold the same view of the reasons for the averaging decision.[138] If that decision was a concession, it was not made in response to the auto industry.

As late as 1967, the MVPCB continued to defend its position on the ground that averaging, whatever its legislative history, was a reasonable approach. Averaging was "the only means of achieving any degree of success in emission control"; any other approach would be "unfair," "arbitrary," "capricious," and "illegal."[aa] Admittedly, the exhaust devices were not perfect, but they were significantly effective, "doing an immense amount of good and . . . worth the money that is being spent on them."[139] But these claims were simply met by new assertions from Los Angeles

---

manufactured by the Detroit auto makers"), p. 42 (statement, quoted earlier, of auto industry's opposition to any requirement that would increase purchase prices). Wiman has stated that the "Breath of Death" program started as a result of the averaging decision. Interview with A. Wiman, 5 Aug. 1971. The program was highly critical of the MVPCB. Perhaps it was for this reason, as well as the contradictions in evidence mentioned above, that one former MVPCB member held the opinion that the program lacked integrity and honesty. Interview with J. Havenner.

[aa]From statements by the MVPCB executive director, quoted in Wiman, "A Breath of Death," pp. 20–21. The basis for the statements was that a standard requiring compliance by *each* vehicle could not in fact be met—that such an approach was not feasible and would not be sustained in a judicial test. Ibid., p. 21. In this light, it is fascinating to wonder whether the MVPCB had been influenced by the earlier legal opinion implying that infeasible standards were not legally enforceable—an opinion rendered by the counsel for Los Angeles County, the body now complaining so bitterly about averaging!

The executive director changed his position later in 1967, saying that it would be "desirable" to abandon averaging and that the MVPCB was "working in this direction." Wiman suggested that this was in response to the "Breath of Death" campaign. Ibid. Other evidence suggests that the change of heart had come about considerably before that program. See *California Motor Vehicle Pollution Control Board Bulletin,* Vol. 6, No. 2, February 1967.

County. Its air pollution control officer alleged that averaging had "effectively undermined both the standards and the intent of the Legislature." Contrary to the MVPCB's claim that any approach other than averaging would likely be unlawful, he argued that averaging was illegal as beyond the agency's authority.[140]

The Los Angeles APCD as much as admitted that it had directed a strong attack at the MVPCB. The reason for the attack, one representative of the APCD said, was the belief that the MVPCB was operating "an inadequate program and trying to cover it up"; accordingly, the APCD "blew the whistle."[141] But less pure motives may also have been at work. Los Angeles County had been engaged in a long feud with the state over the relative contributions of vehicular and stationary sources to Los Angeles's air pollution problem (the county had jurisdiction over the latter).[bb] As others have observed, Los Angeles had "an unarticulated interest in passing off some of the public pressure for cleaner air," and tried to blame "the close relations" between the MVPCB and the automobile industry "for the lack of greater governmental pressure on the auto manufacturers." Los Angeles, in short, needed a "political scapegoat," and the MVPCB was it.[142] A former MVPCB member summarized this view nicely. Los Angeles, he said, "played both sides of the fence" on the averaging issue, first asking for approval of a device even if it were not perfect, and then—when the MVPCB adopted averaging "with the knowledge of Los Angeles"—"screaming" that 50 percent of the vehicles were over the standards, in a "grab for political capital."[143]

Yet the MVPCB must take some responsibility for bringing the rain of criticism on itself; if nothing else, it at least had adopted practices and postures that lent the appearance of credibility to allegations of an unduly cooperative relationship with the industry. There is no question that the agency actively sought industry cooperation; indeed, it talked of teamwork and a division of labor with the industry, for a considerable period announced that the auto companies "constantly demonstrated complete cooperation," approved auto company laboratories for purposes of testing, relied on automobile club surveillance programs as a means to monitor devices in operation, and invited industry engineers to serve on technical

[bb]See Wiman, "A Breath of Death," pp. 25-26. Thus the county claimed that *automobile* controls were not "good enough to rid Los Angeles of smog." G. Getze, "Criticism of Smog Devices Mounts," *Los Angeles Times,* 10 May 1967. The MVPCB, on the other hand, suggested that "our friends downtown [meaning Los Angeles] . . . should pay some attention to [stationary sources] instead of just the motor vehicle." One of the MVPCB's former members publicly singled out Los Angeles as "the first indication of a political body as having a potency against air pollution control." Quoted in Wiman, "A Breath of Death," pp. 25-26, 27.

advisory committees. A former MVPCB executive director accepted a high-salaried position with Ford during the peak of the averaging controversy, and a chief engineer took two trips to Europe at the expense of foreign automotive manufacturers. The leading critics of the MVPCB found considerable capital in all of these events.[cc]

There are sound explanations for some of the MVPCB's practices (some were simply poor judgment). A number of its former members observed that, in light of the agency's slim staff and budget, cooperation with the industry was essential in order to maintain a "line of communication" over which needed technical information could travel. The MVPCB was aware that some of this information would be "biased," and screened it as best it could. But it was "necessarily close to the auto industry," and this put it in a "delicate position."[144] On this count, Los Angeles County's APCD was enviably situated. "As the District did not have to engage in continuous negotiation with the auto manufacturers, it could afford to proceed with less tact, and its leaders often questioned the integrity of the less outspoken [MVPCB] officials."[145] Given the role of Los Angeles in air pollution control and the MVPCB's openly cooperative approach, one can reasonably conclude that the opportunity of the former to question the integrity of the latter was too inviting to forego.

The averaging and pink-slip controversies are tales of a body caught in the middle by conflicting demands, conflicting roles, limited resources, and a measure of its own ineptitude. To pinpoint blame for the affairs is not so important as to note their effects. The averaging debacle surely helped to bring about the death of averaging itself—a net loss to the extent that averaging, if properly formulated and overseen, could be an efficient approach to the emissions problem.[dd] Beyond this small point, the debacle of averaging cemented the poor relations between state and local control

[cc]See, e.g., Wiman, "A Breath of Death," pp. 22, 25, 37–38. See also interview with D. Jensen, 8 Sept. 1971. With respect to laboratories, the MVPCB resolved in 1963, after Los Angeles had donated its own laboratory to the state, to conduct all future testing in that facility. *California Motor Vehicle Pollution Control Board Bulletin,* Vol. 2, No. 8, August 1963. But in August 1964 the MVPCB adopted procedures to approve outside laboratories, and in October of the same year approved automobile company laboratories. Ibid., Vol. 3, No. 8, August 1964; ibid., No. 10, October 1964. There is some evidence that the Los Angeles laboratory was little used; "speculation" had it that this was because it was "housed in the same building with the Los Angeles County Air Pollution Control District." Wiman, "A Breath of Death," p. 38.

[dd]Since it is the total amount of emissions that is important, averaging seems sensible. However, even a well-designed averaging approach has one shortcoming. Because it is inherent in the concept that not *every* vehicle must meet applicable

authorities and, together with the pink-slip affair, undermined a considerable amount of public confidence. Finally, and no doubt related to the matter of public confidence, both controversies sent efforts to control used-vehicle exhausts into a trauma from which they would not even begin to recover until 1970. Considering that in the intervening years such control was considered essential to substantial progress in air quality, this last point is hardly a small one.[146]

FEDERAL ACTIVITIES

The federal government was not so busy as California in the years 1960-1966, but neither was it entirely idle. The 1950s had closed with a substantial consensus that it was appropriate for the federal government to encourage and support research regarding the causes, effects, and control of air pollution. On the issue of federal control, however, there was marked disagreement, even within the Department of Health, Education, and Welfare (Secretary Flemming favored some federal enforcement power, but the Public Health Service opposed it). Concerning automotive emissions control in particular, Congressman Schenck had begun, in the late fifties, to urge legislation prohibiting the use in interstate commerce of motor vehicles discharging dangerous amounts of unburned hydrocarbons. Largely in response to objections from HEW, Schenck's measure was diluted to provide merely for study by the surgeon general of the effects of motor vehicle exhaust on the public health, and it passed the House in this form in 1959. In the next year it passed the Senate with little difficulty, perhaps largely because HEW and the Bureau of the Budget voiced no objections to the bill. The absence of objection was consistent with the view of these agencies that research was an appropriate federal role; by this time, moreover, HEW "recognized the importance of vehicle exhaust pollution and the desirability of keeping Congress informed."[147] President Eisenhower signed the law, the Schenck Act, in mid-1960. Thus the decade opened with Congress's first automotive pollution legislation, legislation that led to increased Public Health Service emphasis on the automotive exhaust problem.[148]

---

standards, the opportunity for an effective inspection program is largely negated by averaging "since there are no standards for the individual vehicle." H. Sullivan, Testimony in Hearings before Subcommittee on Air and Water Pollution of the Senate Committee on Public Works, 90th Cong., 1st Sess. 284-290 (Part I, 1967). See also R. Chass, Statement before Subcommittee on Air Pollution of the California Assembly Committee on Transportation, Los Angeles, Calif., 13 Nov. 1969; Interview with R. Barsky, 23 July 1971.

The year 1960 also saw Secretary Flemming, the primary proponent of federal enforcement powers, continue his campaign for legislation giving the surgeon general power to hold conferences in the case of significant interstate air pollution problems. The Senate passed such a measure in 1960 and again in 1961, but in both years the House Subcommittee on Health and Safety (chaired by Congressman Kenneth Roberts of Alabama) took no action on the bills, perhaps because the Public Health Service continued in its objections to federal enforcement powers (even in the relatively innocuous form urged by Flemming). President Kennedy had given at least indirect support to Flemming's proposal in a February 1961 message calling for a more effective federal air pollution program, but the Public Health Service persisted in its stand that the federal government's job was "provision of technical assistance, research knowledge, and the information on which a State and local government might act."[149] Against this resistance, Kennedy in 1962 became more direct. In a message early in the year, he asked the House to pass the interstate-conference bill upon which the Senate had previously acted favorably. Congressman Roberts introduced the bill, but by midyear concluded that his subcommittee needed more time to study its implications. In its place he proposed an additional two-year extension of the 1955 law providing for research and training, one that would take it through 1966 (a 1959 extension had continued the law, which would otherwise have expired in 1960, through 1964). The extension measure passed.[150]

The 1962 legislation did more than grant the 1955 act an additional two years of life; it also directed the surgeon general to continue permanently the motor vehicle exhaust studies that had been initiated under the Schenck Act. In June the surgeon general had submitted his first report on the subject. *Motor Vehicles, Air Pollution, and Health* discussed the evidence of harmful effects of vehicle emissions on humans, animals, and materials. Research was still in progress, however, and the surgeon general requested more time "to obtain the quantitative research information needed to serve as a basis for equitable and appropriate judgments as to the limitation of discharges of various pollutional substances from motor vehicles."[151] Congress reacted with concern. Its 1962 legislation directed ongoing studies "in view of the nationwide significance of the problem of air pollution from motor vehicles."[152]

The Kennedy administration was disappointed with the paltry congressional response to its request for an expanded federal role in air pollution. The Second National Conference on Air Pollution was upcoming in December 1962, and the secretary of HEW (now Abraham Ribicoff) hoped to publicize the national government's new responsibilities at that time.

Support for some federal enforcement powers was growing, and the national conference could serve a useful function for those who favored such powers. The urban lobby was coalescing around broadened federal authority, and in 1962 one of its spokesmen drafted legislation incorporating that view. The lobby—primarily the Conference of Mayors, the American Municipal Association, and the National Association of Counties—had initially been sympathetic to Flemming's proposals for interstate conferences in certain cases. Its 1962 program implied a more far-reaching attitude, incorporating most of Flemming's ideas but moving beyond them by dropping the traditional reference to the "primary responsibilities and rights of the states and local government," and by adding a policy statement referring to the federal obligation to provide leadership. The lobby, in a successful effort to avoid Congressman Roberts, managed to get its bill introduced by a friendly Democrat from Pennsylvania, but the proposal reached the House too late for action. This had been anticipated by the lobby, but it hoped the measure would nevertheless stimulate discussion at the national conference; in the meantime, it worked on preparation of an even stronger proposal that included explicit federal enforcement powers.

Those supporting increased federal activity had worked diligently to shape the national conference into a platform for urging their program. They received unexpected help from an air pollution episode that hit London shortly before the conference convened. The episode, claimed to have caused up to 700 deaths, was cited at the conference as evidence of the urgency of the air pollution problem. This, plus a speech by Flemming and a telegram of support from the American Medical Association, helped push the move for federal control forward—but not too far. The Public Health Service persisted in its views, and Congressman Roberts maintained his position that abatement and enforcement efforts, with the exception of motor vehicle emissions, should remain in the hands of state and local government. This was the consensus of the conference, despite the fact that only fifteen states had any control program at the time. (Of a total $2 million being spent on state control, over half represented California efforts; local expenditures totaled $8 million, more than half of this attributable to California, and nearly four-fifths of that to Los Angeles.)[153]

While 1962 closed on a note of victory for those opposed to federal enforcement, the tune was soon to change. The division of opinion within HEW remained a marked one, the Public Health Service (backed up by the Bureau of the Budget) continuing its objections to even such a limited federal "enforcement" role as the calling of conferences in the case of interstate pollution problems. Others in the department wanted that and more. In late December 1962 Wilbur Cohen, HEW's assistant secretary for legislation, convinced President Kennedy of the political wisdom of supporting strong federal enforcement. (In the past, Kennedy had given at least halfhearted support to the interstate conference approach originally suggested by Flemming. But Flemming, in the course of his speech at the National Conference on Air Pollution earlier in December, had gone further, arguing that the federal government should have abatement authority in the case of certain interstate air pollution problems—authority it already had with respect to water pollution. The American Medical Association had endorsed this view, and Cohen convinced Kennedy that he could not adopt a position to the right of the association and the Republican Flemming.) In February 1963 Kennedy recommended legislation to provide a more intensive research program, project grants to stimulate state and local control activities, studies on air pollution problems of interstate or nationwide significance, and abatement authority for interstate air pollution similar to the federal authority in the water pollution area.

The administration drafted no specific legislation of its own, but Ribicoff—now a senator and having decided to make air pollution his major project—had already introduced a bill. It drew on earlier proposals made by the urban lobby and added the federal abatement authority subsequently endorsed by Kennedy. Congressman Roberts, in the meantime, had experienced a change of heart. The recent London smog, he said, convinced him of the need for federal abatement authority in the case of especially severe problems. He introduced a bill roughly parallel to Ribicoff's. Debate thereafter centered on the abatement issue. The mayors and conservationists supported a strong federal role; the states and industry opposed it. The opposition relied, in particular, on the past tradition of state and local control and voiced its concern that a stronger federal role would quell state initiative. But the support for federal enforcement power had become too strong. A conference bill compromised on measures already passed in the House and Senate, and became law with President Johnson's signature in December 1963. Instrumental in its passage were Roberts, Ribicoff, Schenck, and a fresh figure—Edmund Muskie of Maine, who had recently been named chairman of a new Senate Special Subcommittee on Air and Water Pollution.[154]

The Clean Air Act of 1963[155] expanded the research and technical assistance programs begun in 1955, provided for development of air quality criteria by HEW, and created federal investigative and abatement authority similar to that already in effect for water pollution. These provisions were more bark than bite, however. Criteria, which were to reflect the most recent scientific knowledge on air pollution effects, were simply advisory guidelines to be used or ignored by state and local government. The investigative and abatement powers were hedged in by a set of complex and time-consuming procedures. As to intrastate problems, HEW could not act at all without a state request. For interstate pollution—but only if it endangered health and welfare—HEW could take the initiative, so long as it first consulted with local officials, then made recommendations, then held public hearings, then made more recommendations. If these were ignored, HEW could request the attorney general to begin abatement proceedings.

Regarding motor vehicle emissions, the act directed HEW to encourage the automotive and fuel industries to develop devices and fuels to aid in prevention of pollutant discharges, and, toward that end, to appoint a technical committee (with representatives from HEW, the automobile industry, the petroleum industry, and exhaust-control device manufacturers) to make recommendations on research programs and to evaluate progress. In addition, HEW was to report to Congress within one year, and semiannually thereafter, on measures taken and still needed to resolve the vehicle pollution problem.

With passage of the Clean Air Act, congressional attention focused more closely on automotive emissions. The credit for this belongs partly to Senator Muskie, partly to the state of California, and partly to the automotive emissions problem itself. By the end of Senate subcommittee hearings in 1963, Muskie had developed a considerable interest in air pollution. To act on it, he decided to hold a series of public hearings across the nation, both to increase congressional knowledge and to generate public concern. The first hearing, held—quite appropriately—in Los Angeles in January 1964, heard Governor Brown say that despite a valiant battle, California was only keeping even. The reason was the automobile. Stationary sources had been largely brought under control, but the resulting gains had been offset by growth in the population of motor vehicles. While California had attacked the automotive emissions problem, accomplishments to date had been modest, and the state wanted help. "The automobile industry," the governor testified, "is in interstate commerce and the Federal Government clearly has jurisdiction."[156] The other hearings repeated the story, and Muskie concluded that motor vehicle exhaust was a national problem calling for further federal intervention. The California experience, he said, should be used as the basis for a national program of automotive pollution control.

The automobile industry did not agree. In its testimony during the hearings, the Automobile Manufacturers Association took the position that automotive emission controls should vary according to the needs of each state. Muskie was surprised; a uniform requirement, he thought, would be the most advantageous to the industry. In any event, he considered it the best approach, and his subcommittee concurred. Its October 1964 report, *Steps Toward Clean Air,* pinned half the blame for the air pollution problem on the automobile and recommended, among other things, national standards for vehicular emissions.[157] Support came from a late 1964 report by HEW that documented the major national importance of automotive emissions. In January 1965 Muskie introduced legislation to establish exhaust emission standards for gasoline-powered vehicles. There had been implications of administration support for the measure but, when presented with a concrete proposal, President Johnson balked. He was concerned that the automobile industry, which had shown some signs of a willingness to consider voluntary action, had not been consulted. Rather than supporting the legislation, Johnson presented a February 1965 message stating he would initiate discussions with the industry designed to lead to elimination of the auto exhaust problem. Two months later no discussions had been started. An HEW assistant secretary, appearing in April before Muskie's subcommittee and bound by the president's decision, testified that the administration was opposed to regulatory legislation until the president's discussions got off the ground. Muskie was openly appalled, and the newspapers came down hard on the administration's stand. Editorials in the *New York Times,* the *Washington Post,* the *Wall Street Journal,* and the *Los Angeles Times* all hinted at a "love affair" between Johnson and the automobile industry.

If so, the industry was a fickle partner. Up to this time its position had been that the automotive pollution problem was not so severe across the nation as to require federal intervention, and that emission standards should vary with location. Now, immediately after the administration's statement of opposition to Muskie's bill, it changed that position in a subtle but significant way. Repeating that automotive pollution was not a national problem, it nevertheless announced that it was prepared to equip new vehicles with exhaust controls should Congress deem that necessary (the industry had already decided to install crankcase controls on all models beginning in 1963). But adequate lead time (two years) would be required, and the industry could only comply if a national standard rather than diverse state standards were established; moreover, those standards could be no more stringent than California's and both the standards and their effective dates should be determined administratively rather than by the legislature (Muskie's bill proposed the latter). The industry's change of heart is not difficult to understand. The argument in favor of state standard-setting had been a tactic to forestall federal intervention; it had not been contemplated that the states would exercise initiative. But an emission control bill had been introduced in Pennsylvania, and a bill in New York proposed standards tougher than California's. The diverse standards the industry feared most appeared to be imminent; in that light, uniform federal standards, hardly better than nothing, were nevertheless the easiest road.

Developments from here on came rapidly. The administration testified that it had been misunderstood and offered to work with the subcommittee to draft effective legislation. The price of cooperation was that standards and deadlines not be written into the legislation but rather left to administrative determination, and this Muskie was ready to do. In the Senate the standards in his bill were deleted in favor of administrative discretion, and the effective date of the standards—originally 1966—was moved back to 1967; in the House the 1967 deadline was deleted in favor of administrative discretion as to this as well as to the setting of standards. In October 1965 the bill became law.[158]

The Motor Vehicle Air Pollution Control Act required HEW to set emission standards for new vehicles, taking into consideration the technological feasibility and economic cost of compliance. A performance-standard approach was employed; the means of compliance were up to the manufacturers, and HEW would simply test submitted prototypes and certify those that met the standards. The law did not say whether the federal standards were to be the *only* ones for new vehicles—whether, that is, the federal government had preempted state standard-setting. But the legislative history seemed more than clear. The House report stated that standards were necessary on a national scale and that, states' rights not withstanding, federal standards were the preferable approach; the Senate report said the whole nation should benefit from control and that state standards could result in "chaos" for manufacturers, dealers, and users.[159] Nevertheless, the important matter of preemption had not been explicitly resolved. It would be, in two years time.

A year after passage of the 1965 act, HEW prescribed standards for emissions from new motor vehicles. The standards, roughly the same as those that became effective in California in 1966, were to become applicable beginning with the 1968 model year.[160] The federal program of automotive pollution control was underway. The accomplishment was not a small one, for the auto industry could be a mean opponent when its dearest interests were at stake. Ralph Nader discovered this when, in 1965, he indicted the industry for its neglect of safety hazards. The reaction of the industry was to embark on a disgraceful investigation designed to produce (in one way or another) information to discredit the young Nader.

The industry handled the matter with laughable ineptitude and ended up making the most public of apologies. The event revealed the industry's guile and at the same time diluted its proud image and political power. Another event in 1965 eventually served the same two ends. In January of that year the Los Angeles County Board of Supervisors requested the United States attorney general to pursue evidence that members of the industry had conspired to delay the development and introduction of pollution control techniques. (Antitrust charges were filed four years later. As mentioned earlier, they resulted in a 1969 consent decree whereby the industry agreed, without conceding the existence of a conspiracy, to cease any unlawful activities.)

11

STATE REORGANIZATION AND THE BEGINNINGS OF FEDERAL DOMINATION: 1967–1969

CALIFORNIA'S 1967 LEGISLATION

Considerable dissatisfaction with the makeup and functions of the MVPCB had developed by the close of 1966. The agency's political credibility and popular reputation had been seriously eroded by the pink-slip and averaging controversies, making it difficult to command compliance from the auto industry or cooperation from the public. There had been too many fiascos and too much fighting with Los Angeles County. The MVPCB's staff spent more time in internal discussions and public hearings than it did in carrying out a program of certification and surveillance. The MVPCB had no authority over stationary sources, no power to coordinate state and local programs, no ability to engage in significant research or carry out long-term planning. There was a general feeling in the air that it was beholden to the auto industry and, growing from this, a developing view that the decision to establish a vested-interest body had been a mistake. All these factors suggested the need for a new board with broader powers and with members who had no axes to grind.[1]

The 1965 federal legislation also encouraged reform, if for no other reason than to preserve a significant role for California's auto control program. The view among MVPCB members was that the federal law clearly promised preemption of automotive emissions if it had not, indeed, already accomplished it. The MVPCB's executive director accepted a position with Ford Motor Company because he thought his agency would now be abolished for lack of a job; he told staff members to look for other employment.[2] As a result of the federal legislation, the MVPCB itself appointed a committee to consider abolition in favor of a new agency with a broader mandate to conserve the total air resource.[3] It actively supported measures to create a body with statewide authority over stationary sources, as did the press.[4] Legislation introduced during 1966 proposed to extend authority to stationary sources,[5] and by the beginning of the next year there was a bill to create a new Air Resources Board with just such broadened powers. The new governor, Ronald Reagan, had in mind an even more encompassing approach, submitting a message calling for an integrated program of waste management to deal with air, water, and solid waste problems.[6]

It was the legislation providing for an Air Resources Board (ARB) that became law in 1967.[7] The new ARB was to be composed of fourteen members, nine appointed by the governor with Senate consent, and five consisting of the heads of specified state agencies (the directors of the Departments of Public Health, Motor Vehicles, Agriculture, and Conservation, and the commissioner of the Highway Patrol). The appointed members were to have "demonstrated interest and proven ability in the field of air pollution control," and would hold office, after an initial staggering of terms, for four years. The Department of Public Health would recommend air quality standards to the ARB, but aside from this its functions were limited ("emasculated," according to one observer).[8] The ARB was to divide the state into air basins and adopt air quality standards for them (the standards could vary from basin to basin). In some senses, at least, control of stationary sources was still the "primary responsibility" of local government (local authority over motor vehicles was preempted). Local government could establish ambient air quality standards more stringent than those set by the ARB; moreover, stationary control was in local hands so long as carried out successfully. (If local programs failed to meet state standards, the ARB could step in.) The initial conception had been for a stronger state role with respect to stationary sources, but local governments had ultimately succeeded in retaining a good measure of jurisdiction.[9] The ARB had authority to conduct research on the effects of air pollution; to inventory sources; to investigate specific air pollution problems; to monitor air pollutants; and to review local rules and regulations and provide local assistance.

As to motor vehicles, much of the existing program was transferred intact to the ARB, though it was also given the new authority (previously in the Public Health Department) to establish vehicular emission standards. Existing provisions by which counties could opt out from vehicular controls were largely retained. Finally, the legislature established for the new agency what the old had had to provide for itself—a technical advisory committee. The committee was to be composed of twelve members appointed by the ARB—physicians, scientists, biologists, chemists, engineers, and meteorologists with professional or technical experience in air pollution. The committee's job was simply to advise, on request or on its own initiative. Committee members received a per diem fee and expenses; ARB members, who were not expected to serve on a full-time basis, received only the latter.

Louis Fuller, the Los Angeles air pollution control officer who had been such a vocal critic of the MVPCB, was named chairman of the ARB. But Fuller also wished to keep his county position, and he eventually resigned the chairmanship on the ground it would create a conflict of interest.[10] Governor Reagan then named another figure well known in the business of air pollution control in California—Haagen-Smit.

THE FEDERAL AIR QUALITY ACT OF 1967

Just as the California program was showing promise of new momentum, developments on the federal level were threatening to constrain state initiative.[11] Pressure for increased federal control of the air pollution problem had been growing between 1963 and 1966. Both Congress and the executive branch were dissatisfied with state and local progress, and with some reason. The federal approach in the past had been to recognize air pollution control as primarily a problem of state and local responsibility, but state and local efforts had been relatively scant. By 1966 little more than half of the country's urban population enjoyed local air pollution control; of "nearly 600 counties with a population greater than 50,000, less than [90] had control programs, and most of these programs [were] far from adequate."[12] Allegations of the shortcomings of state and local control were given strong support by a November 1966 episode in New York City that led to a declaration of emergency by Governor Rockefeller and resulted in a reported eighty deaths.

The situation was more than ripe for administration announcements—

at the Third National Conference on Air Pollution held in Washington in December 1966—of strong new federal initiatives in the air pollution area.[13] Arguing that state and local governments had not taken advantage of the opportunities afforded by earlier federal legislation, HEW used the conference as a platform to urge a comprehensive federal approach to air pollution control that featured national emission limitations for major stationary sources and a system of regional agencies with broad powers to set and enforce standards. Senator Muskie, who also addressed the conference, agreed that further federal action was necessary, but opposed national emission standards for other than vehicular sources. In Muskie's view, priority should go to those areas with the most serious problems; he was concerned, moreover, that national emission standards would pose too large a job for the federal government and in any event impact adversely on the less prosperous areas of the country.[14]

But the administration largely ignored Muskie. In January 1967 President Johnson delivered a message to Congress asking for legislation—an Air Quality Act—the central features of which would be national emission standards for major industrial sources, and regional air quality commissions with authority to establish and carry out control programs in regional airsheds defined without respect to state and local boundaries. Johnson's proposal would also require federal registration of motor fuel additives and provide federal assistance for state automotive pollution control inspection systems and increased support for federal research. In the next month Muskie's subcommittee began a long series of hearings on the administration's proposals and others. The hearings revealed strong industry objection (especially from the coal and oil lobbies) to national emission standards, but rather surprising support from state, county, and municipal organizations.[15] Muskie himself continued in his reservations, a posture that tended to align him with industrial interests and expose him to charges (especially from Ralph Nader's Study Group on Air Pollution) that he had yielded to their influence.[16]

National emission standards were not the only subject of dispute. Johnson's proposal for powerful regional commissions—commissions to be staffed and financed by the federal government and completely under the authority of HEW—also generated considerable debate, as did the role of HEW in formulating the criteria documents that described the effects of various pollutants. HEW had been given the job of establishing criteria by the 1963 act, and by the time of Muskie's hearings had issued a strong document on sulfur oxides. The coal lobby in particular attacked the criteria, taking the position that their application threatened to eliminate use of coal in major cities. The lobby questioned the accuracy of the HEW document, arguing that "HEW's sulfur research program was most limited and still in an incipient stage." Coal interests enlisted the aid of Senator Jennings Randolph of West Virginia, chairman of the Public Works Committee under which Muskie's Special Subcommittee on Air and Water Pollution held its charter. Randolph demanded "complete and irrefutable proof of necessity" of the health standards contained in HEW's criteria document.[17]

The most heated conflict in the course of considering the Air Quality Act, however, had to do with the motor vehicle—more particularly, with the important issue of preemption that had been left open by the 1965 legislation. The auto industry, it should be recalled, much preferred no emission standards; but if standards were to be established, it also much preferred ones set on a uniform basis by the federal government, with no room for state authority. That preference had become markedly stronger since 1965, inasmuch as several states had subsequently enacted or begun to consider motor vehicle emission standards. The original draft of the Air Quality Act contained no preemption provision, but the issue came up early in Muskie's hearings. Los Angeles testified in favor of minimum federal standards that could be made more strict by the states where conditions required them. HEW said such an approach would produce utter confusion and, in any event, would result in wasteful duplication of federal standards. The Automobile Manufacturers Association's position, familiar by now, was that multiple standards would create chaos in the industry and the economy and harm the consumer. Muskie himself had earlier expressed sympathy for the uniform-standards approach; and when the bill came out of his subcommittee and subsequently passed the Senate, it reflected his, HEW's, and the industry's views—with a concession for California (authored by Senator George Murphy). California, but no other state, could get an exemption from federal standards upon showing a problem that required more stringent controls. The measure was an obvious compromise of the industry's desire for uniformity and California's claim that it needed stricter standards. If in the future other states should develop needs as compelling as California's, it was the subcommittee's view that the scope of the exemption could be expanded. California could be a useful testing ground in the meantime.[18]

California had won a victory in the Senate, but a new battle began when the bill reached the House. Congressman Charles Dingell, from Detroit, submitted an amendment that deleted the provision granting California authority to set stricter standards, and the amendment passed in committee. Californians were outraged. Dingell had not consulted any California air pollution control officials before submitting his amendment, and it was clear that his move had been prompted by the automotive industry; indeed, the amendment was reportedly drafted in the Washington offices of the Automobile Manufacturers Association.[19] Control officials in California found it most annoying that they, who had pioneered in efforts to control air pollution, were now in danger of being preempted by the federal government that had so long followed in their footsteps. The Dingell amendment was especially threatening because the state was in the process of considering adoption of stricter standards.[20]

The House committee's action united old foes. A member of the California Assembly wrote California's congressional delegation urging it to restore the Murphy amendment on the floor of the House. California's representatives coalesced for the first time in years, one saying "Air pollution is a bigger issue than Vietnam in California, and every Democrat and Republican in the delegation will fight to the last ditch on this."[21] Decrying the auto industry's lobbying tactics, the delegation immediately began lobbying of its own. States' rights were stressed in talks with other congressmen, photographs of a once-proud Los Angeles now wrapped in a brown haze were displayed without shame, and cans of "smog" marked "poison" were passed around. The delegates were assisted by state and local air pollution control officials from California, who met with congressmen in Washington to provide technical information.[22] The officials had "agreed . . . to forget past differences over enforcement and join in an allout effort against restrictive federal legislation," legislation that would "be an outright violation of the concept of states' rights and an apparent recognition by Congress that the interest of Detroit auto makers prevails over the interests of California's 20 million people." In the view of Californians, "the profit motive has been placed above the protection motive."[23]

In waging their war against preemption, California's delegates and control officials could point to the volume of mail from constituents—a third of a million letters and cards according to one report, more than 400,000 according to another.[24] Most of that mail had been stimulated by the radio program so critical of state air pollution control efforts in California, the "Breath of Death" series.[25] (The reporter conducting the series had uncovered the auto industry's role in the Dingell amendment; he referred to smog as "Dingell's dust.")[26] The California campaign was successful; the state's authority to set stricter standards was restored on the floor of the House.[a] Indeed, it was broadened. The Senate version of the

[a]Senator Murphy said the radio series was instrumental in the success; one observer comments that the series "demonstrated the great potential for superficial mass mobilization." On both points, see T. Roberts, "Motor Vehicular Air Pollution Control in California: A Case Study in Political Unresponsiveness" (Honors thesis, Harvard College, 1969), pp. 55–56.

relevant provision provided in essence that California could set stricter standards only if it proved the need for them; now the section was changed to put on HEW the burden of showing that the federal standards would, in fact, be adequate.[27] This and several other minor differences between the Senate and House bills were resolved (favorably to California) in conference, and the Air Quality Act passed the Congress and was signed by President Johnson in November 1967.[28]

As finally enacted, the Air Quality Act revealed a mixed bag of victories—for Senators Muskie and Randolph, for the auto companies and other large industries, for California—but not for President Johnson.

The act did not provide the national emission standards for major stationary sources that the president had wanted, nor did it give much endorsement to his wish for strong regional controls. Like its predecessors, the legislation announced and manifested that the primary responsibility for air pollution control was to be left with state and local government.[29] National emission standards were to be studied, not established. And while a regional approach was to be used, it rested largely on state initiative rather than, as Johnson wished, federal involvement. HEW would designate air quality control regions consisting of communities with common air pollution problems. Next, HEW would designate air quality criteria—and issue documents discussing the costs, feasibility, and effectiveness of various control technologies—for each pollutant. (Here Senator Randolph's influence was apparent. HEW was directed to reconsider the sulfur oxides criteria previously issued.) Once criteria and control documents were issued, each state was to adopt for its regions air quality standards consistent with the criteria, and then draw up implementation plans setting forth ways of achieving the air quality standards.

The regional approach contemplated by the act was limited and time consuming. Each state was free to establish its own standards and plans, despite interstate regions. The process could occupy as much as a year-and-a-half and would have to be repeated as each criteria document appeared. Standards set by the states were subject to HEW approval, and if a state failed to act, or acted inadequately, HEW could intervene to set or enforce standards. It could do so, however, only after pursuing a long series of procedural steps. The cumbersome abatement provisions of 1963 were carried over to the 1967 legislation, expanded slightly to provide for emergency situations. To help state and local governments carry out their new responsibilities, the act authorized added funds to support and encourage regulatory programs.

Regarding motor vehicles, the Air Quality Act added very little to previous federal legislation. It expanded the federal research role, provided for federal registration of fuel additives, and authorized grants for state motor vehicle inspection programs. More significantly, it clarified the preemption issue. Uniformity was to be the rule, California the exception. States could adopt standards for used, but not for new vehicles. California, however, could establish new-vehicle standards more stringent than the federal ones, unless HEW found they were unnecessary, and so long as they took due account of technological feasibility and economic costs.[30]

CALIFORNIA'S PURE AIR ACT OF 1968[31]

California's victory on the federal preemption issue was an important one, for just as the Air Quality Act of 1967 was passing through Congress, moves for a dramatic new approach to vehicle emission standards were underway in California. The state already had the most advanced air pollution control program in the nation, and smooth operations by the ARB created in 1967 would result, by the end of that year, in at least some degree of emission controls on half of the state's eleven million vehicles.[32] Despite these accomplishments, not all legislators were completely satisfied. They wondered whether greater and faster progress might not be possible, either through conventional approaches or through development of alternatives to the internal combustion engine. The matter was pressed in the form of an Assembly resolution calling for study of the questions. Routine referral of the resolution to the Assembly Transportation and Commerce Committee was the first of a series of coincidences that led to passage of what were to be the most far-reaching controls on automotive emissions for some time.

It was pure coincidence that the resolution was referred to the Transportation and Commerce Committee at all; the prevailing practice had been to send most air pollution measures to the Committee on Public Health. It was coincidence that the Committee on Transportation and Commerce was already more than occupied with other business, such as problems involving the Bay Area Rapid Transit District. This led the committee chairman, who was not enthusiastic about the study but nevertheless considered that there were advantages in expanding his committee's role in air pollution, to ask for assistance from the Assembly Office of Research. In particular he asked the office to organize a series of hearings on the subjects in question. Finally, it was further coincidence that the office even existed; it had been established only shortly before, at the urgings of Speaker Jesse Unruh, to make studies and recommend solutions regarding major social problems in California.

Coincidences aside, the Assembly Office of Research went about its new task with diligence. It organized the hearings that had been requested by the chairman of the Transportation and Commerce Committee, but with the strategy of casting the automotive industry as the "heavy." (The chairman of the committee had intimated that if legislation were to come out of the hearings, he would favor a strong bill; the idea became to make the case for such a bill.) A large number of potential witnesses were interviewed and, when the hearings were held, those whose testimony would be most favorable to the views of the office and the committee were scheduled first, when the press would be present and interest high. A mailing list of persons and groups active in environmental problems was compiled and employed to disseminate information and urge attendance at hearings; by the same method, the press was kept carefully informed of developments.

At the level of the initial hearings, the strategy was successful. Health experts testified of an imminent air pollution crisis and said the auto industry should be warned that further delay would not be tolerated. State and local control officials testified in unison that stricter controls on motor vehicle emissions were necessary. They also alleged that foreign manufacturers were doing a better job than the domestic firms. The auto companies, in their testimony, argued that further improvements on any mass production basis were then impossible. Their position, however, was weak. The Assembly Research Office had earlier polled domestic and foreign manufacturers on the scope of their research programs and their prognosis for the future; while the latter replied with detailed information, the former were evasive. The domestic firms were criticized for this and for having ignored the problem of controlling oxides of nitrogen. Finally, they were lambasted for their efforts to secure federal preemption of emission controls for new motor vehicles. The auto industry had indeed become the heavy.

By the end of 1967, the initial set of hearings had gathered enough information and generated enough ill will toward motor vehicle manufacturers that the Assembly Research Office and the committee concluded it was time to draft a bill. The matter of drafting, like all that had preceded it, was approached with utmost care. Key interest groups among those generally sympathetic to stronger legislation were invited to submit ideas and participate in drafting; though most declined, the move helped to win their support for the measure that finally resulted. Moreover, the Assembly Research Office itself drafted the legislation, rather than relying on the staff of the legislative counsel, the thought being that the latter would devote less care to important matters of detail. And detail was important, for the idea had developed to put specific emission standards for new vehicles into the legislation. California's victory on preemption made this approach possible, and other considerations made it attractive. It would add the force of the legislature to the standards, and it would avoid any timidity on the part of the ARB. (Governor Reagan, an acknowledged conservative, had appointed members to the ARB, and there was concern that his appointees would not stand up to the auto industry.) The ARB's chairman at the time, Louis Fuller, was initially against legislative standards, believing there were important advantages of flexibility in administrative standard-setting (and no doubt believing, too, that a body under his leadership could not be unduly influenced by interests in Detroit). When Fuller later abandoned the chair, however, he also abandoned his view. He supported legislative standards and, in conjunction with others, helped to determine just what they should be.

A bill containing the legislative standards was introduced in the Assembly in January 1968.[b] Central features of the original measure included specific emission standards, which the ARB could make more stringent, governing emissions of hydrocarbons, carbon monoxide, and oxides of nitrogen from the crankcase, the exhaust, and from fuel evaporation of 1970 models, with stricter standards prescribed for 1972 (later changed to 1974) models;[c] expression of standards on the basis of a grams-per-mile rather than parts-per-million basis (the latter, which had formerly been used, discriminated against small vehicles); continuation of the existing system of certifying control devices, but raising the cost ceiling for used-vehicle exhaust devices to $150; a provision that vehicle manufacturers were to indicate by window labels the emission characteristics of the model in question; transfer of the control activities of the California Highway

[b]At about the same time, two related measures were introduced. One was intended to encourage manufacturers to develop very low-emission vehicles by providing that if an appropriate vehicle were developed, the state would be required to purchase it for at least up to one-quarter of its fleet. The other measure encouraged the California Highway Patrol to explore the feasibility of steam engines. Since the standards of the first measure probably could not be met by an internal combustion engine, both can be viewed as intended to provide alternatives to the conventional engine; both were to become law. See Ch. 765, § 1, [1968] Cal. Stats. Reg. Sess. 1486, adding to California Health and Safety Code §§ 39200–39205; Res. Ch. 189, [1968] Cal. Stats. Reg. Sess. 3332.

[c]The ARB was given power to ban the sale of automobiles not in compliance with the law (a power retained in the legislation as finally passed). In an effort to convince the auto industry that the state would actually carry out a ban, the Assembly Office of Research and the Committee on Transportation and Commerce attempted to convince California's aerospace industry to enter the automobile manufacturing business. While the efforts ultimately failed, they are claimed to have been a successful bit of psychological warfare.

Patrol to the ARB; and a provision that the ARB was to develop procedures for assembly-line testing.[d]

The bill in this form enjoyed general support from a majority of the Assembly and from the press—thanks largely to the earlier efforts of the Assembly Research Office and the committee to disparage the auto companies and to keep potential allies informed and solicit their views. And the bill earned the support of another important group, a technical advisory panel that had been appointed by the Speaker of the Assembly to evaluate the standards in the proposed legislation. The panel, composed of nine experts from government, industry, and the universities, was something of a wonderful brainstorm. It had been established for the specific purpose of endorsing the legislative standards as technologically and economically feasible, an endorsement that was important for several reasons. First, it would tend to foreclose claims by the auto industry and recalcitrant legislators that the emission standards needed more study; second, it would assist in convincing the federal government that the standards in the bill—more strict than the existing California and federal standards—were feasible, a condition precedent to any waiver by the federal government.

The panel was cooperative. In April 1968 it issued a report reaching favorable conclusions as to the feasibility of the standards and recommending only a few small changes.[33] One of these became quite controversial. The panel was of the view that there should be provision for a hearing board that could give variances in particular instances where a company demonstrated failure to comply despite the best of efforts; it was thought that such a procedure was needed to avoid undue hardships and in any event might be necessary for the proposed standards to pass muster in applications for a waiver of federal standards. Without such a procedure, a waiver might be denied for the reason that the California standards were not feasible. One member of the Assembly concluded that the variance procedure would drastically dilute the bill and opposed it on that ground.

[d]Early in 1968 the legislature did away with averaging, enacting a provision that the ARB was not to "approve any motor vehicle pollution control device unless each motor vehicle tested with such device for purposes of such approval meets the emission standards set" by the ARB. Ch. 49, § 2.5, [1968] Cal. Stats. Reg. Sess. 189, 191, adding to California Health and Safety Code § 39083.3. The Pure Air Act, passed later in 1968, repealed this provision, but replaced it with one that appears to be of the same import ("emission standards applied to new motor vehicles and to used motor vehicles equipped with emission control devices are standards with which *all* such vehicles shall comply") (emphasis added). See Ch. 764, §§ 7-8, [1968] Cal. Stats. Reg. Sess. 1463, 1467, repealing and adding to California Health and Safety Code §§ 39080-39096.

Only considerable pressure persuaded him to withdraw his opposition in exchange for a provision that the procedure was to be employed only if necessary to obtain a federal waiver.

The variance procedure aside, the bill passed the Assembly without substantial amendment. There were, to be sure, a number of minor changes, but even these may have been part of a grand strategy. The California Highway Patrol, for example, strongly objected to losing control jurisdiction. Its role was reinstated (though it was required to coordinate with the ARB), but reinstatement came in exchange for Highway Patrol support of the bill. Apparently just such an exchange had been the idea of the bill's authors all along. Passage in the Assembly was aided in a number of ways. A public interest group, Stamp Out Smog (SOS), comprised almost entirely of women from Southern California, helped with mailings to urge attendance at hearings and stood in reserve to pack hearing rooms. Conservation interests were not otherwise drawn upon. The approach at the hearings was to elicit supportive testimony from experts, including members of the technical advisory panel, the notion being that the legislators were already familiar with the views of clean-air zealots. And well they should have been. After introduction of the bill, a public forum had been arranged and the reporter of the "Breath of Death" series had been engaged to urge his listeners to participate. The outcome was a meeting attended by hundreds of citizens bent on stricter automotive controls, a turnout that apparently influenced a number of legislators.

The automobile industry chose to adopt a low profile at the Assembly hearings on the bill. Rather than testify, the industry engaged in concerted lobbying; indeed, it sent a full-time representative to Sacramento for the first time in history. When the lobbyist took a hard line, this served only to convince legislators that a large measure of compulsion was necessary. Apparently aware of its mistake, the industry sent a replacement, a man respected in legislative circles for openness and integrity. Largely the new representative appeared too late, although he did convince assemblymen that control of oxides of nitrogen by the 1970 model year was impossible. The 1970 requirement was set back a year in exchange for bimonthly reports from the industry on progress relating to oxides of nitrogen control.

On the Senate side, industry lobbying resulted in more, but only a bit more, softening of the bill. The requirement of window labels disclosing a model's emission characteristics was stricken on the industry's argument that the labels would lead to unfair competition. Proponents of the measure defended it but conceded early in the realization that the requirement was hardly essential. The industry was unsuccessful in attempts to strike provisions relating to assembly-line testing. Its efforts first resulted in deletion of the requirement in the Senate Transportation Committee, but a large amount of mail from constituents coupled with attacks from the press, television, and radio put it back. Nevertheless, there were other important changes in the bill before it passed the Senate, though they were not instigated by the auto industry. The automobile clubs (which, unlike the industry, testified openly about their dissatisfaction with particular provisions) succeeded in reducing the cost ceiling for used-vehicle exhaust controls from $150 to $65; earlier efforts to accomplish the same end in the Assembly had failed. Moreover, rural pressure resulted in continuation of a familiar provision: all but the populous counties of the state could opt out of the control program for used vehicles. But these matters aside, the original conception of a strong control measure containing explicit emission standards for new vehicles had survived. When Governor Reagan signed the Pure Air Act in July 1968, he cited it as another example of the "lead" taken by California. He was "gratified to see that the concern for improving our air quality is recognized as being of equal importance with the conservation of our other natural resources."[34]

SOME SUBSEQUENT DEVELOPMENTS

Following passage of the Pure Air Act, California began the process of seeking the federal waivers necessary to put the act's strict standards for new vehicles into effect. The first waivers were granted in mid-1968, and by 1969 had been obtained for various standards applicable through the 1974 model year.[35] The waivers were important; progress with the state's new-vehicle program was essential because the used-vehicle controls continued to confront obstacles. By the end of 1969 there still had been no certification of an exhaust device for used cars.[36] Thus significant air quality improvements in the state were heavily dependent on strict and effective controls for new vehicles.[37]

Yet, despite the federal waivers, the new-vehicle program was developing some problems of its own. A surveillance program conducted in California in 1969 revealed that controlled vehicles in general had emissions of hydrocarbons higher than the applicable standards.[38] In response, the ARB in 1969 adopted procedures to test a representative sample of new vehicles on the manufacturers' assembly lines.[39] Later in the same year (and pursuant to legislation amending the Pure Air Act), it began discussions with the manufacturers to consider means for 100 percent assembly-line testing.[40] Concurrently, the ARB's technical advisory committee began studying the question of more stringent emission standards for new vehicles subsequent to 1974 models. The study, completed in 1969, concluded that tighter standards were both necessary and technically feasible.[e] The committee's recommendations would be taken up in 1970.[41] In further response to the surveillance data, the ARB in 1969 began planning a study to develop a practical program for inspecting vehicles on the road.[f]

All of these developments concerned vehicular sources, but there were in addition activities of a more general character. Under California's 1967 legislation, the ARB was to divide the state into air basins and establish air quality standards for each. By the end of 1968 it had accomplished the first task and was in the course of considering air quality standards for six pollutants.[42] In 1969, however, these efforts were diverted by federal law. The Air Quality Act of 1967 directed HEW to designate air quality control regions and issue criteria and control technique documents. Once those steps were accomplished as to a particular pollutant, the states were to establish air quality standards and a plan to implement them—all subject to federal approval. Early in 1969 two of California's eleven air basins (previously established by the ARB) were designated as air quality control regions, and criteria and control technique documents for sulfur dioxide and particulates were issued. The ARB established air quality standards for these pollutants and at the end of 1969 was at work on an implementation plan scheduled for completion in mid-1970.[43]

Mention of the Air Quality Act introduces an important development on the federal side in 1968–1969. The act was beginning to draw severe criticism, a primary complaint being that the federal procedures for moving from air quality control regions to implementation of air quality

[e]The statutory standards in the Pure Air Act became increasingly strict only through 1974 models, the most stringent standards being applicable to subsequent model years unless changed by the legislature or the ARB. The ARB had authority under the act not only to adopt more stringent standards for later years, but to establish standards for models through 1974 that were more stringent than the statutory standards, so long as necessary and technologically feasible. See California Health and Safety Code § 39052.5 (1968).

[f]California Air Resources Board, *1969 Annual Report* (1970), p. 26. New Jersey had begun studies to develop an inspection program somewhat earlier, and a program was later put into operation. See F. Grad, *Environmental Law* (New York: Matthew Bender and Co., Inc., 1973), chap. 3, p. 353; E. Angeletti, "Transmogrification: State and Federal Regulation of Automotive Air Pollution," *Natural Resources Journal* 13(1973):448, 464, 471–472.

standards were unduly cumbersome and time consuming. Though the act was passed in 1967, by the end of the decade no completed implementation plans had been filed with HEW.[44] Critics claimed that the federal government was slow in providing criteria documents and the state governments slow in acting once documents were provided. Ralph Nader predicted, based on a report by his Study Group on Air Pollution, that the implementation phase would not be completed until well into the 1980s.[45] Nader's group was critical of the Air Quality Act in general, and Senator Muskie in particular. Criticisms of the act, made by Nader's group and others, included (in addition to that mentioned above) time-consuming enforcement mechanisms for both intrastate and interstate air pollution problems; inadequate funding for research, local assistance, and the training of enforcement personnel; failure by HEW to adopt assembly-line testing of new motor vehicles (HEW asserting that it had authority to test only prototypes, but in the face of a legislative history that indicated otherwise to the critics); and a general failure of the federal vehicle emission control program (studies indicated that 50 percent of new vehicles were not meeting the applicable federal standards).[46]

At the same time, California's program was receiving criticism of its own. The ARB was said to be too large and uninformed; its members, many of them allegedly political appointments by Governor Reagan, worked on only a part-time basis and were not compensated, and their ignorance was said to be multiplied by inattention. The technical advisory committee was said to have stepped into the breach, so that it—rather than the ARB—was making policy though it had no mandate or authority to do so; the ARB was nothing more than a "rubber stamp." An additional important criticism was that the ARB, like its predecessor, had failed to implement an effective program for control of used vehicles.[47]

THE PREVAILING SITUATION AT THE END OF THE 1960S

Despite significant state and federal legislation in 1967 and 1968, the 1960s closed on a note little less gloomy than had been common for decades before. An important source, the motor vehicle, had come under state and federal control, but control was largely limited to new vehicles and only partially successful as to those. California's most stringent standards for hydrocarbon control would not be applicable until 1972, and those for nitrogen oxides not until 1974. (Moreover, by the end of the sixties no system to control oxides of nitrogen to any level had been approved.)[48] The control of exhaust emissions from used vehicles was little further along than it had been years before, and the control of used vehicles nationwide was insignificant. In California, carbon monoxide and hydrocarbon totals were going down despite an increasing number of vehicles, but oxides of nitrogen were increasing and would continue to do so until they came under initial control in 1971.[g] The ARB could project air quality improvements until 1985, when growth in the number of automobiles would begin to eat up the gains of technology.[49] Some observers, however, regarded the projections as too optimistic,[50] and in any event they left a gloomy outlook for the shorter run. Days in excess of the air quality standards by 1975, according to one prediction, would be "discouragingly high."[51]

California's response to the much criticized MVPCB, the ARB, was itself being subjected to familiar complaints, and the federal response to charges of state and federal lethargy, the Air Quality Act, was also said to be lethargic. Considering the many criticisms aimed at state and federal efforts, it is surprising that one went unsaid: those efforts continued to reflect the tight technological fixation that had prevailed since the 1940s. California's Pure Air Act was claimed to be a significant breakthrough inasmuch as it "moved from following technology to leading it,"[52] this because application of its standards was not conditioned on prior approval of control systems. (The same compliment, of course, should have been paid to the federal legislation of 1965 and 1967.) But it would be misleading to carry too far the claim that dependence on the development of technology had been escaped; all legislation existing at the end of the 1960s was limited by constraints of "technological feasibility." True, standards became applicable to new vehicles despite the fact that a technology might not have been approved, but the standards themselves were to be technologically feasible—meaning, of course, that there must be some technology to achieve them. It is difficult to discern some deep difference of principle between the two approaches, though on the level of application there were differences indeed. The latter approach provided incentives to manufacturers that the former did not. Nevertheless, the tie to technology remained. Moreover, it remained in its original form regarding used vehicles, as to which exhaust controls (for 1955 through 1965 models)

[g]California Air Resources Board, *1969 Annual Report,* p. 5. The carbon monoxide and hydrocarbon controls actually aggravated nitrogen oxide emissions by increasing engine combustion temperatures. See the discussion at p. 21. Failure to provide against such a result by mandating simultaneous controls on nitrogen oxides, carbon monoxide, and hydrocarbons, or by providing controls on carbon monoxide and hydrocarbons that did not increase nitrogen oxides, is regarded by most experts as having been a fundamental flaw in California's early approach to vehicle emissions.

were required only upon certification of a control device.[53] Pending such certification, there would be no controls.

Yet, with respect to the technological fixation, there were by 1968–1969 some indications of a change in attitude. Increasing attention was being directed to fiscal incentives, and to some extent this attention was coming to be reflected in law. California had passed considerably earlier a system of subsidies (in the form of tax incentives) to encourage employment of pollution control technologies, but the federal government had been reluctant to act in similar fashion despite repeated attempts in the 1950s and 1960s to enact subsidy legislation.[h] By 1969 this reluctance had largely disappeared. The Tax Reform Act of that year allowed rapid amortization of a portion of the cost of investment in qualified pollution control facilities. The provision was the ultimate result of a "proliferation in the number of proposed methods of providing financial assistance for combatting air and water pollution."[54] The tax incentives were, to be sure, very modest steps away from sole reliance on technical solutions to pollution. They were, first of all, tied in their own way to technological solutions, for they provided incentives to employ "pollution control facilities," not incentives to reduce pollution by whatever means, technological or not. Moreover, to some extent they were not "incentives" at all (and at any rate most likely not effective ones) because in many instances they did nothing more than subsidize part of the cost of pollution controls that were required in any event by other law (such as the controls on stationary sources that had existed in California as early as 1947). In this light, the subsidies appear to have been little more than transfer payments designed to quiet industry objections to pollution control, rather than to serve as a positive system of incentives.[55]

Yet more far-reaching steps were also coming to be proposed. As early as the mid-sixties President Johnson had shown interest in employing such market mechanisms as effluent fees in the control of water pollution.[56] Despite some success with fees in Germany,[57] however, a federal study of the matter concluded in 1967 that effluent fees and emission charges were not yet feasible—the institutions, monitoring techniques, and reliable damage estimates necessary to such programs had first to be developed.[58] Nevertheless, by the end of 1969 suggestions to incorporate some system

[h]With two small exceptions. In 1966 the investment credit was suspended, but continued for pollution control facilities; in 1968 the tax-exempt status of industrial development bonds was repealed, except as to bonds for pollution control facilities. See P. McDaniel and A. Kaplinsky, "The Use of the Federal Income Tax System to Combat Air and Water Pollution: A Case Study in Tax Expenditures," *Boston College Industrial and Commercial Law Review* 12(1971):351.

of pollution charges into the attack on environmental problems were growing in the literature and taking on more concrete form.[59] On the federal level, Senator William Proxmire of Wisconsin, a respected conservationist, proposed "that one answer to water pollution would be to tax industrial polluters in proportion to the noxiousness of their emissions."[i] At about the same time in California, there were related proposals more pertinent to the discussion here. A series of television editorials in October 1969 urged a "smog tax" on "pollutants spewed out by industry and others, . . . including jet planes and automobiles."[j] "If offenders are taxed," the editorials reasoned, "the odds are their experts will soon come up with some solutions. Cash out of pocket provides a stronger incentive than mere warnings in breaking down barriers to meaningful research and retooling." As to the automobile, "owners taxed on pollutants from their cars will develop more selectivity in their buying habits. Resultant pressure on automotive manufacturers in a competitive market promises prompt and positive action in Detroit."

Here again the stress seemed to be on technological solutions, the tax being merely an incentive to induce them. But the proposal of a young California economist, made two months after the editorials, emphasized the advantages of a pollution charge in encouraging other than simply technological responses.[60] If prices are charged for pollution, he argued,

> Firms and individuals are then free to adjust to the prices any way they choose. Consumers will switch from those goods which increase in price to other, less polluting and therefore less costly substitutes; there will be incentive to drive smaller cars, take the bus, or form car pools. Price differentials will encourage plants to locate where their pollution does less harm, and to produce less during the smog season, when the prices are higher.
>
> In short, pollution will be reduced by an unlimited number of subtle adjustments, so that any given reduction is obtained at the lowest possible cost to society.

But this was not to say that pollution charges had no relation to advances in technology. Indeed, the economist suggested that "probably the greatest

[i]G. Hill, "Objections to a Tax on Pollution," *New York Times,* 10 Dec. 1969. At this point in time, conservationists were generally opposed to any notion of employing pollution fees, seeing the approach as a "license to pollute." The attitude was to change.

[j]Editorial, KABC-TV, Los Angeles, Calif., 30–31 Oct. 1969. The idea was presented as a new one. "Whoever heard of a tax on smog?" the editorials asked. Californians had, in 1959. See the discussion at p. 96.

advantage the price system has over its alternatives is the incentive it gives for technological improvements." And here he spoke to a problem that had been plaguing California and the nation for some time:

> . . . under the present system, it would be suicidal for any auto manufacturer to introduce a cleaner, but more expensive car, since no individual consumer is anxious to pay more for something which is only of the smallest direct personal benefit. However, if a cleaner car meant lower registration fees and fuel taxes for the owner, there would be strong, continuous pressure on the manufacturers to be ahead in the clean-car race.
>
> No "conspiracy" among manufacturers to maintain the polluted status quo could last long in the face of consumer demand for clean engines; and the positive incentive of profit will bring results which can never be obtained with lawsuits or legislative bluffs.

Mention of a "conspiracy" among auto manufacturers introduces another significant change in attitude that had developed by the close of the 1960s. As we have seen, the Justice Department's investigation of antitrust conspiracy charges against the auto industry concluded in late 1969 with a consent decree providing for an end to any conspiracy without determining that one had ever existed.[k] The hullabaloo surrounding the investigation and the decree, combined with the earlier fiasco involving attempts to discredit Ralph Nader, had eroded the industry's image to the point that it "had hit rock bottom" by the end of the decade.[61] At the same time, Senator Muskie was feeling the sting of Nader's Study Group on Air Pollution.[62] Both events would contribute to major new federal initiatives in the next year.

But a more overriding change in attitude also developed just as 1969 drew to a close, and it more than anything else portended new directions in the making of pollution policy. "During 1969 there was a groundswell of public interest in environmental issues"; 1969 was the "year of the environment," the year in which a "sense of crisis burst upon us" with a "suddenness" and an "intensity" that were "both perplexing."[63] The National Environmental Policy Act of 1969[64] was the official symbol of the change in attitude and the initial step in a new direction, but it would soon be followed by many others. The first of importance was to be the Clean Air Amendments of 1970[65]—legislation which brings us to the era of present policy.

[k]See the discussion at p. 88.

Part IV

THE ERA OF PRESENT POLICY: 1970–1975

12

THE NEW FEDERAL PROGRAM

The approach to pollution policy adopted by the federal government in 1970 was new, and it was old. There was a new law, the Clean Air Amendments of 1970,[1] but the dramatic expansion of federal intervention that it reflected was simply another step (though a large one) along the path of an old trend. The existing technique of federal preemption of new motor vehicle standards (except for California) was continued, but the new federal standards were to be far more stringent than their predecessors. Moreover, federal emission standards for some stationary sources were now to be established (an idea initially proposed in 1967). It was still up to the states to submit implementation plans, but now those plans had to aim at uniform national air quality standards to be established at the federal level. Beyond this, implementation plans were required to give consideration to land-use and transportation controls as means to meet the federal air quality standards; and if state plans were deemed inadequate, the federal government had the power to draft and implement its own versions (a power that did not clearly exist under the Air Quality Act). Also new was the philosophy of federal standard-setting adopted in the Clean Air Amendments. Air quality and vehicular emission standards were to be designed to protect public health without regard to economic and technological feasibility, though the old theme of feasibility eventually came very much into play.

So much has recently been written about the Clean Air Amendments of 1970, especially the provisions dealing with control of motor vehicle pollution,[2] that it would be redundant to embark on an exhaustive treatment here. Accordingly, this chapter is limited to a brief discussion of the background of the Clean Air Amendments, to a sketch of the law's major provisions, and to consideration of a few federal and state developments that occurred soon after passage of the legislation. The next chapter will describe in detail state and federal efforts to implement the Clean Air Amendments, especially in the Los Angeles area.

THE CLEAN AIR AMENDMENTS OF 1970

Background

The history of federal intervention in pollution control has been one of gradual steps, first toward study, then toward modest intervention, then toward more far-reaching controls. Each step has been accompanied by the announcement that primary responsibility lies with state and local government, but each has eroded state and local authority more than previous measures—and precisely because experience revealed, in the congressional judgment, an inability on the part of lower levels of government to carry out the responsibilities Congress had left to them. The Clean Air Amendments continued this means of making policy, but in more dramatic form. Like its predecessors, the legislation proclaimed the "primary responsibility"[3] of state and local government while at the same time taking some of that responsibility—in this instance, a great deal—away. And as with its predecessors, one strong force behind the new federal incursion was an unfavorable reading by the national government of the states' accomplishments—in this case, under the Air Quality Act. Another important force was congressional skepticism about the good faith of the auto industry.

There was considerable dissatisfaction with federal policy as it existed at the end of the 1960s. Little evidence could be found "that pollution was being reduced on a broad national scale. Progress made in some areas and on some substances was more than canceled overall by population and economic growth."[4] There was increasing opinion that controls more stringent than those imposed under the 1967 act were required. "Scientific evidence" revealed a "pervasive and growing . . . problem," one "more severe than had been thought at the time" of earlier legislation. Experience showed that if public health were to be protected, there was a need for nationwide improvement, and "healthful ambient air quality levels could not be achieved nationwide without drastic additional reductions in automobile emissions."[5] Other factors of a more political bearing were also at play. The "environmental crisis" had emerged in 1969, and by 1970 the press, radio, and television claimed that environmental protection was the issue of the day. Opinion polls in 1970 showed a sharp increase in public concern with pollution problems. Earth Day was proclaimed, and the well-publicized, nationwide activities that celebrated it uniformly called for a massive attack on environmental problems.[6]

Some of the blame for the air pollution situation prevailing in 1969–1970 was laid on the cumbersome procedures of the Air Quality Act, some on the states' inability or unwillingness to fill the roles Congress had given them, and some on the growing problem of automotive emissions. Title I of the Clean Air Amendments reflected congressional reaction to the first two points; Title II related to the last.

Concerning the motivations behind Title I, review in 1970 of progress under the Air Quality Act is said to have "revealed the mistake of relying primarily on the states, which were unable to cope with the burdens imposed by the [earlier] legislation."[7] The job of the states to establish their own ambient standards was seen as diverting resources from the important task of devising state implementation plans, and was in any event regarded as duplicative. Moreover, there was evidence that large industries were able in some instances to pressure states into establishing permissive standards, and that states might compete for industry by setting lax requirements. It appeared that uniform national air quality and emission standards could solve all these problems.[8] Uniform national ambient (air quality) standards appeared to have a number of advantages. Air quality was not improving significantly, and new medical evidence revealed the possibility of adverse health effects given a continuation of the present situation.[9] Ambient improvements were essential, for "protection of health" was seen as "a national priority,"[10] yet it was hardly clear that the states were willing or able to take the necessary steps. Uniformity was required because it was important that standards "protect the health of persons regardless of where such persons reside."[11] "Uniform nationwide air quality standards" would not only guarantee protection of public health nationwide but would also provide "an opportunity to take into account factors that transcend the boundaries of any single state. States cannot be expected to evaluate the total environmental impact of air pollutants, or take it into account in standard setting."[12] (The fact that uniform standards for the nation would themselves ignore factors that *did not* transcend state lines was apparently overlooked.) And if the uniform standards were truly to protect against adverse health effects, then they should be set without regard to the costs of achieving them.[13]

One might suppose that uniform air quality standards necessarily implied variable emission standards (only those necessary in each particular state to achieve the uniform ambient standards), but Congress thought in different terms. To permit state-set emission standards would be to ignore the danger of states competing for industry by setting lenient requirements, or being pressured by industry to do so. The results in either case would be competitive advantages and "polluter havens," and these were to be avoided.[14] In any event, "large stationary sources . . . have adverse effects not only on public health and welfare in their own communities but also on air quality over broad geographic areas. This problem is one that demands national attention."[15]

A final concern reflected in congressional consideration of Title I had to do with the implementation planning process: it needed to be employed both more quickly and more ambitiously. As to the latter in particular, it was thought important that states give more careful attention to land-use and transportation controls, for mere emission controls would in some areas be insufficient to achieve air quality that would protect public health. Thus, there was talk of measures relating to the siting of industrial sources and to "new transportation programs and systems combined with traffic control regulations and restrictions."[16]

The targets of Title I of the Clean Air Amendments were the states and industry in general; those of Title II (labeled the National Emissions Standards Act) were the auto companies and the Nixon administration. A report issued by the Senate Committee on Commerce in 1969 reflected the general state of mind about the automobile. "The automobile is the primary villain in air pollution," the report said. "It accounts for at least 60 percent of the total air pollution in the United States—85 percent of the pollution in some of our sprawling urban areas. . . . If air pollution is to be curtailed, dangerous emissions from automobiles must be substantially reduced."[17] The report found the existing approach to automotive pollution control inadequate in a number of respects. Present controls would not even stabilize, much less reduce, motor vehicle air pollution, given projections of growth in the number of vehicles and the number of miles each would travel. Oxides of nitrogen emissions would actually increase more than would be the case without the existing controls, because those controls—aimed at hydrocarbons and carbon monoxide—tended to raise combustion temperatures, which in turn increased emissions of nitrogen oxides. California now had standards that could keep the situation stable at least for a time, but there was no assurance that the manufacturers could meet those standards. Moreover, there were concerns that control technology tended to deteriorate over time for lack of maintenance. Finally, there were no promising new technologies on the horizon. The report recommended efforts to develop an alternative to the internal combustion engine;[18] others were to propose the same. While the Clean Air Amendments did not end up adopting such a proposal as a primary solution to the vehicular pollution problem, the legislation as passed did reflect attention to many of the worries contained in the Senate document.

As matters developed, Senator Muskie came to have a particular stake in motor vehicle emission control. The report of Ralph Nader's Study Group on Air Pollution had been published, and its criticisms of Senator Muskie threatened a reputation made in the fields of air and water quality.[19] President Nixon was also threatening Muskie. In February 1970 he sent a message to Congress calling for a 90 percent reduction in vehicular emissions by 1980 and proposing a federally supported research program on low-emission vehicles. Senators Gaylord Nelson of Wisconsin and Henry Jackson of Washington constituted yet another threat, taking aggressive positions on environmental protection issues. Nelson went so far as to propose, in August 1970, a bill that would ban the internal combustion engine by 1975.[20] But Muskie, as chairman of the subcommittee with primary jurisdiction over pollution legislation (the Subcommittee on Air and Water Pollution of the Senate Committee on Public Works), was in a position to exercise considerable command of the situation. When a bill that had already passed the House reached his subcommittee, "Muskie took Nixon's standards and Nelson's deadline and fashioned his own program."[21] A 90 percent reduction in vehicular emissions of hydrocarbons and carbon monoxide would be required by 1975, and a similar reduction in nitrogen oxides a year later. Muskie's move put the Nixon administration on the defensive. In November 1970 it asked that the proposed Senate deadlines be relaxed, and that the newly created Environmental Protection Agency be given full authority to extend the deadlines. But the Senate standards and deadlines prevailed.[22] Events that helped were a "major air pollution crisis which arose along the entire East Coast just as the bill was being considered," and the 1970 Collegiate Clean Air Car Race, a "stunt" that "nevertheless indicated that development of a clean car was indeed possible."[23]

There were a few compromises by Muskie reflected in the legislation as finally enacted, but they were relatively minor. The EPA was empowered to grant a single one-year extension of the 1975 and 1976 deadlines, and the National Academy of Sciences was directed to study the feasibility of the standards and deadlines. Despite these points, the ultimate product was a strong one: The environmental movement had made it important that Congress take a tough stand on some pollution issue, and the auto industry—with its recently eroded public image—was a vulnerable target for attack. The Nixon administration was also vulnerable by virtue of a sluggish record of enforcement, and this gave Congress an opportunity to embarrass and scold the administration by denying most of the discretion it had sought with respect to enforcement of the new law. "The overall tone . . . [was] that of an impatient Congress forcing a reluctant administration and a resistant industry to act promptly."[24] Others agree with the characterization.[25]

Central Provisions[26]

As finally enacted, the Clean Air Amendments reflected all of the concerns expressed above. Its major provisions (those regarding ambient air quality standards, state implementation plans, and emission controls for new vehicles) revealed a sharp rejection of the old philosophy of control and a sharp turn to a very new philosophy. The old way was to pay heed to economic and technological feasibility in designing controls. The new way was simply to protect public health, and feasibility was "not to be used to mitigate against" this goal. If technology were inadequate or uneconomic, it would have to catch up.

> The first responsibility of Congress is not the making of technological or economic judgments—or even to be limited by what is or appears to be technologically feasible. Our responsibility is to establish what the public interest requires to protect the health of persons. This may mean that people and industries will be asked to do what seems to be impossible at the present time. But if health is to be protected, these challenges must be met.[27]

The new federal philosophy appeared first in the provisions on air quality standards. As was the case under the 1967 act, the federal government (through the EPA) was to designate air quality control regions and then to publish criteria documents for harmful pollutants. But next, and very much unlike the situation under the old law, the EPA was to publish two types of uniform national ambient air quality standards that the states (at a minimum) were to achieve: primary standards, to protect public health allowing an adequate margin of safety; and (generally, more stringent) secondary standards, to protect public welfare (vegetation, wildlife, materials, and so forth).

Within nine months of EPA promulgation of primary or secondary standards, the states were to submit implementation plans to achieve them—in no more than three years from the date of plan approval with respect to primary standards, within a reasonable time as to secondary standards. Provision was made for extension of the deadlines governing submission of plans (but only regarding secondary standards), and the EPA was empowered to extend up to two years the date by which compliance with a primary standard would be required. Implementation plans were to include, among other things, emission limitations; monitoring provisions; preconstruction review of major new sources; emergency measures; and, if necessary, provisions for periodic inspection and testing of motor vehicles, and for land-use and transportation controls. If a state failed to submit an implementation plan, or if the plan were considered inadequate, the EPA then had the power and responsibility to design a plan for the state. If the EPA learned that a plan, state or federal, was being violated, the agency could initiate direct enforcement action unencumbered by the conference and hearing procedures of the 1963 and 1967 legislation. It could issue orders requiring abatement action by a state, and it could proceed directly against sources by way of abatement orders or civil suits. Stiff fines, up to $25,000 per day, were provided for violation of the act's provisions.

As the legislative history intimated would be the case, the 1970 law provided not only for federal air quality standards but for uniform federal emission standards as well. As to stationary sources, the emission standards were of two types: those for classes of new stationary sources found by the EPA to contribute significantly to air pollution that endangers public health or welfare; those for pollutants that were, in the EPA's judgment, especially hazardous. As to each type, the states could set standards more stringent than the federal ones, and were permitted in any event to enforce the standards, state or federal.

The theme of public health protection dominating economic and technological feasibility was even more pronounced in the vehicular emission standards of Title II than in the provisions of Title I. This is not to say Title II concerned only emission standards; its provisions touched upon many other aspects of the motor vehicle pollution problem as well. Title II provided, for example, for grants to the states to help them bear the costs of developing and maintaining inspection programs for vehicles in use; it had provisions to encourage and support research and development regarding low-emission vehicles; it granted the EPA authority to regulate fuels and fuel additives in order to assure that these would not impair emission-control systems. But the provisions pertaining to emission standards for new motor vehicles were the most far-reaching.

As originally enacted, the Clean Air Amendments required hydrocarbon and carbon monoxide emissions to be reduced 90 percent below 1970 federal standards by 1975, and nitrogen oxide emissions to be reduced the same percentage below 1971 emission levels by 1976. (As we shall see in the next chapter, these requirements were later changed by Congress.) The standards were preemptive as to all states but California, which (as before) could be granted waivers under specified conditions. The 1975–1976 deadlines could be extended by the EPA—but only once, only for one year, and only upon findings that extension was essential to the public interest, that the auto companies had made good-faith efforts, and that the required technology was not available to meet the deadlines. To aid in the last determination, the National Academy of Sciences was directed to study the technological feasibility of the standards and deadlines, and to submit semiannual reports. No extension could be granted if the studies indicated the availability of technology.

The basis for the 90 percent reductions mandated by Congress was a report prepared by the predecessor of EPA (the National Air Pollution Control Administration).[28] It concluded that to provide ambient air quality conditions that would protect health with a reasonable margin of safety in the worst areas of the country, reductions ranging from 92.5 to 99 percent of 1967 vehicle emission levels would be required by 1980. Muskie's subcommittee moved the dates up to 1975–1976, apparently on the basis of programs and studies underway at the time indicating that those were the earliest possible dates for application.[a] The percentages themselves were rounded off to 90, perhaps because this was the figure recommended by the Nixon administration, perhaps because of the five-year advance in the deadline used in arriving at the higher numbers, or perhaps because

[a]The acceleration of the deadlines may have been for other reasons: "Rapid reduction in the pollution level of the ambient air was required under Title I of the law before the impact of dramatic reductions in emissions of motor vehicles required by Title II was to become fully efficacious. Perhaps in recognition of this anomaly, the law required the stringent reductions in automobile emissions in a time frame which even the proponents of the amendments conceded might be unrealistic." F. Grad et al., *The Automobile and the Regulation of Its Impact on the Environment* (Norman: University of Oklahoma Press, 1975), p. 325.

those higher numbers were based on 1967 (not 1970-1971) emissions. While there was some notion, based on the studies and reports mentioned, that the standards and deadlines were feasible, Congress's mandate hardly turned on feasibility. The evidence was that "past reliance on tests of economic and technological feasibility had resulted in standards which compromised the national health. The 1970 Amendments explicitly rejected these tests and relied exclusively on considerations of public health and welfare."[29] Senator Muskie in particular stressed that point, adding that if the auto industry could not satisfy the new requirements, it could appeal to Congress for relief.[30]

In order to implement and enforce the requirements of Title II, the EPA was to prescribe specific emission standards for new automobiles. Until 1975-1976, the agency's regulations were to take appropriate account of economic and technological feasibility, but beginning with those years the reductions mandated by Congress were to be applied without regard to such considerations. The EPA was to test vehicles and systems submitted by manufacturers and certify those meeting the specified requirements. The law provided for assembly-line testing, and any certification could be suspended if testing revealed that production vehicles were not conforming to standards; provision was made for hearings and appeals in the event of suspension. The manufacture, sale, or importation of nonconforming vehicles was prohibited by the legislation, and civil fines up to $10,000 could be assessed for each vehicle manufactured or sold in violation of the law; in addition, the federal courts were given jurisdiction to restrain violations. To guard against deterioration of control systems, the EPA could monitor in-use performance and recall vehicles if a substantial number failed to conform to the prescribed standards even with proper maintenance. To enlist consumer participation in the enforcement effort, the law required manufacturers to warrant that each vehicle as designed and produced would meet the applicable standards, and—under certain limited circumstances—further warrant conformity for five years or 50,000 miles.[b] Any costs associated with bringing vehicles into conformity under the warranties were to be borne by the manufacturer. Finally, the EPA could move to compel the manufacturer of any successful technology or

[b]The latter warranty could be required only after the EPA determined the availability of inspection and testing facilities and equipment that were sound from an engineering standpoint and that yielded results capable of reasonable correlation with the original certification tests. Even after such determinations, vehicle owners could avail themselves of the warranty only upon showing proper maintenance and operation and, further, that noncompliance of the vehicle subjected them to penalty or sanction under state or federal law.

system to share its control methods with the industry through mandatory licensing.[31]

EARLY DEVELOPMENTS

At the Federal Level

On April 30, 1971, the EPA promulgated uniform national air quality standards, generally stated in such a way that the prescribed ambient concentrations were not to be exceeded in any region more than one day per year, nor during more than a limited period of time within that one day.[32] The concentration values themselves were very low. As mentioned earlier, the EPA had been instructed by Congress to establish air quality standards that would protect public health with a reasonable margin of safety. The agency carried out that mandate with a vengeance. The standards were based on a series of worst-case assumptions and set so as to protect the most susceptible part of the population from adverse effects, with a considerable margin for error.[33] As an official of the EPA later said, the law was "interpreted to require that emissions be controlled to a point at which, for example, the sickest emphysema victim on the second worst inversion day of the year should be able to spend eight hours at the busiest street corner of the most polluted city without suffering any ill effects."[34] The stringency of the standards is made especially clear by considering the requirement that generally the ambient concentrations were not to be exceeded more than one day annually. To give an example, the Los Angeles area in 1970 equaled or exceeded the earlier California standard for oxidant (.10 parts per million) on 241 days.[35] Under the 1970 legislation, the area was required by 1975 (or, if the deadline were extended, by 1977) to exceed the more demanding federal standard (.08 parts per million) no more than one day per year! The dramatic degree of improvement required is painfully obvious.

The standards having been established on April 30, 1971, the provisions of the legislation required that states submit implementation plans by the end of January 1972—nine months after promulgation of standards. The EPA issued regulations to govern the preparation of implementation plans, the most striking feature of which was the agency's insistence on a creative approach to the problem. In particular, state efforts at implementation were not to be limited to technological "fixes" (that is, mechanical controls). The regulations specified that in devising a control strategy, the states were to consider, among other things: emission standards for stationary and vehicular sources; emission charges or taxes or other economic incentives or disincentives; measures to shut down polluting facilities during times of emergency or otherwise; changes in the schedules or methods of operation of commercial, industrial, and transportation systems; periodic inspection and testing of motor vehicle emission control systems once the EPA determined the feasibility of such measures; measures to reduce motor vehicle traffic, including commuter taxes, gasoline rationing, parking restrictions, staggered working hours, and promotion of mass transit.[36]

Finally, in mid-1971 the EPA established automotive emission standards, expressed in grams per mile, to implement the strict 1975–1976 requirements for new vehicles set out by Congress in the Clean Air Amendments.[37]

In California

By 1970, if not earlier, it was obvious to most observers that reduction of the air pollution problem in California to a manageable point was very heavily dependent on motor vehicle emission controls.[38] The Los Angeles APCD was especially anxious to make this clear, for it meant that air quality improvements turned "entirely upon the effectiveness with which the State and Federal governments enforce the programs blueprinted for them by law." The APCD could add that its responsibility, stationary sources, "are already, with few exceptions, well controlled today, and will be even more stringently controlled under additional regulations that are now in preparation."[39] In the Los Angeles area in particular, control of vehicle emissions was of crucial importance. In the fourteen years since 1956 (see tables 6 and 7 in chapter 7), population had grown by over 1.5 million to a total of almost seven million persons, the number of motor vehicles had increased to four million (an addition of about 1.2 million since 1956), and gasoline consumption and the total number of vehicle-miles traveled had increased by even larger factors.[40] In 1970 the county's "more than four million motor vehicles [were] responsible for 90 percent of the total pollutant tonnage, and for 64 percent of all pollutants other than carbon monoxide. Vehicles contribute[d] virtually all of this latter pollutant."[41] Oxides of nitrogen had become an especially serious problem; the emissions of this pollutant had increased considerably due to the methods used to control carbon monoxide and hydrocarbons from motor vehicles. Control of nitrogen oxide emissions was scheduled for 1971 models, however, and would gradually result in substantial reductions.[42]

To some, the outlook was not considered entirely gloomy. The Los Angeles APCD believed that if the state program could go forward smoothly, "marked improvement in the effects of photochemical smog should be apparent by 1980," and it predicted that by 1990 the atmosphere would meet air quality standards.[43] Others (including the technical advisory committee of the ARB, the Environmental Quality Laboratory of the California Institute of Technology, the EPA, and the Rand Corporation) were much less optimistic, concluding that the standards would not be met by 1990, and that in any event growth would lead to declining quality once again after 1990 (or perhaps even earlier).[44] Even the Los Angeles APCD made its projections with reference to the state standards, not the new, more stringent federal ones. Moreover, even the APCD estimates depended upon steady progress under the state program. "Should anything interfere with the orderly administration of [California's] program . . . predictions . . . would be invalidated." Yet with the passage of the Clean Air Amendments, interference had become a clear possibility, the APCD noting that "impending conflict between the Federal program for vehicular pollution control and that of the State" could result in "slowing the prosecution of California's program."[45]

The California program acquired several new features in 1970–1971, each of which would play a part (or at least was supposed to) in efforts to devise an implementation plan to meet the federal requirements. In 1970 the legislature enacted provisions requiring the creation of basinwide air pollution control coordinating councils for those air basins consisting of more than one county (the same legislation activated air pollution control districts in all counties in which they did not yet exist). The councils were mandated in order to aid in regional planning for each basin; a central idea was that they would play an important role in devising implementation plans.[46] Subsequently, a council for the South Coast Air Basin—made up of all or part of six counties, including Los Angeles, and the scene of California's most serious air pollution problems—was formed. As matters developed, the council played little part in devising the basin implementation plan. Another piece of legislation in 1970 directed the ARB to study the costs and benefits of a program for periodic inspection of vehicles in use.[47]

In 1971, and in response to the increasing problem of nitrogen oxide emissions, the legislature instructed the ARB to set standards for devices to control these emissions from 1966 through 1970 models (earlier law had provided for control of the emissions from new vehicles beginning with 1971 models). Once a technologically feasible device was available and certified, the ARB was to schedule its installation upon registration or transfer of ownership of the covered vehicles.[48] Finally, and also in 1971, the ARB was reorganized to consist of five members appointed by the governor. Unlike the members of the earlier ARB, these were to be paid and were expected to devote at least sixty hours monthly to official duties.[49] The time had come, and the means had been provided, to implement the new federal requirements in California and across the nation.

13

PROBLEMS OF IMPLEMENTATION

None of the various projections for air quality in Southern California, even the most optimistic, would result in a situation remotely in compliance with federal standards on photochemical pollution. Policymakers knew from the outset that if the improvements called for by the Clean Air Amendments were to be realized, enormous reductions in vehicle miles traveled would be necessary. As matters developed, some of the most controversial and telling aspects of the implementation process dealt with efforts to achieve those reductions through transportation controls in the Los Angeles area. Accordingly, this chapter gives them considerable attention.[1] It also discusses developments in the federal program to control new vehicles nationwide, for these too are revealing. Discussion of the new-vehicle program, however, is largely by way of summary, because the material events have been detailed at length elsewhere. The chapter then briefly describes some moves to amend the Clean Air Amendments, and concludes with a closing note on the California scene. These matters bring the formal historical narrative to an end, but they and the final chapter show that the process and problems of making pollution policy have hardly drawn to a close.

TRANSPORTATION CONTROLS IN THE LOS ANGELES AREA

The California Plan

California officials were far from optimistic that they could devise a plan that would meet federal requirements, especially in Los Angeles. As early as June 1971 (six months before the plan was due to be submitted), Haagen-Smit, chairman of the ARB, and others took the position that some aspects of the federal requirements were "unachievable."[2] The general posture of the responsible officials was that if the federal government desired a plan to achieve unrealistic goals, it would have to draft the plan itself; the state was determined to stick to its own program of increasingly stringent emission controls on vehicular and stationary sources. The Los Angeles APCD, which tended to be the most optimistic about the future promise of the California program, wanted nothing to do with transportation controls. The ARB itself "concluded very early in the game that most transportation controls afforded such limited air pollution control potential as to not merit serious consideration in the state implementation plan." Accordingly, it did not contact many of the "transportation-related agencies in the Los Angeles area . . . about possible solutions."[3]

California's lack of enthusiasm for the seemingly impossible job the federal government had handed it was in some ways encouraged by the federal government itself. While the EPA "attacked the ARB in public hearings for deficiencies in early versions of the proposal," complaining that it "had to push" because "the state wasn't taking us seriously enough," still its "air quality experts admit[ted] that the standards the state must meet by 1975 are not worth taking seriously." The agency's director for the southwestern United States stated flatly that in the Los Angeles area "the standards cannot be met by 1975, . . . 1985 is the best we can hope for." Even that date was considered unduly optimistic.[4]

Despite general agreement that the undertaking was futile, California submitted an implementation plan on January 30, 1972, the due date. If all of the measures proposed in the plan were carried out, the federal standards would still not be fully met in Los Angeles and some other areas by the 1975 deadline; beyond this, a number of crucial provisions in the plan were considered "unworkable."[5] Accordingly, submission of the plan was quickly followed by Governor Reagan's request for the two-year extension (until 1977) allowed by the 1970 legislation under appropriate circumstances.

The request covered several air quality standards in California's most populous air basins.[6]

The plan submitted by California proposed to achieve the largest emission reductions through a continuation and expansion of state and local programs governing stationary and vehicular sources. As to the latter, the major elements of the strategy for control of new vehicles consisted of emission standards to become increasingly stringent through the 1975 model year, fuel evaporative controls that began with 1970 light-duty vehicles and that would apply to heavy-duty vehicles beginning in 1973, and assembly-line testing of all light-duty vehicles to be sold in the state after January 1973. For used vehicles, the plan contemplated continuation of the state program. Now, however, that program would include not only crankcase controls, but exhaust controls as well.[7] One exhaust control device for 1955–1965 models had been approved shortly before submission of the state plan, and others were approved later in 1972; they would be required in several areas at the time of transfer of ownership or initial registration in California.[8] Moreover, the plan anticipated additional approval of exhaust devices to control oxides of nitrogen emissions from 1966–1970 models.

In addition to the direct controls on emissions from new and used vehicles sketched above, there were other elements of the California plan that also related to the motor vehicle pollution problem: limitations on the degree of unsaturation of gasoline sold in the South Coast Basin (the basin in which Los Angeles is located), and on the volatility of gasoline sold throughout the state; fuel tax exemptions for gaseous fuels when used in connection with approved conversion systems; gasoline sales tax funds to support public transportation; testing of incentive programs to encourage commuter use of public transportation and car pools; improvement of traffic flow on freeways; a general air pollution research program; exhaust emission testing of samples of vehicles at two Los Angeles locations; and a program of enforcement (certificates of compliance, prohibitions of excessive smoke from vehicle exhausts, random roadside checks of exhaust emissions from 1966 and later model vehicles).[9]

Finally, the plan proposed investigations to determine the feasibility of a number of more far-reaching measures for control of vehicular emissions in areas with particularly acute problems—areas like Los Angeles. Measures to be considered included control of emissions from currently exempt vehicles (for example, those with small engine displacements); retrofitting of pre-1970 vehicles with evaporative emission control devices; prevention of spillage and vapor losses at service stations; transportation control measures (such as car pooling, changes in work schedules) to reduce vehicle use by as much as 20 percent; mandatory vehicle inspection and maintenance;[a] conversion of up to one-third of the vehicle population to gaseous fuels; and requirements that local planning agencies develop land-use plans to reduce air pollution through growth limitations and regulations, industrial location controls, coordinated transportation systems, and provisions for open space.[10]

Despite the ambitious nature of some of these measures, still the total package would not yield compliance with the federal standards. Moreover, state officials believed many of the measures could not be implemented. Both the 20 percent reduction in vehicle use and the one-third conversion to gaseous fuels were considered "unattainable"; there was "no way" they would be realized. The proposal for evaporative controls was regarded as

[a]As indicated in the last chapter, the legislature in 1970 directed the ARB to study the costs and benefits of a periodic inspection-maintenance program. See the discussion at p. 210. The agency responded by commissioning the Northrop Corporation to undertake the study. Northrop concluded that an inspection program was feasible and would be beneficial in terms of costs and emission reductions. See Northrop Corporation with Olson Laboratories, Inc., *Mandatory Vehicle Emission Inspection and Maintenance, Part A—Feasibility Study* (May 1971, Vol. 1, Summary). During the time of the Northrop study, the technical advisory committee of the ARB undertook an investigation of its own and recommended development of an inspection program. California Air Resources Board, Technical Advisory Committee, *Emission Control of Used Cars; Available Options: Their Effectiveness, Cost and Feasibility* (June 1971). Subsequently, in July 1971, the ARB recommended a pilot program. California Air Resources Board, *A Report to the Legislature on Vehicle Emission Inspection* (1 July 1971). Not much later, the California legislative analyst made a similar proposal. Report of the Legislative Analyst to the Joint Legislative Budget Committee, *Analysis of the Budget Bill of the State of California for the Fiscal Year July 1, 1972 to June 30, 1973* (California Legislature 1972 Regular Session, 1972), p. 456. Bills introduced in the 1972 session of the legislature would have provided for inspection programs, but none was enacted. In February 1972 Governor Reagan established his own task force to study the matter; it too recommended a program. Governor's Task Force on Periodic Vehicle Inspection and Maintenance for Emissions Control, *Task Force Report and Recommended Program* (October 1972). In fall 1973 the legislature provided for a pilot inspection program in the South Coast Air Basin that would expand eventually to cover all vehicles in the basin. Ch. 1154, § 1, [1973] Cal. Stats. Reg. Sess. 2392, adding to California Business and Professions Code, Health and Safety Code, and Vehicle Code. For further background, see E. Leong, "Air Pollution Control in California from 1970 to 1974: Some Comments on the Implementation Planning Process" (Doctor of Environmental Science and Engineering thesis, University of California, Los Angeles, 1974), pp. 86-91; D. Fisher, "Car Inspections: Cost Vs. Results Is the Big Issue," *Los Angeles Times,* 18 April 1972; L. Pryor, "Legislature Passes Car Inspection Bill," *Los Angeles Times,* 16 Sept. 1973; J. Gillam, "Statewide Energy Board Vetoed, Vehicle Smog Inspections OKd," *Los Angeles Times,* 3 Oct. 1973.

"awfully shaky." Effective land-use planning measures would be hard to obtain against the objections of local officials, would in any event be slow in developing, and were unpredictable in terms of the benefits they might yield. Quick control of uncontrolled vehicles was "highly unrealistic." Nevertheless, federal approval was anticipated. One official of the EPA pointed out that " 'if we don't accept the state's proposal then we'll have to write our own impossible plan.' Chances seem . . . good that the EPA will accept it more or less as submitted."[11]

But the EPA did not. At the end of May 1972 the administrator of the agency disapproved California's plan for the Los Angeles area on the ground that it did not provide for attainment of the ambient air quality standard for photochemical oxidant. He noted further, however, that attainment of the standard in Los Angeles and other areas with severe vehicle-related pollution problems depended heavily upon transportation control measures. Because of the complexities involved in such controls, and because neither the state nor federal government had had much experience with them, he extended the time to submit plans for such controls to February 15, 1973. He determined further that in many areas transportation controls would not be available soon enough to permit attainment of federal primary standards by 1975. Accordingly, he granted a number of these areas, including Los Angeles, an additional two years in which to meet the air quality standards for vehicle-related pollutants. Thus, a transportation control plan for Los Angeles would be due February 15, 1973, and it was to contain controls that would achieve the ambient standards by mid-1977. In the meantime, the agency and the states would have time to study the design and implementation of control measures.[12]

Actually, the EPA had started study of alternative measures for transportation control as early as 1971.[13] The so-called *Six Cities Study*[14] was designed to evaluate and recommend measures that could reduce vehicular emissions within the deadlines of the Clean Air Amendments; it was to focus in particular on the technical and institutional feasibility and probable costs of alternative approaches. The study was completed in March 1972, but it did little more than identify potential measures in general terms; closer evaluation was necessary with respect to particular problem areas. Realizing the need to provide additional assistance, the EPA initiated a second inquiry—the *Fourteen Cities Study.*[15] This study, undertaken by a number of consulting firms, was to provide technical support to states and cities confronted with the need to develop transportation control measures. It would be the foundation for specific proposals in implementation plans. The study was scheduled for completion in December 1972, leaving some two and one-half months to convert its learning into plans by the February 15, 1973, deadline.

That portion of the *Fourteen Cities Study* pertaining to Los Angeles was undertaken by TRW, Inc.[16] The study was rather unusual in several respects. Although it was intended to assist both local and state agencies in devising a plan for the area, TRW was explicitly directed by the EPA not to contact the Los Angeles APCD. In view of the APCD's distaste for transportation controls expressed earlier, it was considered likely that contact with the agency would be more a hindrance than a help. The South Coast Air Basin Coordinating Council established in 1970 was also ignored until very near the end of the project. The opinion of the ARB and the EPA appeared to be that local control districts, represented by the council, could make few useful inputs.

As the *Fourteen Cities Study* was underway, the ARB requested the California Department of Public Works to conduct a similar inquiry to identify means to reduce vehicle-miles traveled in the South Coast Air Basin.[17] The effort was significant not so much for its findings but because it resulted in the first active involvement of state and local transportation planning agencies in the implementation problem. The ARB itself had long been reluctant to propose controls over matters in which it had no legal authority, and this included transportation measures. Yet it was not until the Public Works study that the transportation agencies that did have control authority were brought into the process on any significant scale. This was not the only instance of a relevant body being excluded from the planning process. For example, the Southern California Association of Governments—a logical focal point for organizing efforts to develop transportation controls for the South Coast Air Basin—had been asked by the Los Angeles APCD early in the game to "stay out" of local air pollution control matters. Since the organization was young and had little authority, it was dependent on the faith and goodwill of member governments. One observer concluded that there was "not enough faith and good will to go around." As a result, the organization's participation was for some time "peripheral."[18] By 1973-1974, however, it had begun to take an active role in transportation control development, coordinating the efforts of a number of agencies in order to arrive at sound local plans for short-term transportation controls.[19]

In the course of the *Fourteen Cities Study,* events occurred that were to make development of a transportation control plan for the Los Angeles area more urgent. In September 1972 the California cities of Riverside and San Bernardino, in conjunction with several citizens' organizations, brought suit against the EPA to compel compliance with the statutory duty to promulgate an implementation plan for the Los Angeles area. The basic assertion of the suit was that, inasmuch as the Los Angeles plan had not been approved, the EPA had no alternative under the clear language of the Clean Air Amendments but to draft a plan for the area. The agency had no authority for extending to February 1973 the deadline for submission of the plan, nor—until an adequate plan finally existed—for extending the deadline for attaining air quality standards. These actions, the suit claimed, had been unlawful. The court agreed. It ruled in November 1972 that the EPA had acted illegally in extending the time for submission of a transportation control plan. The EPA was ordered to prepare, by January 15, 1973, a plan to achieve the federal air quality standards. The plan was to include "all necessary transportation controls and land use controls."[b]

The lawsuits against the EPA put the agency into the planning business. The *Riverside* case in particular did so because it called not for state sub-

[b]City of Riverside v. Ruckelshaus, 4 E.R.C. 1728 (C.D. Cal. 1972). A subsequent suit, brought by the Natural Resources Defense Council, involved the same points but was of more general application. Natural Resources Defense Council, Inc. v. Environmental Protection Agency, 475 F.2d 968 (D.C. Cir. 1973). In the *NRDC* case the court held that although the EPA had acted in good faith, it had no authority to extend the time for submission of implementation plans; moreover, it could not grant extensions for meeting federal air quality standards until after plans had been submitted. The court ordered the EPA to rescind the extensions of time for submission of transportation control plans and attainment of air quality standards previously granted. Transportation control plans were to be submitted by April 15, 1973, and were to provide for attainment of federal standards by 1975. Subsequently, a large number of states were notified that extensions were canceled and that they were required to submit by April 15, 1973, transportation control plans that would achieve the air quality standards by 1975. The EPA was to approve or disapprove the plans by June 15, 1973, and, in the case of disapproval, propose federal regulations to be promulgated finally by August 15, 1973 (later extended to October). The June and October dates were to become those governing the proposed and final plans for Los Angeles. At the time of submission of plans, state governors could request an extension of up to two years in which to meet the standards. If extensions were granted, they would have to be accompanied by the imposition of all interim measures reasonable under the circumstances. On the two lawsuits, see U.S. Environmental Protection Agency, *Transportation Controls to Reduce Automobile Use and Improve Air Quality in Cities* (November 1974), p. 23; F. Grad et al., *The Automobile and the Regulation of Its Impact on the Environment* (Norman: University of Oklahoma Press, 1975), pp. 366-367; T. Bracken, "Transportation Controls Under the Clean Air Act: A Legal Analysis," *Boston College Industrial and Commercial Law Review* 15(1974):749, 752-53; D. Fisher, "Air Cleanup Delays Rescinded by Court," *Los Angeles Times,* 2 Feb. 1973.

mission of plans, but for agency development of a plan for the Los Angeles area. The EPA had never been eager to get into implementation planning, expressing from the outset its distaste for the task of writing an "impossible plan." Yet it no longer had any choice. State and local officials had reflected a closed mind with respect to transportation controls; they knew that attempts to meet the federal standards in the Los Angeles area were foolhardy, and they as much as told the government so. In their view, the "problems were of such an overwhelming nature that initiatives would have to come—if at all—from the Federal Government. There was much skepticism expressed . . . about the willingness of federal officials to get involved or their ability to arrive at workable solutions."[20]

But willing or able or not, the EPA had now been mandated to take the initiative. The challenge was not a pleasant one. Earlier federal review of the Los Angeles situation had led to the conclusion that it was "difficult to be sanguine about solutions to the automobile air pollution problem in Los Angeles, at least over the short-to-medium-term." Los Angeles had a more serious problem than other areas, and resolution was "made exceedingly difficult by practical and political problems of an entirely different order than elsewhere."[21] No doubt for reasons like these, the EPA had argued in the *Riverside* case that a transportation control plan for Los Angeles could not be available before February. Inasmuch as the *Fourteen Cities Study* had not yet been completed, more work was necessary before transportation controls could be proposed. Additional study was especially important because those controls were likely to "have a severe impact on the public." The judge was unmoved; "none of the excuses for the delay are valid."[22]

Thus it came to be that the EPA prepared a transportation control plan for Los Angeles. Before it finished the task (through a series of proposed plans and subsequent revisions), it would see stalling tactics, finger pointing, and inactivity. Complaints about federal intervention would become common, and the agencies with the responsibility to draft a plan were to be those most critical of federal planning efforts; they would lead the charge that federal bureaucrats should not intervene in local problems. Their default proved to be politically and legally adroit. To a considerable extent it shielded them from what was to become a very heated issue, and to a considerable extent it was endorsed by federal courts.

The First EPA Plan

The *Riverside* "court decision pushed EPA staff members into a string of 10- and 12-hour days in search of a way to meet the tough . . . standards" and deadlines mandated for the Los Angeles area. The agency "took the best available statistics and estimated" the emission projections, and next considered "their possible reduction through various means of control available through the EPA. . . . When the EPA exhausted every means of control it could implement (except gasoline rationing)," it learned that still the federal air quality standards would not be met.[23] That of itself was no surprise. Analysis in the early stages of the Los Angeles portion of the *Fourteen Cities Study* had been directed to finding solutions that were politically and institutionally acceptable and would minimize social and economic impacts. It quickly became apparent that the requirements of the law and the objectives of the study were diametrically opposed. Acceptable control measures would not come close to realizing the federal standards, while the measures that would achieve them appeared totally unacceptable in social and political terms. The Los Angeles study, for example, concluded that no technical measures could assure compliance with the standards; only gasoline rationing could do so. Yet rationing was not recommended, on the ground that it would require too radical disruption of the Los Angeles life-style. Beyond this, the study concluded that no "incentive-type" measure (for example, parking and driving taxes) was adequate to deal with the Los Angeles problem.[24]

Some observers disagreed, at least in part. In November 1972, for example, the Environmental Quality Laboratory of the California Institute of Technology proposed a set of measures that it claimed would lead to substantial improvements in Los Angeles air quality by 1976. Included in the proposal, in addition to a number of technical measures, were incentives for mass transit and car pools and, as a "last resort," increased gasoline taxes amounting to approximately a 25 to 50 percent increase in fuel prices. Later versions of the proposal included rationing that would reduce gasoline consumption by about one-third over a period of several years.[25] While the recommendations of the Environmental Quality Laboratory were not so drastic as to appear totally infeasible, the fact remained that they would not result in compliance with federal standards for photochemical pollution. They would reduce the days in excess of those standards in the Los Angeles area from about 250 to 40 annually, but federal law permitted no more than one day per year. More dramatic controls were necessary, and rationing appeared to be the only available alternative. Consistent with the Los Angeles study, the EPA concluded that "'carrot and steak' [*sic*] approaches, like high parking fees, increased gasoline prices, freeway bus lanes, and car-pooling won't work well enough to do the job. Our studies show that the public will simply pay for the convenience of using its cars." Thus, while gasoline rationing was a "last resort" that went into the plan only after considering "every other means open to the EPA," it went into the plan nevertheless. EPA officials said "it was the only avenue open to our agency if we are to fulfill the court's mandate."[26]

That the plan devised by the EPA staff called for gasoline rationing was not of itself the most striking feature of the proposal; rather it was the amount of rationing—86 percent during the six most critical months of May through October. (By way of contrast, rationing during World War II cut consumption by less than one-third at the most extreme.) In this form the proposed plan went to William Ruckelshaus, administrator of the EPA, for review. While he had three alternatives, "yes, no, or maybe," in fact only the first response was legally available. Either of the last two would amount to defiance of the court order and could result in a contempt citation.[27]

Ruckelshaus acted accordingly. In a widely publicized news conference held in Los Angeles on January 15, 1973, he proposed a plan for the Los Angeles area. It contained a number of measures: reduction of evaporative losses from dry cleaning plants; elimination of highly reactive hydrocarbons used in degreasing for electroplating and other industrial operations; other stationary source controls; aircraft emission controls; gasoline marketing vapor controls; used-vehicle retrofits; inspection and maintenance programs; and gaseous fuel conversion. But the most noticeable element of the proposed plan was that calling for gasoline rationing (now 82 percent) during the months of May through October.[28] These measures were not, of course, a plan, but only a proposal, a "proposed rule making" to comply with the earlier court order. Hearings were scheduled, comments were solicited, and revisions were anticipated. A two-year extension for California was still being considered, and it was expected that the state would continue to proceed in accord with the original timetable allowed by the EPA. Thus state proposals for transportation controls would presumably be submitted in mid-February, and it was "the Administrator's policy to be guided in his final promulgation by approvable segments of the State plan."[29]

Given the background, it could not be expected that the EPA's proposed rule-making would be enthusiastic in tone, at least as regards its rationing proposals. The agency's announcement stated that rationing would "have a great economic and social impact" and probably "cause serious dislocations" to the Los Angeles area. Rationing was regarded as an "extreme measure"; the EPA had "serious reservations as to [its] feasibility and desirability . . . [but] legal requirements placed on the Agency" left no alternatives. The proposals themselves consisted not of concrete provisions, but rather vague language that "the sale of gasoline shall be controlled by the Administrator," in amounts to "be determined" subsequently.[30] The introductory section of the rule-making, however, made clear the amount in mind, though it also disclosed that the means of achieving it had not been settled upon:

> The amount of rationing may be different for different months, up to an expected maximum of 82 percent. The rationing system may be enforced at the individual vehicle operator's level, with gas coupons required to purchase gasoline. Or it may be enforced at the manufacturer's level only, with price controls at the retail level (to prevent windfall profits) and all gasoline sold on the basis of first-come-first-served. Public comment is invited on these options, and on questions of implementation, including eligibility for coupons, and transferability of coupons.[31]

The mention of "first-come-first-served" made it especially difficult to believe the EPA was serious in its proposal. There was some thought that the rule-making was designed not to comply with the court order, but "primarily to encourage public pressure on Congress to extend the deadline for achieving the clean air standards. . . . The tone of Ruckelshaus' remarks [at the press conference] indicated he does not believe the drastic gasoline rationing he proposed will ever occur." The administrator said he did not " 'believe that the final plan . . . will achieve reductions in that neighborhood [of 82 percent].' " He saw the rule-making as an " 'important step that we are attempting to take today in forcing people to pay attention to the seriousness of the (smog) problem.' "[32]

People tended instead to pay attention to the foolishness of the rationing proposal. Several days after announcing the rationing plan, the EPA appointed a special task force to conduct public hearings and give a technical critique of the various aspects of the rule-making. Even before rationing had been officially proposed, people in the Los Angeles area had begun to speak out against it. One observer saw it as "a lesson in over-reaction. . . . Gasoline rationing as a solution to today's commuter and smog problem is about as sensible as smashing all drinking glasses in the area to conserve the Los Angeles water supply."[33] The public hearings produced more of the same. Jesse Unruh, a candidate for mayor in Los Angeles, testified that a 20 percent reduction might be possible, but he predicted "quite possibly a revolution" if the 82 percent plan went forth. There would be, he thought, "armed raids on gasoline stations." The mayor himself, Sam Yorty, said the federal plan was "shocking" and "economy destroying." In his view, "the cost of meeting the federal air quality standards should be 'borne by the authority mandating the changes'—meaning the federal government." Senator John Tunney of California found the plan "drastic and nonsensical." One member of the Los Angeles County Board of Supervisors proclaimed it "unrealistic" and "unnecessary."[34]

To be sure, the EPA found a bit of support. In the view of the Coalition for Clean Air, the fault was not with the agency but "was the direct result of gross failure by the state to meet its commitments." But even the Coalition's spokesman admitted that "82% rationing is 'unrealistic.'"[35] The Sierra Club as much as conceded the same point, proposing a 30 percent reduction through, among other measures, an emissions tax on automobiles.[36] All in all the tone was hostile, and the hostility, by and large, was directed not at the Congress, not at state and local agencies, but at the EPA. "Most witnesses attacked the prospect of drastic gasoline rationing," said an account of one of the public hearings.[37] The rationing proposal "brought an immediate negative reaction from wide sectors of the community," said another.[38] An editorial summed it up: "The consensus has it that the Environmental Protection Agency's proposal to ration gasoline sales in Los Angeles during the six smoggiest months of the year is economically infeasible and politically unrealistic."[39] Ruckelshaus could only agree. He "indicated that it may result in an extension of federal clean air deadlines. . . . An extension of the timetable 'makes sense,'" he said.[c]

The Second EPA Plan

As mentioned earlier, the EPA, shortly after announcing its "hastily drafted gas rationing proposal," had established a task force to hold hearings and provide a technical critique of the January rule-making. Subsequently, the task force "began a three-pronged reappraisal" that led to

[c]L. Dye, "Failure of State Blamed for Gas Rationing Plan," *Los Angeles Times*, 19 Jan. 1973. The EPA subsequently submitted to Congress a proposed amendment to the Clean Air Act that would permit extending "for not more than five years the deadline for attainment of national primary ambient air quality standards where transportation control measures are necessary for the attainment of such standards, and where the implementation of such control measures would have serious adverse social or economic effects." The amendment was submitted because at least ten metropolitan areas could not achieve the federal standards by 1977 except by gasoline rationing, and the EPA did "not consider gasoline rationing to be a reasonable or desirable approach to emissions reduction." U.S. Environmental Protection Agency, *Transportation Controls*, pp. 26, 69, Appendix C. See also P. Houston, "Delay in L.A. Clean Air Program Sought," *Los Angeles Times*, 24 May 1973; D. Fisher, "U.S. Proposal Could Ban Autos in L.A.," *Los Angeles Times*, 16 June 1973; Bureau of National Affairs, *Environment Reporter, Current Developments*, Vol. 4, No. 8, 22 June 1973, p. 266; ibid., No. 21, 21 Sept. 1973, p. 824; ibid., No. 48, 29 March 1974, pp. 1975, 2004.

some discouraging conclusions. First, new calculations concerning "what it would take to clean up the air in Southern California" suggested "that the task may be much more difficult than originally thought." Second, public hearings indicated that in Southern California "motorists would tolerate a cut in auto use of 30% at the most."[40]

The first point requires elaboration. The January 1973 proposal for 82 percent gasoline rationing rested on certain technical assumptions about the relationship between emission reductions and air quality improvements. Since the time of that proposal, there had been developed an approach to the emissions-air quality relationship considered more technically sound than the technique used for the original rule-making. The rule-making had used the approach of linear rollback, which proceeds on the simple (and scientifically unjustifiable) assumption that maximum air quality concentrations will change directly in proportion to changes in emission levels.[41] But the task force employed a statistical approach based on relationships between actual morning hydrocarbon levels and the observed air quality conditions that accompanied them. The approach took into account, in a way the rollback technique did not, the known nonlinearities in emission levels-air quality relationships. It was not only a more rigorous method in scientific terms, but in terms of results as well. That is, it indicated the need for more stringent controls than did the rollback approach.[42] Had the statistical approach been used in the original January rule-making, it would have been necessary to reduce *entirely* (that is, by 100 percent) the gasoline consumption of light-duty vehicles, for the emissions from uncontrolled stationary sources, trucks, aircraft, and motorcycles would of themselves exceed the maximum allowable. Compared with this, the original proposal for 82 percent rationing was wonderfully permissive!

The implication of the new statistical approach, of course, was that it would be even more difficult than originally thought to achieve the air quality standards in Los Angeles (a mean realization considering that the original measures were considered impossible). The EPA was painfully aware of the point. It announced in May 1973, a month before the revised proposal for Los Angeles was due, that it would ask Congress to give the area additional time—from two to four years—to meet the federal standards. But even four years would be inadequate. An EPA official "who asked not to be named" said that "short of creating economic and social chaos, meeting the standards will take a lot longer than that—perhaps a decade or more." The agency had no intention of creating chaos. It was "considering an alternate 'transportation control' plan for Los Angeles that would abandon last January's suggested emphasis on massive gas rationing and switch to heavy reliance on bus transit and car pooling." The new acting administrator of the EPA, Robert Fri (Ruckelshaus had gone to serve as acting head of the FBI), "conceded that the gas rationing proposal . . . was made solely to satisfy a court order." There would be a new proposal in June, and even it would be something of a charade. The "main body" of the proposal "should be considered just a 'formality . . . a technicality' to follow the letter of the law. He said the real meat will be in the preamble, where EPA will present a 'sensible' plan for reducing air pollutants in Los Angeles. . . ." That plan, however, "probably would enable the Los Angeles area to move only halfway toward achieving federal air quality standards by 1977."[43]

The revised proposal for Los Angeles was announced less than a month later, on June 15, 1974. A final plan was to be promulgated on August 15, and the purpose of the June 15 revision was "to solicit public comment on the current thinking of the Agency." That thinking was as follows: Reduction in hydrocarbon emissions "on the order of 90 percent" would be necessary to achieve the federal standards. Much of the reduction would be achieved through measures submitted by the state and already approved, but still the standards would not be met. The EPA concluded that an extension to 1977 was justified; in the meantime, "reasonable interim measures" were to be imposed. These would include bus and car-pool lanes on freeways, reductions in off-street parking, limitations on the construction of new parking facilities, limitations on motorcycles, mandatory inspection and maintenance of light-duty vehicles, "and a lid on further increases in gasoline consumption"—all to become effective by 1975.[44]

The proposal to put "a lid on further increases in gasoline consumption" revealed that the gasoline limitations of the original January 1973 rule-making had been at least temporarily abandoned. In that rule-making the EPA had indicated that it was considering two approaches—rationing, and limitations "at the manufacturer's level." Now it opted for the latter, proposing that "distributors not be permitted to deliver to retail outlets any more gasoline than was distributed during the 'base year' of July 1972-June 1973." Gasoline consumption had been increasing 4.5 percent annually. The new approach would hold it constant, and would be "much less severe and far more easily administered than a rationing system of great magnitude as was earlier proposed." Unfortunately, however, it was "quite possible, indeed probable," that even with full implementation of all the newly proposed controls, the national standards would not be achieved by 1977. Thus, the EPA had "no choice" but to propose further gasoline limitations "of whatever degree necessary" to meet the standards in that year. The "degree necessary" appeared ambiguous, but a table setting forth the "compilation of control strategy effects" made it perfectly clear. The air quality standards would be met in 1977 only if there were

"100% Gasoline and 60% Diesel Fuel Reduction."[45] Thus the new proposal was quite accurately seen as one that "could result in banning all automobiles from the streets of Los Angeles by 1977." Since such a ban was obviously "unreasonable," the acting administrator concluded that "we're going to have to go back to Congress and get this situation corrected." The general view was that "Congress had little choice but to give the Southland more time."[46]

State and local officials were nevertheless disenchanted. The provision for a 100 percent reduction in gasoline consumption aside, the revised proposal was still unsatisfactory. Its parking restrictions "could mean additional traffic tieups in work areas while motorists search for parking places—and consequently more pollution," according to the Los Angeles APCD. The executive officer of the ARB was concerned with "how the public will react to such measures." He "questioned the feasibility and effectiveness of the proposed measures, the adequacy of the state's and the EPA's authority to enforce the plans, and the availability of the funds necessary" to carry them out.[47] He "also said that if Congress legislates clean air, they also should be obligated to vote the funds necessary to reach the standards."[48] As to the new gasoline limitation, Los Angeles County at least regarded it as "nothing more than gasoline rationing with a sugar coating around it and it will not work here."[49]

It is not surprising that California officials were unhappy with the EPA's revised plan, for they had never been enthusiastic participants in the process that led up to it. The January rule-making had stated the hope that a state transportation control plan would be submitted by the original February deadline, the administrator saying that he would be "guided" by it. There was no state submission by that deadline, however, nor was there one by the April deadline set in the *NRDC* case (though there was a partial proposal by April 18, 1973).[50] Indeed, no new plan was forthcoming from the state until July 1973—after the EPA had proposed its own revisions—and even then the state deleted its earlier proposals pertaining to a 20 percent reduction in vehicle use on the ground that controls on vehicle use were now to be developed by local and regional transportation planning groups.[51] As a result of the state's reluctance to get involved in transportation controls, the regional office of the EPA was forced to take most of the initiative. That situation was to change little in the course of EPA's revision of its June rule-making into a final plan.

The Third ("Final") EPA Plan

The EPA's June rule-making declared that if California would submit adequate control measures, the administrator could approve these "rather than promulgate his own." Timely submissions would be required, however, so it was "suggested that planning to comply with these measures not be delayed in hopes of approval of substitute measures." Public hearings were scheduled, and the agency would thereafter draft or approve a final plan by the court-mandated deadline of August 15, 1973.

California officials tended to view the timetable as too constrained, but for reasons that were now somewhat more positive: state and local agencies were ready to participate more heartily in the planning process. Haagen-Smit wrote the acting administrator in early July, requesting an extension in order that local, state, and federal officials could "develop jointly a plan that is feasible to implement." This would take time, but would ensure a workable product in the long run.[52] A letter from city, county, and regional officials to the EPA's regional office expressed the same sentiments. Efforts to develop feasible transportation controls, it said, were now beginning. In July 1973 a local task force had been formed to work on specific alternatives for EPA consideration in order to ensure local inputs to the planning process. The Southern California Association of Governments, which a year-and-a-half before had been directed by the Los Angeles APCD to stay out of air pollution planning activities, was now playing a central role in task force activities.[53]

The EPA's reluctant proposal for drastic reductions in gasoline consumption had had some positive impact. State and local planning efforts were underway that would not, by the admission of the agencies themselves, have come about without the federal threat.[54] But defenses to the threat were also being prepared. In its June rule-making the EPA had warned that "the Agency will require State and local governmental entities to take actions wherever possible, and will not involve the Federal government in the administration of local programs and in direct enforcement against individual citizens."[55] The ARB, at least, saw that language as implying that lawsuits might be brought against it should it fail to comply with federal requirements. It asked the California attorney general's office for an opinion on the liability of state and local officials should they fail to comply with provisions in an implementation plan or orders from the EPA. The opinion, set forth in a "confidential" memorandum, concluded that there was no liability. In case of a "State failure," it was up to the EPA to enforce the plan "either through an order requiring any violator in such State to comply or by bringing action against any such violator." It was possible, of course, that "court action may be commenced . . . but we are of the opinion that any such action would not be successfully maintained."[56]

The opinion made the ARB more comfortable for a time. The agency continued to remain fundamentally at odds with the EPA, especially with respect to assumptions about the relationships between emissions and air quality. The EPA tended to employ conservative assumptions, thus concluding that very severe controls were necessary; the ARB and the Los Angeles APCD leaned toward the opposite approach.[57] There was no objective way to resolve the debate, for many of the uncertainties involved were not capable of resolution given the state of knowledge. Meetings among the best technicians from all agencies involved—meetings to discuss differences in assumptions and calculations—might have been helpful, but few if any such meetings were held.[d] The EPA went forth on its assumptions, and the ARB on its. The ARB would propose for implementation only measures for which technology existed "and which have been proven to be feasible and effective to implement." Careful consideration was given to "the impact of the measures, and enforcement difficulties, costs, and method of funding."[58] By the due date for announcement of a final plan, October 15, 1973 (the August deadline had been extended), the ARB had not submitted a control strategy that would achieve the 1977 standards. With respect to transportation controls in particular, the necessary steps had not been "spelled out in regulatory form."[59] As had been the case in the past, the EPA was forced to rely largely on its own judgments, and on opinions expressed in public hearings, to come up with a final plan for Los Angeles.

That plan was similar in many respects to the June rule-making. The extension to 1977 was retained, as were most of the control measures to be applied in the meantime.[60] There were a few significant changes, however. The EPA's June proposal to reduce public parking space (by 20 percent) "drew almost universal adverse comment during" the public hearings. "At the same time," however, "the use of regulatory fees to discourage pollution-causing activities was widely supported. In particular the use of fees to control parking was mentioned." As a result, the agency withdrew the proposal to limit public parking and substituted "a regulatory fee to increase the price of parking." (Limitations on the construction of new parking facilities were maintained largely intact.) A second change was related to the parking surcharge; it provided "for employer-paid mass transit fares and special parking privileges to those who travel by carpool." Finally, the provisions for gasoline limitations to be effective by 1975 were abandoned. The 100 percent reduction for 1977 was retained, however; the administrator had "no alternative."[61]

[d]There were a few such meetings later, though they appear to have resolved nothing. See, e.g., L. Pryor, "U.S., County Air Experts Argue Issues in Private," *Los Angeles Times,* 22 Aug. 1974.

Aftermath

Reaction to the EPA's final plan was mixed. Federal and local officials were generally pleased with the result, claiming it represented "a fundamental turning point in the war on smog." There "was an emerging cooperation between federal, state and local governments." There had been clashes during the three previous years, but now Mayor Bradley saw "a new direction. . . . EPA is working with us and listening to us at the local level." Important "concessions" had been made—the provisions relating to gasoline rationing and parking surcharges in particular.[62] The EPA noted that a number of its final transportation controls had been "recommended by State and local task force[s]."[63]

The controls, however, were not really "final," for further concessions and revisions were on the way. Not one month after publication of the plan for the Los Angeles area, the EPA deferred the parking surcharge control measures pending further study. "Public reaction" to the measures had "been tense." The surcharges had been "widely criticized as arbitrary, illegal, administratively burdensome, and economically disastrous. A great many petitions for judicial review have been filed."[64] Shortly thereafter, in early 1974, the agency withdrew the parking surcharges as well as the employer incentive measures; at the same time, it deferred implementation of regulations on the construction of new parking facilities. The state was free to submit surcharge and incentive measures of its own, but it was not required to do so.[65]

In taking these actions, the EPA was proceeding in accord with "firm congressional guidance." The Clean Air Amendments had been born out of the environmental crisis; now the energy crisis had developed, and Congress was considering further measures to respond to it. In December 1973 the House Committee on Interstate and Foreign Commerce had reported out its version of the Energy Emergency Act; the bill forbade (for reasons that could not possibly be related to the energy problem) imposition of parking surcharges without the consent of Congress. Amendment on the floor of the House resulted in a further provision forbidding special lanes for buses and car pools, and review of new parking facilities without congressional approval. In conference, the surcharge prohibition was retained, the special-lanes prohibition dropped, and the provision regarding review of new parking facilities altered to permit the EPA to suspend such a requirement. Though the bill was not enacted, "for reasons completely unrelated to the amendments to the Clean Air Act which it contained," Russell Train (the EPA's new administrator) did not believe he could ignore its "strong expressions of intent." He pointed out that the bill reflected the views of a number of members of the House and Senate who sat on committees with jurisdiction over air pollution legislation.[e]

Unlike federal and local officials, those at the state level were not at all happy with the EPA's final plan for the Los Angeles area. The ARB considered that the agency had overstepped its boundaries and intervened in matters that properly belonged to state and local government. One week after promulgation of the plan, its executive officer had written Train to remind him of the congressional intent "that state and local agencies were to retain the primary responsibility for the control of air pollution. . . . EPA's policy during the last few years has, however, not reflected such a policy."[66] Train's subsequent compromises on parking surcharges were apparently insufficiently soothing to the ARB, for it remained dissatisfied with the federal government's aggressive posture. In setting out its final plan, the EPA had announced that the "measures are legally enforceable, and EPA will enforce them unless and until alternatives are suggested and found to be preferable."[67] Despite the soothing legal opinion rendered earlier by the California attorney general, the possibility of a lawsuit continued to nag the ARB. Its counsel recommended a counterattack, a suit against the EPA to challenge the agency's authority.[68] Not long after, in late November 1973, a suit was filed. Haagen-Smit expressed to Train his reluctance to take the action; it had been "compelled . . . to protect our right . . . and prevent civil and criminal actions by you against us."[69] The suit alleged in essence that the United States Constitution precluded the federal government from compelling states to commit funds and manpower to carry out federal mandates.[70]

Other state-level officials in California were also disenchanted with the EPA's plan for Los Angeles, and—like the ARB—they took steps to register their feelings. In November 1974 a task force consisting of representatives from several state agencies (Business, Resources, Commerce, Transportation, and Employment Development) completed a study of the plan and issued a report. According to the task force, the EPA plan would lead to "complete paralysis" of commerce in the Los Angeles area, and produce "social and economic disruption of staggering proportions."

[e]39 *Federal Register* 1848 (1974). Train was speaking in early January 1974. An Energy Emergency Act of 1974 was passed by Congress little more than a month later. S. 2589, 93d Cong., 1st Sess. (1973). President Nixon vetoed it in March. S. Doc. No. 93-61, 93d Cong., 2d Sess. (1974); *Congressional Record* 120 (6 March 1974):2883. In June Congress passed another bill—the Energy Supply and Environmental Coordination Act of 1974—very similar to the earlier measure, and this became law. Pub. L. No. 91-190, 22 June 1974. The legislation ratified the EPA's earlier actions pertaining to parking surcharges and construction of new parking facilities.

To achieve the federal standards by 1977 would require mass transit, yet a sufficient transit system could not be developed in time. As a result, millions of persons would be without the means to get to work—they could not drive (because of the gasoline limitations), and there would be no satisfactory alternative means of transportation. Moreover, the provisions of the final plan relating to the construction of parking facilities would result in the loss of millions of dollars worth of development and thousands of jobs. Controls on new stationary sources could prevent the construction of major industrial facilities, including power plants and oil refineries, "further aggravating an already critical statewide energy situation and greatly restricting the economy's ability to expand." The task force concluded by recommending a number of amendments to the federal law. Cost-benefit studies should be required of the EPA before it promulgated any regulations, and the EPA should be precluded from "implementation of regulations that would cause unreasonable social and economic disruption." Moreover, the EPA should be granted authority to extend the statutory dates for achieving federal air quality standards. Finally, the gasoline limitations and parking management regulations of the transportation control plan should be nullified.[71]

The EPA responded quickly. The task force report, it said, gave "a misleading impression of potential social and economic impacts." The assumptions of the report were "unrealistic" and had "the effect of a scare tactic." The report "greatly exaggerate[d]" matters by considering only the "worst case" situation rather than the "realistic situation." Specifically, the report was based on the gasoline limitations scheduled for 1977, yet even the EPA considered these "an unreasonable burden on the people of the South Coast Air Basin"; it would not enforce the limitations and had even moved to amend the law to remove the need for them. As to regulations on construction of parking facilities, these had been deferred. Finally, the EPA noted that the report neglected the "severe health effects" anticipated if the air pollution control program did not go forth, and failed to take account of the jobs and income that would be generated by that program.[72]

Following this exchange, relations between California and the federal government deteriorated further. In May 1975 the ARB openly challenged federal authority by voting to repeal certain provisions of the final plan relating to emergency controls during periods of especially heavy pollution. The EPA had "sought to implement the emergency provisions of the Clean Air Act by getting California to shut down businesses, offices and recreational facilities during high smog episodes." The ARB rejected the measures "because of the economic and political problems associated with a wide-spread disruption of daily affairs." Even an avid environmental lawyer on the ARB voted against the federal plan, finding it "not only a hoax but . . . a meaningless gesture because it isn't even a plan." The agency's chairman said it was "hard to know if EPA is serious or whether they [*sic*] have been making these outlandish proposals in an effort to destroy the Clean Air Act."[73] The EPA, however, claimed to be in complete earnest, saying it would "intervene to force state implementation of plans to cope with air pollution emergencies."[74]

It was, of course, the fear of just such intervention as this that had prompted the ARB lawsuit against the EPA. In August 1975, three months after the EPA's latest threat to force state action, the federal Court of Appeals for the Ninth Circuit reached its decision in that case.[75] The opinion reflected a dramatic victory for state government, and a potentially disastrous setback for the federal program. The court distinguished between the "pollution-creating activities" of state and local government on the one hand, and those of private citizens on the other. Both could be appropriately regulated by the federal government,[f] but *implementation* and *enforcement* of regulations in the two cases was subject to an enormously important difference. If actual polluting activities of state and local government (for example, pollution from state-owned facilities) violated federal requirements, sanctions could be brought to bear directly on state and local officials to encourage compliance. The EPA could, for example, seek an injunction or civil contempt order against the responsible officials. If, on the other hand, violations of federal requirements occurred because state and local officials simply refused to carry out EPA regulations (contained in the federal plan for the area) directing enactment and enforcement of measures to control private pollution-creating activities, the EPA had no recourse but to promulgate and implement controls itself and enforce them by proceeding directly against violators. The agency could not compel state and local officials to perform those tasks. "Tersely put," the court said, "the Act, as we see it, permits sanctions against a state that pollutes the air, but not against a state that chooses not to govern polluters as the Administrator directs."[76] Regulations providing for the latter were held invalid.

This interpretation of the Clean Air Amendments made it unnecessary

[f]Thus in a later, related suit, the court upheld the authority of the EPA to promulgate and implement the regulations providing for gasoline limitations, notwithstanding that they might result in "substantial economic loss and social disruption." City of Santa Rosa v. Environmental Protection Agency, 534 F.2d 150, 155 (9th Cir. 1976).

for the court to confront constitutional issues—in particular, those arising from California's claims that any broader enforcement authority in the EPA would reach beyond the Commerce Power and into areas of responsibility reserved to the states by the Tenth Amendment. But it hardly regarded these assertions as "frivolous"; indeed, it had been "induced" to adopt its interpretation of the Clean Air Amendments in order to avoid "constitutional difficulties" that would otherwise be posed by the legislation.[77]

Three other federal courts of appeals decisions on the question of EPA enforcement authority reflect mixed views. One upheld the EPA, one reached conclusions essentially the same as those in the California case, and one recognized some authority in the EPA to force state action, but not so much as the agency would like to have. The decisions presented the federal government with a considerable dilemma—whether it should seek review in the United States Supreme Court to resolve the differences of opinion among the lower federal courts. The attorney general of the United States "expressed his doubts about the constitutionality of forcing states to take specified actions."[78] If the Supreme Court shared those doubts, review would not only prove fruitless, it would preclude the EPA from relying—in the areas in question—on the one opinion upholding its views and on the other concluding that the agency has at least some authority to force state action.

These considerations argued against review, but others argued, urgently, in favor of it. In those areas where the federal court decisions left the EPA with only very narrow enforcement power, areas like California, the agency had but one alternative—"direct federal enforcement." But the "EPA has determined that this would be impracticable." It is not difficult to see why. Instead of having to bring only one or a few suits against recalcitrant state officials to force them to administer federal regulations, the agency would have to implement the regulations in each area itself. If violations of the regulations occurred, it would have to bring enforcement actions—potentially thousands of them—against citizen-polluters to realize compliance. Thus the EPA said that the narrow judicial decisions "may virtually paralyze our present efforts to reduce auto pollution in metropolitan areas," and "make it virtually impossible to implement those measures which have already been developed."[79] Faced with such a prospect, the government finally did opt in favor of review, and the Supreme Court agreed to hear the cases. Unless the EPA's views are ultimately upheld, transportation controls under the Clean Air Amendments are likely to be much more a matter of state pleasure than federal mandate. If nothing else, the states will have a strength in bargaining that Congress (at least according to the EPA) had intended to deny them. There will be cooperation rather than cooptation, federalism rather than federalization.[g]

THE NATIONAL PROGRAM FOR NEW VEHICLES

By the end of 1975 the federal program for transportation controls—concerned with regulating the emissions and especially the use of all vehicles, new and old—was in serious jeopardy. One could say more, but not a great deal, for the program to control emissions from new vehicles. Here Congress had demanded large accomplishments in a small time. Not surprisingly, setbacks and delays quickly developed.[80]

As mentioned in the last chapter, the EPA in mid-1971 had set emission standards to implement the Clean Air Amendments' strict 1975-1976

[g]On the courts of appeals decisions, see "The Clean Air Amendments of 1970: Can Congress Compel State Cooperation in Achieving National Environmental Standards?", *Harvard Civil Rights-Civil Liberties Law Review* 11(1976):701. On the Supreme Court's decision to grant review, see Bureau of National Affairs, *Environment Reporter, Current Developments,* Vol. 7, No. 5, 4 June 1976, p. 199; 44 U.S.L.W. 3681-3682 (1976). Subsequent to the government's decision to seek and the Supreme Court's decision to grant review, the Court rendered a decision that jeopardized the EPA's position. National League of Cities v. Usery, 44 U.S.L.W. 4974 (1976), in holding unconstitutional the 1974 amendments to the Fair Labor Standards Act, "sharply limits the power of the Federal Government over states pursuant to the Commerce Clause of the U. S. Constitution. The authority of the Federal Government under the Commerce Clause to control air pollution has been cited as the basis for the Environmental Protection Agency's authority to require states to enact and enforce specified transportation controls. . . ." Bureau of National Affairs, *Environment Reporter, Current Developments,* Vol. 7, No. 9, 2 July 1976, p. 399. Still later (and after this book was in press), the Supreme Court rendered a decision (of sorts) in the transportation control cases. The federal government's petition for review in the Supreme Court had challenged the decisions of the courts of appeals only insofar as they invalidated EPA regulations requiring state inspection and maintenance programs. In other words, the government had narrowed its claims of authority, and it narrowed them further in its Supreme Court brief, admitting that the regulations remaining in controversy were invalid unless modified to remove requirements that the states submit legally adopted regulations as part of their transportation control plans. Given this concession, the Court declined to pass on the EPA regulations; it would not review regulations not yet promulgated, and it implied that when promulgated the regulations should be reviewed first by the courts of appeals. Thus the Court vacated the judgments of the courts of appeals and remanded to them for further proceedings, including consideration of whether the cases had become moot. Environmental Protection Agency v. Brown, 45 U.S.L.W. 4445 (1977).

requirements for new vehicles.[h] Subsequently (January 1, 1972), the National Academy of Sciences issued its initial report—called for by the Clean Air Amendments—on the feasibility of the standards and the deadlines. The report did not recommend suspension of the 1975-1976 standards, but it lent support to such a position, and the auto companies seized the opportunity and applied for suspensions shortly thereafter. When the EPA denied these, to the surprise of some people, in May 1972, the companies appealed the action to the Court of Appeals for the District of Columbia. The court's decision marked a success—for the companies; it held that in denying the application, the EPA administrator had failed to support his conclusion that the technology necessary to meet the standards would be available. The matter was remanded to the agency for further consideration.[81]

Following the court's decision, the National Academy issued a second report suggesting that delay in the 1975-1976 emission standards might be "prudent." The EPA apparently recognized a losing battle. In April 1973 it granted the auto companies a one-year suspension of the 1975 standards and at the same time promulgated a set of interim standards. California was given a waiver; its 1975 standards were to be more stringent than those for the rest of the country. The effect of the waiver would be to require catalytic converters in California, and this was seen as a means to phase in that technology. In July 1973 the EPA granted the automakers another suspension, this one of the 1976 statutory standard for nitrogen oxides. Again an interim standard was established.[82]

These early suspensions resulted from (reluctant) administrative action by the EPA; the next ones were initiated by Congress, but followed by further EPA suspensions. The Energy Supply and Environmental Coordination Act of 1974, passed in response to the "energy crisis," had ratified earlier EPA actions on parking facilities and surcharges.[i] It also did something else. For reasons related to energy efficiency, it continued through 1976 the interim standards for new vehicles that the EPA had earlier been forced to set for hydrocarbons, carbon monoxide, and nitrogen oxides, and set a maximum standard for the latter through 1977. It postponed the original 1975-1976 standards of the Clean Air Amendments to 1977-1978, and gave the EPA authority to suspend for an additional year the 1977 (hydrocarbon and carbon monoxide) standards.[83] Subsequent to the legislation, California was granted waivers that permitted it to carry out its plans for higher standards in 1976.[84]

[h]See the discussion at p. 209.

[i]See the discussion at pp. 229-230, footnote e.

In early March 1975 the administrator of the EPA (Russel Train) exercised the authority given him under the 1974 energy legislation and suspended for one year the 1977 standards for new vehicles.[85] The reasons for the action are reminiscent of the earlier discovery that methods of hydrocarbon and carbon monoxide control were increasing emissions of oxides of nitrogen. The EPA had discovered a new health hazard that could well result were the 1977 standards not suspended. The oxidizing catalysts that would be used to meet those standards (catalysts that were already in use on many 1975 vehicles in the United States, and on virtually all such vehicles in California) encouraged the production of sulfates, and these could pose a considerable threat to health. "Some EPA researchers contend that as more catalyst-equipped cars are sold, the resulting increase in sulfate emissions will be more hazardous to human health than the hydrocarbon and carbon monoxide emissions the catalysts are designed to reduce."[86] Not only did Train suspend the 1977 standards, he also recommended that the projected control schedule for later years be slowed down; the standards he proposed would be even less stringent than those already applicable in California. The California standards had been established before the severity of the sulfate problem had been recognized,[j] and had been designed to encourage the introduction of the very technology—the catalysts—now subject to question.[87]

Train's actions made a number of people unhappy. President Ford had proposed, shortly before, a five-year moratorium on auto standards in exchange for a promise from the industry that it would improve gasoline mileage 40 percent by 1980. The president's moratorium would have amounted essentially to freezing the national standards at the California 1975 level. The sulfate developments embarrassed the president inasmuch

[j]At least, this appears to be the common understanding. In fact, as early as October 1973, the EPA was considering whether "they should do away with the catalytic converters" because tests revealed the sulfate problem. See "Auto Pollution: EPA Worrying That the Catalyst May Backfire," *Science* 182(1973):368; D. Fisher, "Smog Control Snag—Threat to Air Cleanup," *Los Angeles Times,* 26 Oct. 1973; idem, "Ecology Agency Backs Catalysts on Cars Despite Pollution Threat," *Los Angeles Times,* 7 Nov. 1973. By 1949–1950 the Los Angeles APCD had recognized a problem of sulfate development in the atmosphere. The agency, however, viewed sulfates not as a health hazard but as a factor that accounted for as much as 30 to 60 percent reduction in visibility. Los Angeles Air Pollution Control District, *1949–1950 Annual Report* (1950), pp. 7–8.

The difference between the information known to the EPA in 1973 and that spurring the EPA administrator's 1975 decision was not "enormous," though the later information was "more specific." Previously the administrator was "not conas the California level was based on the use of catalysts; when he learned of the controversy, he "chastised Train for failing to bring it to his attention."[k] Environmentalists were also unhappy, believing the "new health concerns have been vastly overplayed."[88] Finally, the ARB was unhappy. The decision of the administrator was "outrageous"—"an obvious capitulation to the auto and oil industries." The ARB regarded it as " 'possible' . . . that the alleged hazard was used 'as a convenient propaganda vehicle' to justify an essentially political decision."[89]

The EPA's actions were especially threatening in California because they put "pressure on the State's Air Resources Board to separately order the first rollback of California auto emission standards in history."[90] If the state did roll back its standards, there would probably be an increase in hydrocarbon emissions; moreover, there would be that distasteful step backward by an agency that prided itself on always going only in the other direction. If the rollback were not effected, there would likely be more sulfates. "It begins to sound like you're trading the bubonic plague for typhoid," said one ARB member.[91] Ultimately, the ARB decided not to back down, voting instead to adopt more stringent emission standards for 1977; at the same time, it postponed decisions concerning a sulfate emission standard and a low-sulfur gasoline requirement.[92] (It is too early to judge the wisdom of these actions. As of this writing, the state has received a federal waiver for its standards,[93] and the EPA has found new evidence suggesting the problem may be less severe than thought at first.[94] Like the ARB, the EPA is considering a sulfate emission standard for automobiles.[95] It is also concerned with some new problems, having found that "noncatalyst cars might represent equal or greater hazards because of emissions of other strong acids." But the agency is capitalizing on the sulfates experience, "seeking the earliest possible evaluation of a potential health hazard" from the newly discovered acids in order to "avoid the kind of uncoordinated situation in which the EPA found itself on the [sulfates] issue.")[96]

---

vinced there was a problem." While there was still considerable scientific debate at the time of his 1975 decision, he considered that decision the only "prudent" course. See Bureau of National Affairs, *Environment Reporter, Current Developments,* Vol. 5, No. 48, 28 March 1975, p. 1866; D. Fisher, "State under EPA Pressure to Roll Back Emission Rules," *Los Angeles Times,* 6 March 1975.

[k]D. Fisher, "Auto Makers to Be Given Year's Delay on Emissions," *Los Angeles Times,* 28 Feb. 1975. As a result of the administrator's actions, the president was led to reconsider. See Bureau of National Affairs, *Environment Reporter, Current Developments,* Vol. 5, No. 48, 28 March 1975, p. 1864.

MOVES TO AMEND THE LAW

One could expect even in 1970 that there would soon be moves to amend the Clean Air Amendments. The history of federal air pollution legislation, at least since 1963, had established a trend in that direction; the fact that the 1970 legislation brought heavy federal intervention into the affairs of powerful interests promised its continuation; early troubles with the legislation made the moves all but inevitable. Indeed, the law had already been amended once—by the Energy Supply and Environmental Coordination Act of 1974—and the EPA had been seeking other amendments for some time. But more was in the air by 1975, and not simply because of recent energy problems. "Well before the current fuel crunch hit the world economy," one source has noted, "efforts were astir to ease statutory mandates for reduction of air pollution in the United States."[97] While amendments other than those in the 1974 energy legislation have not been enacted as of this writing, a large number of them, touching virtually every important aspect of the present federal legislation, were pending at the end of 1975. (Developments in 1976 are discussed in chapter 15.) They ranged from such obvious matters as new-vehicle emission standards, transportation control plans, and air quality standards, to issues relating to the use of coal by stationary sources, preconstruction review of "indirect" sources (shopping centers, parking lots), proposals for emission charges, new standards for hazardous pollutants, and prevention of significant deterioration.[1] Of most interest to us, however, are proposals pertaining to the first three items.[98]

A number of bills, introduced by the Ford administration and others, proposed to "delay, relax, or restrict the applicability of the new motor

[1]In Sierra Club v. Ruckelshaus, 344 F. Supp. 253 (D.D.C.), affirmed, 41 U.S.L.W. 2255 (D.C. Cir. 1972), affirmed by an equally divided court, 412 U.S. 541 (1973), it was held that areas with air quality already better than federal secondary air quality standards could not permit "significant deterioration" in the ambient air. After considerable debate, the EPA promulgated regulations on the matter. They automatically designated all areas as "Class II," in which moderate air quality deterioration would be permitted. A state could change an area to "Class I" (virtually no deterioration in quality permitted) or "Class III" (almost any change permitted so long as national standards are not violated) after it published the reasons for the action and held public hearings. Attention had to be given by the state to anticipated growth in the area; social, environmental, and economic effects of a classification; and impacts of the classification on regional or national interests. The regulations were attacked from both conservationist and industrial perspectives, and upheld, in Sierra Club v. Environmental Protection Agency, 9 E.R.C. 1129 (D.C. Cir. 1976).

vehicle emission standards" in various ways.[99] The administration, for example, wished to freeze the standards for five years (1977–1981) at the levels applicable in California in 1975. After 1981 the standards governing emissions of oxides of nitrogen would be established at levels determined appropriate by the EPA, considering air quality, energy efficiency, availability of technology, costs, and other relevant factors. The purpose of the amendment was to "strike a reasonable balance between air quality and fuel economy" in the "context of giving highest priority to the protection of public health."[m]

Proposals to change requirements for transportation control plans were related to the delay in the effective dates of very stringent new-vehicle emission standards. At least twenty-seven regions of the country were known to require transportation control plans, with the likelihood that more would be added to the list. Ten areas, Los Angeles included, would be unable to meet 1977 air quality standards without gasoline rationing.

> There is general agreement by the affected parties . . . that additional time is needed to attain the standards. Moreover, because of relaxation in the schedule for cleaning up new car emissions, many argue that it is unfair for even greater transportation control to be imposed by 1977. Finally, it has been argued that curtailment of private auto use should not be required until an adequate public transit alternative is available. An additional problem is raised by transportation control plans. Many cities feel aggrieved by the fact that these plans impact on the cities, but responsibility and authority for their adoption and implementation is [*sic*] vested in the States.[100]

[m]"Letter from Environmental Protection Agency Administrator to Senate Public Works Committee Chairman Supporting Proposed Amendments to the Clean Air Act," in Bureau of National Affairs, *Environment Reporter, Current Developments,* Vol. 5, No. 41, 7 Feb. 1975, pp. 1570, 1571. See also "House Interstate and Foreign Commerce Committee Print on Clean Air Act Amendment Proposals," in ibid., No. 49, 4 April 1975, pp. 1929, 1930.

Subsequent to the proposed amendments, the National Research Council issued a report challenging efforts to ease the restrictions on new-vehicle emissions. The report recommended sticking to the 1978 timetable established by the Energy Supply and Environmental Coordination Act. At least as to hydrocarbons and carbon monoxide, that timetable "is both feasible and worthwhile," the report said. The EPA was claimed to have exaggerated the sulfate problem, and in any event relaxing the standards was considered by the report to be unlikely to reduce the problem; controls on the sulfur content of gasoline would be a better approach. "The report also said that the recommended emission improvements could be done while improving fuel economy." See "Easing Auto Pollution Rules Is Opposed By Scientists; Current Timetable Favored," *Wall Street Journal,* 5 June 1975.

The administration bill would have authorized the EPA to grant up to two five-year extensions of transportation control plan requirements if all reasonable measures had been implemented and standards still could not be achieved. Other proposals suggested shorter extensions if accompanied by strict interim control measures.[101]

Finally, the air quality standards set pursuant to the Clean Air Amendments came "under attack from a variety of perspectives." A particular criticism was that the "standards should take costs and technology into account and should be based on cost-benefit analysis." While the administration was silent on the issue, several bills responsive to the criticism were introduced.[102] In chapter 15 we shall consider a realistic approach to setting and achieving air quality standards—one that would respond to the criticism made above—and raise several other issues that amendments proposed thus far have ignored.

A CLOSING NOTE ON THE CALIFORNIA SCENE: 1965 REPLAYED

By the end of 1975, federal air pollution policy was in a state of anxious review; many of its initial assumptions had been called into sharp question. As to internal California pollution policy, we should not leave our story with the impression that all was quiet, for that is not the case. In 1974–1975 the state replayed a series of events markedly similar to those of a decade earlier—a time when policy was made more by bumbling than anything else. History does indeed repeat itself on occasion, perhaps because its initial lessons went unlearned.

California's pink-slip controversy of 1965 involved futile efforts to require motorists to have crankcase devices installed on their vehicles on a staggered schedule. The installation program was repealed six months after it became effective. Subsequent to the repeal, devices were required only upon transfer of ownership (a phase of the original program that had gone forth without controversy). Ten years later a similar series of events reached their culmination. In April 1975 the legislature passed, and the governor signed, a bill repealing a requirement that owners of 1966–1970 vehicles install control devices on their vehicles according to a staggered schedule. Pursuant to the repeal, devices would be required only upon initial registration in California or at the time of transfer (again a phase of the original program that went forth with little controversy). The only apparent differences between the two series of events are their separation by ten years; the fact the controls in question were aimed at different sorts of emissions; and the fact the controversy in the second series dragged out a bit more, and became a bit more complicated, than did that in the first. Had policymakers in California been more sensitive to the past, perhaps they would have foreseen and been able to deal with a predictable set of events; indeed, perhaps they might simply have opted for a time-of-transfer approach in the first place to avoid the aggravation their program was likely to produce. But they did not, and the results were what one would expect.

To recapitulate slightly, in 1971 the California legislature had enacted a requirement that 1966-1970 vehicles be equipped with a device—as soon as accredited by the ARB—to control oxides of nitrogen emissions. Under the legislation, devices would be required upon initial registration or transfer of ownership, and, beyond this, would also be required for all covered vehicles upon renewal of registration in 1973.[103] But even as the ARB was searching for a device suitable for accreditation, there were doubts—within and without the agency—about the wisdom of the new requirement. "Top engineers in the state's air pollution program are having second thoughts, fearful that the whole thing might backfire. They are afraid people will be angered over the cost and disappointed over the results, thereby weakening public support for the state's antipollution programs."[104] In fact, direct costs would not be high: the legislation specified that devices were to cost no more than $35, were to last for 50,000 miles, and were to require no more than $15 worth of servicing every 12,000 miles.[105] But there could be high indirect costs—engine overheating, or incompatibility between the device and the vehicle, depending upon the sorts of controls approved. As one engineer put it, "You are asking the car to do something it was not designed to do, . . . and when you do that, you're asking for trouble."[n]

Regarding results, there was a consensus they would not be startling. "Air pollution statistics will improve slightly, but experts agree that the

[n]Quoted in L. Dye, "Smog Device: Engineers Take a Second Look," *Los Angeles Times,* 28 Dec. 1971. One device under consideration by the ARB at the time would bypass the vacuum spark advance in an engine's timing system (the spark advance increases power during acceleration), but would reconnect the advance if the car overheated. The man "credited with discovering the principle behind the development" (McJones) contended overheating was not a problem and that it was sufficient simply to disconnect the spark advance. Some engineers disagreed, believing overheating was likely in the summer months. McJones discovered his control technique "almost by accident." He was working as a consultant to a corporation interested in converting its fleet to natural gas. In trying to determine the best spark timings for natural gas, he disconnected the spark advance. The vehicles ran just about as well, and testing revealed that they also produced "considerably" less nitrogen oxide emissions. The ARB confirmed the fact. "It was great news." Ibid.

air will look just as dirty as ever." Haagen-Smit, chairman of the ARB at the time, backed the requirement nevertheless; he believed "that victory in the war against air pollution will be measured in inches, not miles."[o] The devices would reduce oxides of nitrogen by about 10 percent, and though this would not lead to noticeable improvement, "every little bit counts." Haagen-Smit's support had been instrumental in passage of the law. Engineers on the ARB were reluctant to express disagreement with his views. Other opponents of the measure—such as the Los Angeles APCD—were not inhibited by the chairman's prestige and authority, but did feel other constraints. "Nobody dared oppose [the requirement] because it will reduce nitrogen oxides, and a certain group of environmentalists will tear you to shreds when you oppose controls," according to the head of the APCD.[106]

In August 1972 the ARB approved two devices despite a recommendation against approval by a special committee, and against staff advice that at least one of the devices would "cause substantial adverse effects"—overheating and valve damage. One or the other of the devices would be required upon initial registration or transfer of a vehicle, and also (effective February 1973) at the time of renewal of registration,[107] unless the ARB exercised its authority under the legislation to extend the effective date of the renewal-of-registration requirement "for extraordinary and compelling reasons only."[108] Less than a week after its initial approval, the ARB indicated it would put off that requirement "by at least a year." The announced reason, plausible enough, was exactly the same as one consideration that had led to lag in implementing past programs: without more time, manufacturers of the devices could not "possibly build and distribute enough control systems to take care of the estimated 4 million vehicles affected by the program." Moreover, mechanics would be unable to install the devices by the deadline assuming they were available; in fact, they could not do so "even if the deadline [were] extended a year. . . ." But there was more in the background—negative "reactions from motorists' organizations, auto makers and the auto repair industry." If these were "indicative, the looming controversy over the program may make for further delays."[109] That prediction proved to be accurate.

[o]Ibid. Haagen-Smit later said that while he did not recommend against compulsory installation upon renewal of registration, he also did not endorse it; remembering the pink-slip debacle of 1965, he favored a voluntary program. But several members of the Senate and the Assembly considered a mandatory program essential, and when legislation to that effect was enacted, Haagen-Smit felt that he could not "sabotage" the program. Interview with A. Haagen-Smit, 26 July 1976.

The ARB acted officially in September 1972 to delay the phase of the control program requiring installation upon renewal of registration. It set no new timetable, but several alternatives were under consideration—a year's delay to February 1974, or a staggered system based on license plate numbers or county of registration.[110] The next month it approved a general plan "in principle," but withheld formal adoption out of concern that a pilot program was necessary to check the time, costs, and skills required for installation. If the pilot program went smoothly, vehicles in the South Coast Air Basin would be required, beginning in February 1973, to have devices upon transfer. Later, installation would be required of all vehicles according to a staggered schedule extending to February 1974. The balance of the state would be covered by February 1975. The program was becoming sufficiently complicated that Haagen-Smit was "seriously thinking we should rescind this whole business and start all over again."[p]

In subsequent months the ARB approved several new devices and finally arrived, after several changes of mind, at a firm compliance schedule. Installation would be required on virtually all vehicles between January and October 1974, depending on the vehicle's license plate number.[111] The program, however, was becoming more controversial. Complaints about the possibility of engine damage had continued, and both the Senate Transportation Committee and Governor Reagan were planning reviews of the entire effort; it was "possible the program [would] be delayed or even scuttled as a result." The senator planning the Transportation Committee hearings said it was a "legislative mistake" that 1966–1970 vehicles had been permitted to emit large amounts of nitrogen

[p]Quoted in D. Fisher, "Used Car Smog Device Schedule Initially OKd," *Los Angeles Times,* 26 Oct. 1972. The program was also becoming sufficiently complicated that it was difficult to keep track of the various installation schedules being proposed from time to time by the ARB. The discussion in preceding and subsequent portions of the text relies on news accounts. The California Supreme Court, in an opinion discussed shortly, suggests that the various schedules were somewhat different from what the newspapers indicate. The court refers to an initial delay (caused by a shortage of mechanics and devices) followed by the first schedule: installation upon transfer or initial registration to take effect between February and June 1973, depending upon geographical area; plus a license plate schedule to become effective between June 1973 and April 1974. The latter portion was subsequently delayed until renewal of registration for the year 1975. In June 1973, according to the opinion, the ARB again deferred the program and, shortly thereafter, adopted a new schedule requiring installation by the end of 1974. We assume the last schedule to be that discussed in the next paragraph of the text. See Clean Air Constituency v. California State Air Resources Board, 11 Cal. 3d 801, 523 P. 2d 617, 114 Cal. Rptr. 577 (1974).

oxides in the first place; he did not "want another colossal error."[112] At the end of October 1973 the ARB determined to stick to its revised schedule, "despite a recent flurry of about 600 letters protesting the program." The installation requirement would be enforced through random checks by the California Highway Patrol and through 1975 license plate renewal procedures (which would catch those vehicles that had not complied with the staggered installation schedule).[113]

Governor Reagan was unhappy with the new program. He was also in a position to do something about it. The 1971 legislation that had reconstituted the ARB into a five-member body[q] had accomplished something else as well: by neglect or otherwise, the members of the new agency, as opposed to those of the old, had no stated term of office.[114] They could be replaced by the governor, and they were. In mid-December he appointed four new members in order "to take a fresh look at the air pollution problems confronting California, particularly in light of the energy crisis." Many observers believed, however, that the action "was aimed at killing [the] controversial program to require [nitrogen oxide] devices on 1966–70 model cars in the state."[115] Subsequent events confirmed that view. Within a week of its appointment the new ARB voted to delay the oxides of nitrogen control program—the license plate schedule by one year, the initial registration and transfer requirements by a few months. The action was justified on the ground that the energy crisis presented the "extraordinary and compelling reason for further delay" that the legislation required.[116]

Several lawsuits followed. One, filed in Los Angeles Superior Court in January 1974 by an environmental group, several manufacturers of control equipment, and one of the ARB members replaced by the governor, charged that the ARB had exceeded its authority and abused its discretion in granting the delay.[117] The court disagreed; the agency had "implied authority to take cognizance of facts in the real world." One of these was the energy crisis—"gas is at a premium"—and, considering this, the ARB "did the appropriate thing. . . . It is uncontradicted there will be a savings in gasoline" as a result of the decision. The judge expressed sympathy for the two manufacturers, who had invested millions in the devices and had hundreds of thousands of them on hand. They were "very unhappy" and the judge could not "say as I blame them." But the ARB had acted legally.[118] The same group of litigants tried again, this time bringing an action under the original jurisdiction of the California Supreme Court.[119] The suit succeeded. In June 1974 the court held that the ARB's limited discretionary authority pertained only to effective implementation of the installation program and did not extend to considerations having to do

[q]See the discussion at p. 211.

with current energy problems; no view was expressed as to the propriety of the earlier delays.[120]

What followed the court's decision was, if anything, more of a circus than the events that had come before. In July the new ARB reinstated the program: installation upon initial registration or transfer of ownership would stand as previously ordered. (The date to which the agency had deferred these requirements—April 1, 1974—had arrived prior to the supreme court ruling. While the issue arising from that action was moot, the court did indicate that the ARB had implicit authority for such a deferral under appropriate circumstances.) The installation schedule based on license plate numbers would run from August 1974 through June 1975. The Highway Patrol would give owners one month's grace, and thereafter (in October) begin issuing citations to offending vehicles.[121]

In August the legislature acted to eliminate the staggered (license plate) phase of the program in all counties but those in the South Coast Air Basin (though the legislation would not become effective until early 1975).[122] In January 1975 Los Angeles County and a member of its board of supervisors brought suit to stop enforcement of the law in the South Coast Basin, arguing that the legislature's partial elimination was irrational, arbitrary, and discriminatory. The suit failed. In the meantime, however, the Highway Patrol—at the request of the legislature—had adopted a policy of not citing motorists who were not in compliance with the new installation schedule; only warning tickets were being issued.[123] Prior to the January lawsuit, a bill had been introduced in the state Senate to repeal the law in the South Coast Basin in the same fashion as in the other counties of California.[124] It and another identical measure survived the Senate, but both were held up in an Assembly committee—one version rejected outright, the other given no action at its author's request. Backers of the measures urged that the "people of Los Angeles should write their legislators and put the pressure on them."[125]

March and April witnessed a frantic program of maneuvering by those on the two sides of the controversy. The Highway Patrol had been directed to begin issuing citations despite a Senate resolution asking for delay in enforcement;[126] proponents of repeal developed new arguments, asserting now that the control devices would add hundreds of thousands of pounds of lead to the air of the South Coast Basin each year;[127] the Assembly committee continued in its refusal to pass a repeal, despite "hundreds of letters and telephone calls from angry constituents who have had trouble with the NOX devices or don't want to install them on their 1966-70 automobiles."[128] New appeals were made to the courts. In mid-March the Los Angeles County Superior Court refused to restrain the Highway Patrol from citing motorists.[129] A few days later a judge in Santa Barbara granted the relief,[130] perhaps because in the intervening period the Assembly committee acted to "revive" the repeal legislation it had previously blocked. The change of heart had been prompted by an "out-pouring of mail."[131] Three weeks later the full Assembly voted against repeal.[132] There was no intention to let the matter die, however. The Assembly vote was only eight short of the number needed for passage, and the "thousands of letters and telephone calls" coming in from unhappy constituents were expected to have some effect.[133] They did. Three days after its initial rejection, the Assembly reversed itself and approved repeal, thanks to "growing public pressure" which one assemblyman said had "been pure hell."[134] The Senate quickly concurred in the Assembly version of the repeal, and the bill went to the new governor, Edmund G. ("Jerry") Brown, Jr., the son of California's earlier Governor Brown.[135]

Brown had already begun a series of consultations with "experts on both sides" regarding the action he should take on the bill. He faced a difficult decision, considering that "an all-out war on smog" had been a central item in his campaign, and that the new ARB chairman he had appointed was strongly against repeal.[136] Despite these considerations, the governor signed the repeal after a series of public hearings. He found no "credible benefit" in the nitrogen oxide controls. Experts might have "grandiose plans," but these had to be "tempered with the human touch and common sense." His ARB chairman claimed to agree with the decision, saying now that the program "should be abandoned if the antismog devices could not be properly installed and adjusted." Brown acted as he did because of contradictory evidence concerning whether the control devices in fact yielded a net reduction in emissions. There were in addition, of course, the assertions of engine damage and reduced gasoline mileage caused by the devices. And another factor might also have figured in the governor's decision: polls indicated that 72 percent of Southern Californians opposed the program.[137]

As finally passed, the repeal was rather odd. All the arguments urged in its support would suggest that the nitrogen oxide program should have been dropped entirely. Instead, the repeal operated only to eliminate in the South Coast Basin that phase of the program relating to the license plate timetable. Vehicles throughout the state were still required to have the devices installed upon initial registration or transfer.[138] But, of course, it was only the former aspect of the program that had generated controversy from the outset. As with the pink-slip affair ten years earlier, the time-of-transfer requirement had been received with relative quiet. A news article written during the height of the 1975 controversy noted the similarity to events of a decade before, and observed that "the same thing might happen again." The article saw the controversy as a test of the willingness of policymakers "to support programs that might generate intense opposition while producing relatively small reductions in overall amounts of smog."[139] As of this writing, the legislature is considering means to reimburse owners who had complied with the law and installed devices, perhaps to atone for its sins of forgetting a lesson and insisting on a program that its constituents simply did not want.[140] One is led to wonder how much would have been saved in terms of public goodwill and state resources had the original requirement been limited to time of transfer only. In any event, by the end of 1975 California pollution policy—like that on the federal level—was hardly marked entirely by smooth, programmed progress.[r]

[r]California experienced setbacks in subsequent years as well. Although events in those years are largely beyond the scope of this study, one in particular deserves brief mention because it illustrates continuing public resistance to transportation controls. In early 1976 the California Department of Transportation instituted the Santa Monica Freeway Diamond Lane program in Los Angeles. The program, "the most controversial project in state highway history," aimed at reducing traffic, air pollution, and energy consumption by permitting only car pools and buses to travel in a freeway lane reserved exclusively for them—the "Diamond Lane." The project angered drivers and local officials, and a federal court order brought it to an end after five months of bitter controversy. A somewhat similar program on the San Diego Freeway (in Los Angeles) was abandoned on the order of Governor Brown early in 1977. See R. Hebert, "She's Still the Driver at Caltrans," *Los Angeles Times,* 4 March 1977.

Part V

POLLUTION AND POLICY: AN APPRAISAL

14

SOME THEMES IN THE POLICY PROCESS

The process of making pollution policy—at least the particular process described in preceding chapters—reveals a story of sense and nonsense, of an approach based more on reaction to circumstance than on careful planning, consisting as much in foolish mistakes and understandable errors as in wise steps. The themes of that story show continuity among disparate events occurring in discrete blocks of time. This, at least, is quite clearly the case up to 1970, when policy appeared to undergo rather dramatic shifts in stride and direction. Even in developments since then, one can see some recurrence of themes that had emerged in earlier years, but 1970 draws a clear enough line between continuity and change, between past and present policy, that the years on each side of it deserve separate treatment. This chapter traces in considerable detail the main themes in events up to 1970, then summarizes what has occurred since that time. The burden of the chapter is to identify general strains in particular developments, to speculate about the reasons behind them, and to assess their significance in the making of pollution policy. That accomplished, we should be in a good position to appraise some central elements in present policy—the task of the next (and last) chapter.

THE YEARS UP TO 1970

Policy-by-Least-Steps

From beginning to end, pollution policy in the years up to 1970 was made by a process of least steps taken down the path of least resistance—steps, that is, quite consciously designed to disturb the existing situation as little as possible. One might regard the federal government's preemption of new-vehicle emissions in 1967, or the quite bold initiatives of California's Pure Air Act of 1968, as exceptions, but these should not obscure the central fact that policy generally came in small, conservative increments. Indeed, in these terms it is well to recall that the two examples mentioned above were themselves but culminations of a process that began in the 1940s. Nor should the occasional panicked reaction to an episode, such as the closing down of the synthetic rubber plant after the Los Angeles "smog attacks" of fall 1943, hide the essential process by which policy was made. Episodes played their part, but more by stimulating the next small steps than by producing sweeping change. As James Willard Hurst has pointed out, "most of life is not a melodrama," and most of "law consists more in policy developed out of rather routinely handled instances than in policy developed in moments of drama."[1] Later we will suggest that episodes were a troublesome product of policy-by-least-steps, but also a helpful way of moving policy along. They were a facet of the incremental process.

The process of least steps down the path of least resistance is revealed by a host of examples over the years in question. At times it was a sensible process, given existing knowledge, but on other occasions it consisted at best of senseless experiments that quickly yielded predictable results. As we suggested in chapter 2, for example, it was sensible enough to rely—as policy did in its early stages—first on the courts and next on local ordinances to deal with particular problems as incidents defined them. There was little ground for believing that more was necessary, and thus little occasion for incurring the heavy costs of more substantial intervention at the state level, much less the federal. Within a very short period of time, however, there was little basis to doubt that such intervention was needed. A few moments thought should have suggested to anyone curious about matters that they would only get worse with time. Trends in population and industry, the behavioral components of pollution problems (see chapter 1), and what was already known about Southern California's meteorology and topography dictated that the problem would grow, and grow quickly. But policy resisted such ideas, developing out of reaction rather than initiative. There was no planning for a certain future.

This was the character not only of adolescent policy, it was also a feature that persisted into relative maturity. As late as 1955, eight years after the first significant control efforts, in a period of painful awareness that matters were indeed increasingly serious, California's Department of Public Health noted "too much emphasis . . . placed upon cure, once smog has occurred, and not enough upon prevention and determination of cause"; this despite the fact that "preventive measures taken before air pollution becomes a problem would undoubtedly be more effective and less expensive for all concerned, including the taxpayer and industry."[2] The path of least resistance, of least *apparent* cost, was to react to problems as they developed, and then by the smallest, least disruptive steps. Thus, the response to the new problem that appeared in Los Angeles in the early 1940s, after the initial foray on the rubber plant, was more local ordinances. There were hopes that other cities in Los Angeles County would voluntarily follow suit, but for the most part they did not. Perhaps their conditions were in fact less severe, perhaps they were regarded as a necessary price of progress (or the war effort), or perhaps pollutants escaped to the city of Los Angeles before they did harm.

In any event, initial efforts at voluntary cooperation failed, as the common property and collective goods effects discussed in chapter 1 perhaps suggested they would. Only then was the more incursive step of state legislation taken. It too reflected a minimalist approach, one "adopted without thoughtful consideration for the possible needs of the other regions of the State":[3] under the 1947 law, establishment of control districts was voluntary. Moreover, taken literally the legislation did not even authorize a district until a finding that the air was polluted and could not be managed by local ordinance. It was this feature that inspired the Public Health Department's lament that the emphasis was on cure, not prevention. Finally, districts were to be countywide, despite the fact that pollution ignored county boundaries as well as more local ones—a stubbornness about the problem that had led to the 1947 act in the first place. But the counties were, after all, conveniently there to be used, and they were at least a better locus of authority than the cities; using them recognized this small physical advantage. More importantly, using them recognized the large political advantage of avoiding the disruption of county sovereignty that would follow from creation of special control districts whose boundaries cut across county lines. The countywide districts were more a response to the political realities of the situation than the physical ones. When the shortcomings of countywide districts became apparent two years later, the reaction was a similarly least-step provision—an amendment to the 1947 act permitting voluntary integration of districts. This measure too overlooked an obvious problem, that a "larger county with more advanced control programs, larger population, and more sources fears a detrimental loss of control to the smaller counties."[4] Thus, the aftermath of the 1949 amendment—and of a later (1967) provision authorizing regional districts, again on an optional basis—was predictably unstartling. Two counties merged in 1968, two more in 1971. Nothing further developed until 1975, when four counties merged "voluntarily" under pressure from the legislature.[a]

Lessons about the fruitlessness of voluntary approaches appear to have come hard. When serious pollution problems developed in the San Francisco Bay Area, the first step (in 1949) was "to formulate a voluntary, co-operative scheme of regulation and control." Despite the fact that subsequent cooperation was "singularly lacking," popular pressure for voluntary control continued, and for clearly stated reasons. "Unless some system of voluntary control is devised and adopted, the smog nuisance will worsen to the great detriment of this region's health and economy. The result may well be a series of rigid, drastic and expensive smog-control laws." It was considered that the failure of a voluntary program "passes understanding,"[5] but of course the reasons were clear; common property resources are overconsumed, collective goods underproduced in the absence of coercion. It took until 1955, the year of creation of the Bay Area Air Pollution Control District, to comprehend these fundamental truths.[6]

[a]On the 1968 and 1971 mergers, see W. Simmons and R. Cutting, "A Many-Layered Wonder: Nonvehicular Air Pollution Control Law in California," *Hastings Law Journal* 26(1974):109, 122. The 1975 merger involved the counties of Los Angeles, Orange, Riverside, and San Bernardino; in July of that year they formed, by "voluntary" agreement, the Southern California Air Pollution Control District. It, in turn, was put out of business with the enactment, on July 1, 1976, of legislation creating the South Coast Air Quality Management District. Ch. 324, § 5, [1976] Cal. Stats. Reg. Sess., adding to California Health and Safety Code §§ 40400-40525. (Apparently, it was the threat of just such legislation as this that had stimulated the voluntary merger in 1975.) The new district consists of the same four counties as the old (with provisions for the possible inclusion of Ventura and parts of Santa Barbara counties); it is governed by ten members including not only representatives from county boards of supervisors (as was the case in the past), but also representatives from city councils plus one gubernatorial appointee. The Southern California APCD formed in 1975 had been considered to be too dominated by county supervisors. Supervisors in Los Angeles, Orange, and San Bernardino counties had urged the governor to veto the 1976 bill creating the new district. See W. Rood, "Brown Signs Bill for New Southland Smog Agency," *Los Angeles Times,* 3 July 1976.

There are so many other examples of the least-step approach to policy-making that the discussion would become unwieldy if it did other than merely call them to mind. In Los Angeles: the APCD's policy of "fair-but-strict"; of taking into account the "practical needs of industry"; of reaching only to the limits of engineering knowledge, and often falling short. On the state level: California's reluctance to become involved in the pollution problem at all; its decision to limit itself to a research role when state intervention first occurred, and to permit local options when state control finally took place (not only in the 1947 legislation discussed above, but again in 1960, in 1967, and in 1968); its requirement that vehicle emission controls be installed only upon certification of satisfactory devices (a practice that stopped for new vehicles in 1968), thus implicitly relying on industry to develop controls when logic suggested it would not. All of these examples illustrate the incremental approach to policy. So do the largely identical steps taken on the federal level: the reluctance to intervene; the initial adoption of a research role; the subsequent decision to control only at the margin, leaving most initiative to state and local government and then very slowly taking more and more of it away.

The process is typical of the making of environmental policy,[7] probably of government policy generally—at least as to social problems similar to those considered here. If there is a note of condemnation in the discussion, it does not entirely belong. While policy-by-increments is conservative (perhaps overconservative), while it necessarily takes considerable time, it has some desirable qualities. Least steps, by being just that, appear to be relatively inexpensive. They disrupt the status quo as little as necessary; they avoid the "rigid, drastic and expensive" controls that concerned the Bay Area, controls that might be rejected by their targets or otherwise backfire, thus requiring embarrassing retreat that erodes public confidence. (The pink-slip controversy is an example.) Least steps are easy to realize because they minimize political opposition and because they tend, as each of them proves inadequate, to make their own best case for the next least step. Moreover, they may mark the wisest path through a state of great uncertainty. That some steps on that path might be foolish—such as relying on voluntary controls, or on the auto companies to develop emission devices—can perhaps be forgiven.

Those "foolish" steps can also be attributed at times to more than simply the shortsightedness of policymakers. To be sure, these people made mistakes. To some extent, however, the process of their decisions was dictated by extraneous factors. Politics, already mentioned, played a heavy role in the making of pollution policy. There was repeated concern that state controls not tread on local toes, and the result was a series of local options for powerful rural counties. There was similar concern at the national level to avoid bruising other participants in the federal system. Thus, as just one instance, Congress was careful to point out that its 1955 legislation did "not propose any exercise of police power by the Federal Government and no provision in it invades the sovereignty of States, counties, or cities."[8] These are just two illustrations of a strain that ran throughout the years up to 1970. One might add, simply to remind, instances where the automobile manufacturers and other major pollution sources exercised their political power.

There were other extraneous constraints on policy; in particular, there was law. Legal considerations contributed much to the shape of decisions at the state and federal levels. It was the constitutional doctrine (and the tradition) of "home rule" that made the 1947 act necessary and that led Governor Pat Brown to be "all for" local control; it was other constitutional law (for example, the prohibition against "special legislation") that dictated the form of the 1947 legislation, a form later subjected to sharp criticism. It was California nuisance law that forbade public suits against sources operating in appropriate zones. (With reference to politics and the tendency toward voluntary controls, recall that efforts to amend California's nuisance law in 1945 failed, thanks to the opposition of stationary sources that had long promoted self-policing and wanted prosecutors to leave industry alone to work out its own problems.) It was regarded as a legally required prerequisite to air quality standards that they have a sound scientific base. It was the force of constitutional law, at least as read by the Los Angeles County counsel, that prohibited motor vehicle emission controls before there were "perfected," "practicable," "available" devices. On the federal level, "legal opinion" was the source of the "feeling that the Federal Government should not enter into policing the State problems of pollution of the air."[9] Federal intervention might confront constitutional problems if it occurred without a showing that the states were "unable to or fail to control" pollution themselves.[10] The "principle . . . observed" for considerable time was "that the Federal Government should not and *cannot* impose or enforce regulations designed to prevent air pollution, regardless of how desirable such activity might at times appear."[11] The foregoing examples illustrate specific ways in which law constrained the course of policy. Some of these involved real constraints—that is, factors that actually limited the initiative of policymakers. Others probably involved excuses and contrivances, instances where policymakers or interest groups manipulated legal doctrine to help make a convincing case that some passive course had to prevail over a more disruptive one. (The constitutional barrier raised against federal intervention in the 1950s is probably the best example.) In both cases, however, law was an important part of the constraining force. Law had another influence as well, a more general one. The formalism of legal procedures, the many check points that must be passed to achieve change through legal process, built inertia into the system. There is some sense to inertia. The change that intervention through legal institutions (for example, courts and legislatures) brings about always involves costs. The mere process of employing these institutions to arrive at and enforce collective decisions is itself expensive. So is an alteration in the ongoing social situation, the status quo that exists precisely because it is perceived as being to the individual benefit of all those participants in a society whose activities make it up. Since intervention is in these senses costly, simple logic suggests that it go forth only when there is the promise of compensating benefits, the prevailing situation otherwise to be maintained.

The general influence of inertia, though surely bearing on events throughout the years in question, is difficult to illustrate—perhaps because of its very generality. The best we can do is consider briefly the play of a particular instance of formal procedure; as to it, there are a number of examples.

Allocating the Burden of Uncertainty

Foregoing chapters provided many examples of the high degree of uncertainty concerning the air pollution problem. "In a situation of imperfect knowledge," a 1969 study by the National Academy of Sciences points out,

> it becomes critical to decide where the burden of such uncertainties should fall. Historically, that burden has tended to fall on those who challenge the wisdom of an ongoing technological trend. The working presumption has been that such a trend ought to be continued so long as it can be expected to yield a profit for those who have chosen to exploit it, and that any deleterious consequences that might ensue will either be manageable or will in any event not be serious enough to warrant a deliberate decision to interfere with technological momentum.[12]

The historical theme of which the National Academy report speaks is a general one, reaching beyond dealings with technology, rooted early in American law,[13] and characteristic of the approach of most institutions, legal institutions (legislative, judicial, and administrative) in particular. The burden of uncertainty is allocated to those who seek change. It is their job to show good cause, and if good cause cannot be made out, the tendency is against intervention.

Consider, as one instance of the point, California's 1947 legislation and its implementation. Under the act, control districts were not even to be established until it could be shown that a problem existed; even then proponents of control carried the burden of showing that a district would be worthwhile. Six years after the 1947 legislation, clean air advocates argued that in view of the spreading pollution problem, counties adjoining Los Angeles "should be urged to establish air pollution control districts without delay." In response, at least one of those counties took "the position that science just doesn't know enough about smog and how to control it to justify setting up a control district here."[14] Once established, districts had little power to control agricultural operations. These had been broadly exempted on the ground of an insufficient showing that control of such operations would yield any benefits in air quality. As to sources not exempted, there were to be "specific anti-air pollution rule[s]" only when "scientific evidence could demonstrate the need" and only when it could be shown that "feasible methods for controlling the output of the pollution were available."[15] The last point conflicts with the legislative history of the 1947 act, but may nevertheless reflect what its drafters had in mind. They claimed that there was no constitutional barrier to abating sources even if control techniques did not exist; sources could simply be shut down. Of course, enforcement never reached this far. The Los Angeles APCD adopted its policy of fair-but-strict; it was "delighted to report" that no business had been put out of operation.[16] Emphasis was on fairness (as the APCD saw it), not strictness. Perhaps this was because major sources, such as oil refineries, could rest on the argument that the hard line considered constitutionally permissible under the law would be "too strict . . . without sufficient proof" that it would "cure the overall smog problem."[17] Thus as late as 1958 the Los Angeles APCD, while acknowledging that conditions "compelled . . . the most stringent rules possible," would enact such rules only when shown to be "technically sound and economically feasible," to be backed by "proper technical substantiation."[b] In the event of substantial uncertainty, there would be no rules.

[b]The APCD added that to act otherwise would be "arbitrary, unreasonable, capricious or confiscatory." R. Chass, "Extent to Which Available Control Techniques Have Been Utilized by Communities (1)—Los Angeles County," in *Proceedings of National Conference on Air Pollution* (Washington, D.C., 18-20 Nov. 1958), pp. 353, 365. The quotation in the text illustrates allocation of the burden of uncertainty; that in this note suggests the effect of more general constitutional constraints. The quotation makes clear that the constitutional reach of the 1947 act was never approached.

As another example of the same point, consider the process that led to automotive controls. By 1947 at least two bodies of research conducted in the Los Angeles area indicated the automobile as a prime suspect in the pollution mystery. Yet in 1949 a Los Angeles official branded any notion that the automobile might be an important source as "folklore" that would "place the onus of metropolitan area atmospheric pollution on the automobile, without proof."[18] It was thought to be improper, of course, to proceed in the absence of such proof. Thus, some six years later another official noted "evidence . . . pointing in [the] direction" that vehicle exhaust should "be an item for control," but this alone was not enough: "the contribution" of the auto "must be clearly demonstrated before any action can be taken."[19]

The automotive industry, of course, was happy to pick up on that theme in its strategy of "minimal feasible retreat" that began about 1953. In that year, Ford Motor Company wrote of its belief that exhaust gases were "dissipated in the atmosphere quickly and do not present an air pollution problem. . . . To date, the need for a device which will more effectively reduce exhaust vapors has not been established."[20] Later in the year the position was that factors other than automobile exhaust were contributing to smog. In 1954 it was admitted that exhaust could be considered a source of smog, but more "confirmatory work" was necessary.[21] In 1955 the industry conceded the auto was the major source of Los Angeles air pollution, but not necessarily of "smog." A year after that, the auto was accepted as a major smog source, but more data would be necessary before a judgment that it was the principal source. The Air Pollution Foundation, in the same year, concluded that motor vehicles were indeed *the* principal source. In 1957, and for some time thereafter, the industry maintained that this was the case only in Los Angeles, which it considered peculiar.

Only in 1960 did California enact legislation to control vehicular pollution—four years after independent scientists had reached the firm opinion that autos were the primary problem, three years after the industry had as much as conceded the point. Part of the delay was due to the "view" that ambient air quality and motor vehicle emission standards should be formulated as a prerequisite to control efforts; these "would provide a legal basis for enforcement."[22] Of course, that view was not a necessary one. Neither legal nor practical constraints demanded such standards before enforcement; indeed, the 1947 act had proceeded without them. Still, standards could be useful—ambient standards as a goal, emission standards as a guide to the degree of control needed to achieve the goal. The difficulty was that there existed little information on which to base standards. Evidence on health effects was uncertain at best. This would not have been particularly serious if the scientists charged to establish standards were ready to make some guesses, but they were not. Largely they demanded evidence that would "meet the expectations of the scientific community."[23] Since there was not a great deal of that, in many instances standards were not set.

One cannot entirely blame the standard-setters. Some believed that unless standards had a sound scientific base, they would not be legally enforceable. Moreover, the standard-setters were scientists being asked to do a politician's job. Understandably, they approached the job as scientists. The politician might sense the urgency to resolve uncertainty out, doing research in the meantime. He works with what he has, and if what he has is insufficient to support a conclusion, he waits for more evidence to develop. Thus, where standards could not be "based on and supported by available reliable data," they were foregone until sometime "in the future."[24] The political approach—"weighing both sides of the question, and arriving at a point of equilibrium of pressure from interested parties"—was explicitly rejected in favor of one involving "decisions based as much as possible on scientific knowledge."[25] Only when the standard-setters were forced by their own ambient standards to establish emission standards to attain them did they at times adopt a more accommodating approach to uncertainty.[c]

Some standards were finally established, of course, and motor vehicle control legislation was passed in 1960. But this was only after similar legislation proposed a year earlier had failed on the grounds that Los Angeles smog was a local phenomenon and that no "guaranteed" control devices existed. The 1960 legislation reflected these dual concerns. It contained a local option as to used vehicles, and required that devices be certified as feasible and available before they could be required. It is not difficult to imagine the roots of these provisions. Opponents of the bill had been "concerned about results and proof," worried the state was "rushing too fast. . . . There was no proof . . . that . . . devices have been developed and no proof whether automobiles are the major cause of smog."[26] The provisions dealing with local options and demonstrated feasibility

[c]The approach of the California standard-setters was hardly unique. Recall that in 1958 the Department of Health, Education, and Welfare successfully opposed a bill to prohibit the use in interstate commerce of vehicles discharging dangerous amounts of hydrocarbons; HEW argued there were no suitable criteria for "dangerous." Four years later the surgeon general noted the harmful effects of vehicle emissions but requested more time to obtain the research information needed for "equitable and appropriate" emission limitations. This was two years after California had adopted standards, and also after the auto industry had announced it would control crankcase emissions on 1963 vehicles.

became essentially permanent features of vehicular pollution control legislation in California.

It should be clear that the allocation of the burden of uncertainty did not pose consistently insuperable barriers to intervention; intervention did, after all, occur. The point is that it came about very slowly. The auto companies were able to employ the burden of uncertainty to the end of considerable delay. The Stanford Research Institute, on behalf of the petroleum industry, was able to make a successful case "that community interest would be best served by withholding action of an enforcement nature pending further research and development of technical remedies"; "no single class of violators could be condemned" and premature judgments might "not eliminate smog and would be extremely costly."[27] On the federal level, the coal industry was able to secure reconsideration of sulfur oxides criteria on Senator Randolph's demand for "complete and irrefutable proof of necessity."[28]

As suggested earlier, the approach to uncertainty reflected in all the foregoing examples was not entirely senseless. Change is costly, and it should not be imposed without reasonable promise of net gain. This attitude characterized much of the thinking about pollution policy up to 1970. The Air Pollution Foundation admitted, for example, that "we cannot in a battle 'research' out the last percent of uncertainty; but we must know what we are doing, for we can kill the patient by too many 'untested cures.' Industry is our livelihood."[29] Public health experts asked for "absolute or definite proof" of health effects; this was "reasonable" because it would be "bad" to cause "the public to worry about things which we do not know to be true."[30] To these experts it was "a matter of great importance that all of the public health effects of air pollution be understood and considered in the establishment of policy as to what level of damage will warrant exercising police power to protect the public."[31] It was equally important to pollution sources; they argued for understanding "in detail" because such detail "matters to industry which is blamed for 90 per cent of the pollution in the air."[32]

Thus "legal controls" had to "await scientific clarification of the subject."[33] But here an element of nonsense entered. Clarification would take considerable time, and there were some forces that would not wait. As to the health effects of air pollution, in 1949 "unequivocal evidence" was "meager."[34] Throughout the 1950s, health evidence was considered insufficient for standard-setting. Even as late as 1962, the surgeon general asked: "Are automotive emissions a potential hazard to human health, and if so, at what concentration in air do they cease to be safe?" The response? "This question . . . will require years for its definitive answer."[35] In 1968 a public health specialist lamented that "if . . . each regulation must await specific evidence of health effects prior to enactment, the situation is hopeless."[36]

What could be expected to occur as policy awaited the resolution of uncertainty? The discussion in chapter 1 of the technical and behavioral components of pollution problems suggests that matters would only get worse, and of course they did. Population trends dictated this; so did the common property and collective goods effects. The latter in particular injected substantial bias into the process of making pollution policy (they did in the past, they still do). They built a system of incentives favoring actions and technologies producing external costs—like pollution—over those producing external (and collective) benefits—like clean air.[37] Beyond this, polluting actions and technologies were self-imposing, employed through a process of voluntary individual decisions each of which appeared beneficial to its maker, none of which needed recourse to the legal system to go forth. Control measures, technologies, and other steps yielding collective goods were not self-imposing in this sense. Acts of individual decision did not lead to the collective good of control; it could be achieved only by collective decision. One did not need the help of law to pollute, only to control polluters.

Here the legal system introduced further bias, or at least provided the means for introducing that already prevalent in the general situation. Its inertia, in particular its allocation of the burden of uncertainty, made intervention difficult to realize. Those who sought it confronted the job of showing net benefits, yet all that was certain was that change would be costly. Bias here was deeper than might at first appear. It was one thing for law to say "show me"; it was another to say this to those who, precisely because they sought a collective good, would have difficulty organizing to support the research and lobbying efforts needed to make the showing.[38] On the other hand, those opposed to control, primarily pollution sources, were organized and able (largely through trade associations) to carry out research in support of their positions, and to hire public relations experts and lobbyists to promote them.[d] Perhaps more importantly, they also had only the easy job of showing that things were not clear. Throughout the years, the approach of industry was not to make an affirmative case for innocence, but rather to argue constantly that blame had not been clearly made out. The approach, in short, was to rely on uncertainty.

[d]Sources could better organize because of their smaller numbers and because the pollution problem was an item of higher priority on their agendas of concern than on those of individual citizens, among other reasons. For them, free-riding was a smaller problem than for those interested in pollution control. See generally M. Olson, *The Logic of Collective Action* (New York: Schocken Books, 1971).

In the face of such a layer of biases, control came about piecemeal, in small steps, the problem growing quickly in the meanwhile (quickly because technology, population, and consumption were all increasing exponentially). Under such circumstances, one would expect serious pollution episodes—what we will call "crises." And just such crises recurred.

Crises and the Making of Pollution Policy

It is not enough to note that episodic crises were (still are) a factor in making pollution policy; one must also account for their role. That such a role exists seems indisputable. J. Clarence Davies, in his study of *The Politics of Pollution,* observes that "the growth of concern over pollution has been stirred by crises, both real and created."[39] Others have noted the same points, as to environmental problems generally and as to the specific problem of air pollution.[40] Indeed, some see the history of air pollution control as bearing out a "catastrophe theory of planning. Much of the legislation can be analyzed as a direct response to dramatic incidents."[41]

Even a cursory review of the years up to 1970 reveals a number of instances in which pollution crises appear to have affected the course of policy. A Los Angeles episode in 1903 resulted two years later in a local ordinance controlling smoke emissions; a subsequent series of local ordinances, all prior to 1940, came in sporadic response to particular problems (such as orchard heaters) defined by specific outbreaks; severe episodes in Los Angeles in 1943 provoked complaints from citizens, public studies, and a new wave of ordinances, this time somewhat more general in character than those before 1940; a 1946 episode led to an "angry march" on city hall and resulted in further studies and strong (and successful) impetus for state legislation; an episode three years later reinforced criticism of the prevailing approach to air pollution and thus played a role in encouraging important changes in direction. There were other instances as well at the state level—for example, the 1954 episode that led shortly to Governor Knight's "all-out smog war." On the federal level, the story was much the same: the famous Donora episode of 1948 focused national attention on air pollution as a health problem for the first time;[42] the London disaster of 1962 was much cited at the National Conference on Air Pollution of that year, and played a part in pushing forward the move for federal control; events in New York City in 1966 resulted in a declaration of emergency by Governor Rockefeller and served as the preface to the Johnson administration's proposals for strong new federal intervention. Looking slightly outside the compass of our study, we can note that the early involvement of the federal government in the problem of water pollution was brought on by a typhoid scare; and the breakup of the *Torrey Canyon* (in 1967), a similar event a year later, and the Santa Barbara spill of 1969 "focused national attention on the problem of oil pollution."[43] One could go on with examples. A serious episode in St. Louis in 1939 is said to have led directly to that city's program of air pollution control.[44] In Japan and England series of episodes also played their part in stimulating response by way of control programs.[45]

One should not, of course, get too carried away with the place of episodic crises in the making of pollution policy. In many instances we know of, episodes have been ignored or, in any event, unacknowledged by action. There was, so to speak, an objective crisis (the episode), but no subjective crisis—no great concern or reaction in the public mind. Thus episodes in Southern California in 1940, 1941, and 1942 (episodes that were, apparently, as severe as those in 1943 and 1946) generated no great clamor and concern. To the extent they were noticed at all, they were attributed to "gas attacks" by a World War II enemy. The war may have contributed to the sense of quiet attending these episodes in another way. The constant demands of a life-and-death struggle probably put the condition of the environment low on the list of private and public priorities. Minds preoccupied with more urgent matters might not have noticed the episodes or, to the extent they did, considered them the price of defense. By 1943, especially by 1946, the domestic war effort may have stabilized sufficiently that attention could be turned to the situation closer to home. We can find analogs for this in other societies.[e]

There are further instances of crises that stimulated no response. For example, the serious incident in London in 1952, unlike its successor of ten years later, had no apparent impact on American policy. It did in England, but the action following the 1952 disaster there stood in marked contrast to the inaction characterizing the British government's response (or, better, lack of it) to earlier events of a magnitude similar to or greater than the Donora episode.[46] In New York City in 1953 a brief period of weather stagnation produced pollution conditions that reportedly resulted in 200 deaths, but "this disaster attracted far less attention than the loss

[e]Quick study suggests that Tokyo's pollution problems first emerged during and after World War I. Significant measures to deal with the problems "were quietly shelved" when "the outbreak of Japanese military action in China in 1937 put the national economy on a war footing, and ways to increase war production took precedence over measures for the public welfare." Environmental problems and responses to them began once again with the prosperity that followed 1945. See Tokyo Metropolitan Government, *Tokyo Fights Pollution* (Tokyo, 1971), pp. 19, 22-23, 25-28.

of a comparable number of lives in the collision of airliners over the same city."[f]

Moreover, the distance between crises and significant action is often great. Donora, for example, "brought the air pollution problem to national attention, but it took fifteen more years until passage of the first permanent [federal] control law."[47] As another example, four years passed between the first serious incident in Los Angeles and the first substantial response; thirteen years more elapsed before the first legislation aimed at controlling motor vehicles, despite a constant and dramatic increase in their number and several intervening episodes. This delay, or "lag," can be explained, of course; it might even be useful (points to be considered shortly). But neither it nor occasional instances of no reaction should hide the part that episodic crises played: they typically set the boundaries in which policy was made by small increments. That crises preceded policy does not necessarily mean, of course, a causal connection. But looking at all the surrounding circumstances, the evidence of a connection in most cases seems too clear to deny. Crises appear to have been a quite characteristic part of environmental policymaking.

It is not difficult to understand why this might be so. Policy-by-least-steps, encumbered by inertia and by the manner of allocating the burden of uncertainty, meant a process of minimal response to an increasing problem. And it was increasing. As pointed out in chapter 1, the total output of pollutants is related to population, technology, production, and consumption; without control, pollution will increase with growth in these factors.[48] The relationship among the factors means that their growth produces a very accelerated increase in the total amount of pollutants. Population grows at compound rates, and historically production and consumption have grown at even higher rates than those applying to population growth.[49] At the same time, technology, which enhances the ability to produce pollutant by-products in large amounts, has enjoyed a similar history of vigorous expansion. Moreover, the effects of growth have been aggravated by the trend toward geographic concentration. A pollution problem occurs, and the potential for a crisis is greatest, when the assimilative capacity of the environment is used up; and the greater the concentration of growth, the greater the strain on that capacity. Given these factors, it is hardly surprising that Southern California began to

[f]Staff of the Senate Committee on Public Works, 88th Cong., 1st Sess., *A Study of Pollution—Air* 13 (Comm. Print 1963). Perhaps the absence of a reaction in New York City owed to the fact that mortality statistics were not disclosed until nine years after the event. Yet the early Los Angeles episodes often led to uproar with no evidence at all of severe health effects.

experience a noticeable decline in air quality by the year 1940 and thereafter. Substantial growth had taken place in the area in the years prior to the onset of World War II; further growth occurred during and immediately after the war (see tables 6 and 7 in chapter 7). Raymond Tucker, the St. Louis pollution expert hired as a consultant by the *Los Angeles Times,* noted that industry in the area had grown 85 percent during the years 1940-1946. All the rates of growth were much higher than those prevailing in the nation as a whole and—considering the absence of much control and the existence of especially aggravating meteorological and geographical conditions in Los Angeles—meant that a large amount of pollutants was escaping into a relatively small volume.

Rapid, concentrated growth need not necessarily have increased pollution, of course. Sources could have voluntarily employed control measures, turning technology toward easing rather than irritating the problem; citizens could have brought lawsuits or agitated for government intervention. But the discussion in chapter 1 and in the last section suggests such events would be unlikely, and the evidence is consistent with that suggestion. Tucker noted that air quality declined markedly after 1940, yet this burgeoning pollution problem was left largely uncontrolled—until the happenstance of especially unfavorable meteorology turned a persistent problem into a pressing episode. The process of policy and the structure of environmental problems literally begged for crises. But this explains only partially why crises have been so characteristic of environmental policymaking; it accounts for their occurrence, but not for their effects.

To some extent those effects seem abundantly clear. Crises stimulated government action by helping to overcome inertia and uncertainty. They did so simply by suddenly making the expense of inaction appear too high. If government was reluctant to act under uncertainty because of the costs of doing so, it would be more motivated when the costs of *not* doing something seemed even higher. There is good evidence of this sort of thinking over the years in question. The first head of the Los Angeles APCD, for example, noted in 1949 that "the attempt to show a direct relationship between ill health and atmospheric pollution has been disappointing to those demanding exacting demonstration." He added, however, that the "health considerations have been greatly strengthened by the Meuse Valley and the Donora disasters." He cited a federal report on Donora pointing out that it was not until that episode "that the Nation as a whole became aware that there might be a serious danger to health from air contaminants."[50] In 1952 a California Assembly Committee reported that "in those areas of the State where conditions of air pollution have not reached the critical nuisance stage . . . , enforcement must be based upon sound technical facts, . . . definite information."[51] The implication of how to proceed if conditions *were* critical is clear. In 1960 two active participants in California's pollution history noted the many uncertainties attending standard-setting and observed: "Nevertheless, the urgency of California's air pollution problem required action on the basis of what is presently known."[52] (Some of the standard-setters must have disagreed.)

But crises did more than directly stimulate government initiative by helping to overcome inertia and uncertainty. They also served to provoke citizen demands for action. Such demands can be essential; a crisis without them might not be enough.[53] As one California legislator put it with reference to the state's pollution problem, the legislature will only respond to a crisis to the extent the public itself believes a crisis exists. As evidence of such public belief, he cited a case in which constituents sent in hundreds of letters.[54] Public pressure is especially important in contexts where strong concentrated interests (like the trade associations of important pollution sources) stand against government intervention. In such instances in particular, citizens need numbers and influence on their side; many "persons . . . must turn their attention to the same matter."[55]

Looking back at the past, we can see examples where crises of themselves did not motivate government action, where such action came about only when episodes provoked citizen demands. There was, for example, no response to the Los Angeles smog sieges of 1940–1942, but there was to those in 1943 and 1946; the only apparent difference between the two sets of situations was the presence of citizen action in the later years. The panicky response in 1943 (the closing down of the synthetic rubber plant) was spurred not simply by the episodes in the fall of that year, but in addition was "prodded by an angry citizenry" reacting to the episodes. The second significant government response, increased efforts to secure state legislation, came when "hundreds . . . marched on City Hall" after several severe episodes in the fall of 1946.[56]

Crises played their part in the making of environmental policy. In a report issued in 1955, California's Department of Public Health observed that "in air pollution, it has been disasters which have arrested the attention of the people and the scientists alike. We have become alarmed about damage from air pollution only after a catastrophe."[57] There have been similar observations as to water resources, one good study noting that with respect to them, "general interests are asleep, to be awakened only by floods and droughts, by catastrophe."[58] How crises might function to awaken, arrest, or alarm in terms of government taking initiative appears, as suggested above, quite obvious; how they stimulate citizen demands is much less clear. Let us consider the roles that crises play in the citizens' behalf.

It is more than coincidence that the two quotations just above both dealt with the effect of catastrophes in the context of provision of collective goods—in the one case flood prevention, in the other air pollution control. It was mentioned earlier that organized citizen demand for collective goods is likely to be understated. Each citizen tends to see the costs to him of preventive action as larger than the expected benefits, especially because he might not view even a substantial improvement in air quality as being all that beneficial in his total agenda of concerns; other problems—a job, salary, the children's schooling—are likely to be regarded as more pressing. Each tends to view any contribution he might make to organized efforts toward preventive action (whether self-help or petitioning government) as so small of itself as to be negligible, thus foregoing the contribution; more importantly, each is likely to be aware that the efforts of others, if successful, will redound to his benefit even though he does nothing to help support them—he can take a free ride.[g] Thus, unless the expected benefits of action for any individual are larger than the costs to him of taking action, the demand for action will be understated, even where the aggregate expected benefit to the group interested in the collective good is larger than the total cost of achieving it.[h] It is for these reasons that the "general interest" in collective goods tends to be "asleep"—precisely because it is general and unorganized, as opposed to a concentrated or organized or "special" interest.[i]

[g]The inclination to free-ride will be reduced to the extent that the individual sees his contribution as making possible more lobbying or more clean air than would otherwise be the case. But if the desire is to have only some fixed amount of air quality or lobbying, the tendency will be to take a free ride on the efforts of others to procure that amount. See R. Posner, *Economic Analysis of Law* (Boston: Little, Brown and Co., 1972), pp. 253-254.

[h]Thus, Davies reports that in "almost all the studies of public opinion on pollution a significant gap was found between those who were very concerned about pollution and those who had done anything about their concern. The latter category was a small percentage of the former." J. Davies, *The Politics of Pollution* (New York: Pegasus, 1970), p. 83. See generally Olson, *The Logic of Collective Action*; J. Krier, "Environmental Watchdogs: Some Lessons from a 'Study' Council," *Stanford Law Review* 23(1971):623, 664-665.

[i]See H. Hart, "Crisis, Community, and Consent in Water Politics," *Law and Contemporary Problems* 22(1957): 510, 524-528. Hart points out that only a tiny fraction of persons polled regarded water resources as a serious problem, and even among this fraction the problem was again only a tiny fraction of the polled persons' total concerns—personal and family matters weighed most heavily. Given

How does an episode, a crisis, awaken the general interest; how does it transform it into an effectively organized one? We can speculate that, in addition to identifying problems and spurring government to take steps on its own initiative, crises might serve in a number of ways to make more effective the demands of environmental interest groups for governmental intervention (or more or different intervention once a control program is underway). Let us first sketch some of these ways here, and then suggest how they help to explain why crises seem so characteristic a part of the process of making pollution policy.

If, as suggested above, it is difficult for concerned citizens to coalesce around a problem like air quality partly because each citizen sees the costs of doing something as higher than the costs of doing nothing, then it is plausible that this difficulty might be overcome to the extent that the costs of passivity go up. A crisis, a particularly severe episode, should have the effect of aggravating the situation in the citizen's eyes. A bad problem suddenly becomes intolerable, and for all the citizen knows, it will persist; action becomes relatively more worthwhile. Beyond this, if the peculiar arithmetic of pollution suggests that the effects of environmental problems are a product of growth in population as well as sources (because there are *more* people suffering a given effect),[59] then we can suppose that just as the conditions for an episode are ripening (just as, that is, air quality is deteriorating), so too there is an increasing number of people among whom the episode might trigger response. Granted, each citizen might see the general problem (and thus the episodic crisis) as only twice as bad as the situation prevailing before a doubling of population and sources, even though the total costs might have increased fourfold (twice as many people suffering twice the effect). But it is also the case that there are likely to be twice as many people who will take some form of action. It is hardly necessary that all citizens, or even most of them, suddenly change their calculation of the costs and benefits of doing something in the face of a crisis; it is often enough that a large handful, those most sensitive or outraged, do so. The key to effective citizen demands is absolute, not relative size. An effective protest seldom requires tens of thousands; a few hundred are

---

this, Hart found "no great mystery . . . about the dominance of special and individual interests . . . over against the more general interest" such as that in water resources. The fact that catastrophes were often central to flood control efforts, and the fact that air pollution episodes played such a regular role over the years canvassed in this study (even if only by way of spurring government to take initiative without prodding from without), are evidence of the somnolence of general interests in collective goods.

enough. (Consider, for example, not only the "march on City Hall" mentioned above but the other instances where organized protest from a small proportion of citizens helped turn the tide: in the case of the pink-slip controversy, only about 2 percent of the population complained about the crankcase requirement; nevertheless, 100 percent of the mail of one legislator was against the requirement. Another example is the public forum organized by backers of the Pure Air Act of 1968; only a few hundred citizens attended, but they were all gathered in one room, and their presence is said to have influenced a number of legislators.) The foregoing becomes the more plausible if we remember that the wealth effect discussed in chapter 1 suggests that higher income people will be those most sensitive to episodes, and that they "are precisely the individuals who tend to be the most active politically and organizationally."[60] A crisis, in short, might trigger organized citizen response more easily than would at first appear, especially in the context of a public mood receptive to action.

There is another fashion in which crises might mitigate the cost-benefit problem of individual initiative. Thus far, we have been viewing citizen action as a "production" decision in which the individual weighs the costs of acting against the results of a successful outcome (the expected benefits). Crises, however, might function to change what is ordinarily a production decision into one about "consumption" where the individual can, at a relatively low price, obtain an immediate, satisfying benefit—the sense of doing one's duty, of being an active participant in grass-roots democracy. The act of voting is a rough analogy. Considering its expected costs and benefits, it is virtually impossible to explain the decision to vote in production terms. Given anything more than a very small number of voters, the costs of voting, however low, are almost certain to be higher than the expected benefits, for the likelihood that one vote will be determinative is infinitesimal. Hence the voter's paradox. But if we view the matter from the other side, we see it in a new light. If the voter attaches positive utility to the very act of voting (as opposed to the outcome of the election), if he sees the time and trouble as a way of "buying" at a low price something rewarding (doing one's duty, participating), then we can see him as a rational consumer rather than an irrational producer.[61]

Turning back to crises, there is evidence of a similar phenomenon. Oates and Baumol have suggested that individuals will take steps in situations of crisis that they otherwise would not. They cite as an example the dramatic increase in voluntary blood donations in New York City when the newspapers announced "a blood crisis in which reserves of blood had fallen to a level insufficient for a single day of operation." People gave who previously had not, standing in line for as long as ninety minutes. They explained their reactions by saying they were "paying a sort of personal guilt complex"; that it was "the least [they] could do for the city." Oates and Baumol give other examples and generalize by saying the behavior in question appears most likely to occur "in times of crisis." But they are uncertain as to the reasons, guessing they have to do with "social pressures and a sense of urgency."[62] But we can speculate that more of the answer might lie in the production-consumption distinction. Crises may give rise to a sense of duty, a sense of a need to do one's part, not so much to abate the crisis as to satisfy the urgings of a conscience untroubled in more mundane times. (Some people—like Ralph Nader—might feel this sense of duty, in this case "to save the environment," even in the absence of the urgings of a crisis. But the evidence suggests there are relatively few of these.)

It is important to note that since satisfaction of conscience can come about only through individual action, crises help to overcome not only the general cost-benefit problem discussed above but its particular aspect of free-riding (where no cost is worth bearing since the benefit will be produced in any event, if produced at all). In the consumption perspective, one must join in to be better off; the acts of others are of no help. By the same token, to the extent crises motivate consumption decisions, they also encourage action by individuals who might otherwise see participation on their part as so negligible a contribution to a group effort as to be totally dispensable. If the decision has to do with consumption, with the desire to fulfill a sense of duty, then of course taking individual action is indispensable.

Regarding the last point, there is an additional way in which crises might play a part. The individual might see his own contributions as dispensable precisely because he sees them as his own, rather than as a small but necessary part of a larger whole. If he could be assured of the need for his contribution, and be further assured that others will indeed contribute so that effective action will be undertaken, he is more likely to become a supporter. (Given assurances to an individual of the need for his contribution, and given assurances that his contribution together with that of others will yield the collective good, there seems no reason to expect free-riding.) What is needed, of course, is some system to give assurances, but it is exactly such a system—an organized means of communication and cooperation—that is so difficult to provide when collective goods are involved. Crises, however, might spontaneously provide a sort of tacit system. Speaking in a related context, Thomas Schelling discusses a situation where the

> problem is to act in unison without overt leadership, to find some common signal that makes everyone confident that, if he acts on it,

> he will not be acting alone. The role of "incidents" can thus be seen as a coordinating role; it is a substitute for overt leadership and communication. Without something like an incident, it may be difficult to get any action at all. . . . Bandwagon behavior, in the selection of leadership or in voting behavior, may also depend on "mutually perceived" signals, when a part of each person's preference is a desire to be in a majority or, at least, to see some majority coalesce.[63]

To the extent that an incident—a severe episode, a crisis—makes many believe that it is obvious that many will take action, and believe further that this must be obvious to many, many are likely to act. The crisis coordinates expectations, and thus actions.

The discussion thus far might be taken to suggest that the main function of crises is to facilitate something like mob action on the part of citizens, but in fact there is a bit more to it than that. Equally consistent with each of the points mentioned above is the notion that crises will facilitate the formation of environmental interest groups consisting of a small core of individuals (the most concerned and committed) supported by contributions raised during crises, contributions from citizens who would otherwise be inclined to withhold them. The formation of Stamp Out Smog (SOS) in Los Angeles is a rough case in point. The group was formed by a few Los Angeles women in 1958. The original impetus behind the organization was a crisis in every sense: a Beverly Hills mother had to rush her young child to a hospital as a result of an asthma attack brought on by smog conditions. Advised by the child's doctor that it would be best for the family to leave Los Angeles, the mother " 'decided instead to stay and do something about air pollution.' " She organized a small group of friends who formed the nucleus of SOS. They in turn invited other organizations—garden clubs, chambers of commerce, health associations—to join and support their efforts. The result was thousands of members, including a small number of active members. "The founders of SOS did not need to recruit other active members; air pollution did the recruiting for them." One active member joined out of anger that her children could not play outdoors during smog attacks, another after "a particularly bad spell of smog."[64] Since that time SOS has regularly participated in the making of California pollution policy, and with some impact (though its members confess to lacking the energy and resources to organize major lobbying efforts).[65]

There are several implications to the foregoing discussion. First, the roles of crisis can be expected to weaken with time. For example, if episodes increase the costs of doing nothing and thus encourage doing something because citizens suddenly find the problem intolerable and fear it will persist, experience might reveal that the conditions are in fact tolerable and do not persist—precisely because they are episodic. Similarly, if episodes become familiar (but not too familiar, not persistent), citizens may undergo a sort of acclimatization, such that the episodes are no longer crises but rather events not out of the ordinary at all. As to both points, the sky is not really falling, it just seems to be every once in a while. Finally, the utility of consumptive behavior may diminish as a problem becomes old hat; the peculiar satisfaction of doing one's duty wears thinner the more often duty must be done. (This suggests that a policy of voluntary controls—of government pleas to "do better"—might well be effective during times of crisis conditions, but only for the short term, and only where compliance by a high percentage of the relevant population is not necessary.)

The evidence in our story is consistent with these points. Putting aside the formation of SOS, it is difficult to find episodes that stimulated much *citizen* response after the 1940s. (And not because there were no episodes. For example, an especially serious "five-day siege of smog" occurred in the fall of 1953. It stimulated response, but mostly from the press.) Other casual evidence also appears to be consistent. For example, driver compliance with government pleas to reduce speed voluntarily on the highways during the so-called energy crisis appeared to diminish gradually with time, to the point that now many drivers exceed even a mandatory fifty-five-mile-per-hour speed limit. Oates and Baumol note from similar casual examples "that the sense of high moral purpose is likely to slip away rather rapidly."[66] Of course, to the extent that crises might contribute to the setting up of more or less permanent organizations, such as SOS, citizen prodding of government might continue even after the usefulness of episodes wears out. The effectiveness of such organizations, however, is to a considerable degree dependent on citizen support, so even in these cases one might expect to see familiarity takes its toll.[j]

The second implication may be that one is more likely to see crises result in citizen response at the local than at the state or federal levels. It is easier to organize local coalitions, as opposed to more broadly based ones, to the

[j]Along these lines, consider "Hard Times Hit Environmental Groups as Interest Wilts, Dollars Dwindle," *Los Angeles Times,* 3 Sept. 1972: "Environmental organizations, especially those that specialize in political and legal battles, have fallen on hard times. The heady days of 1969 through mid-1971, when checks flowed in from new members, are gone. They are replaced by a new reality—that people have drifted away from the fadlike concern with the environment, taking their money with them."

extent that physical proximity and commonness of interests facilitate organization—an extent that is probably considerable.[k] Moreover, it is one thing to organize a local group to march impulsively on city hall; it is quite another to put together a coalition to travel to the state capital, much less to Washington.[l] (Though if the result of a crisis is a standing organization, it might be quite feasible to arrange lobbying trips by its core members to distant seats of government; SOS campaigned in Sacramento and Washington.)[67] Again, the evidence is consistent. To the extent crises provoked citizen action, it was usually action aimed at local officials. The response to episodes in Sacramento and Washington was largely a matter of government taking the initiative (but with the press often spurring matters on). Indeed, after the 1940s it was very seldom that pollution crises stimulated citizen action at all. Perhaps this is because of both points discussed: episodes had become more familiar, and in any event much of the responsibility had been assumed by first the state and then the federal government. Thus, at the same time that crises were becoming a less effective organizing device, the most promising targets for organized action were becoming largely irrelevant. (This is not to say there was no citizen action after the 1940s, but the instances in which it occurred are good evidence of the point made. The citizen forum held in conjunction with legislative consideration of California's Pure Air Act of 1968 was well attended, for example, but this was the result of efforts by the legislative promotors of the bill, by a radio station, and by SOS. There were, in other words, existing groups ready to take on the organizational task; they did not need the help of an episode. The same can be said with respect to the hundreds of thousands of letters sent by Californians to their congressmen during the time of the debate over federal preemption; the letter campaign was directed by the "Breath of Death" radio series.)

[k]Cf. Davies, *The Politics of Pollution,* p. 79 ("The major determinant of public concern with air pollution is the actual level of pollution prevalent in the area of residence"). Hart, in his study of the politics of flood control, notes that in "chronically vulnerable areas" there was a tradition of organized self-help and, beyond this, some success in setting up and supporting associations to lobby in Washington. Hart, "Crisis, Community, and Consent," p. 526.

[l]Perhaps it is for reasons like these that one study of the federal air pollution policy process notes that federal policymaking "was not overwhelmed by episodes of crisis proportion," though it was often stimulated by them. R. Dyck, "Evolution of Federal Air Pollution Control Policy" (Ph.D. diss., University of Pittsburgh, 1971), p. 262. An additional reason might be that federal (and state) legislative processes are characterized by more "official and unofficial veto-points" than those of local government, and that these tend to constrain any quick reactions to crises. Cf. G. Schwartz, "Book Review," *U.C.L.A. Law Review* 19(1971):148, 152.

We should not conclude our speculations about crises without at least brief reference to two quite obvious ways in which they might stimulate lawmaking without regard to moving government to take action on its own initiative (by defining a problem), or to facilitating organized citizen demands for action. One way has to do with the press. Recall, for example, that the *Los Angeles Times* capitalized on the 1946 episode in Los Angeles by starting a campaign of its own. The explicit purpose of the *Times* and other papers covering the smog issue was "a campaign or crusade in the public interest,"[68] but surely there were other ends in mind. The publisher of one Los Angeles paper (the *Daily News*) was candid in admitting as much: "The community has a paramount right to clean air—and the people expect this right to be championed by the press. It is also true, of course, that smog is a prolific producer of issues for dramatic copy."[69] It could not have been simply coincidence that the interest of the *Times* and other Los Angeles papers developed just as the war was ending. The taste for headlines must have been high, and a new war was needed to satisfy it; thus the headlines talked of stepping up the "War on Smog," and publishers spoke of "battles" and "victories."[70] Crises made the war a real one, and helped sell papers.

Subsequent to 1946, the Los Angeles press devoted a great deal of space to the smog issue—by one estimate, 61,500 column inches in the years 1949-1951, by another estimate, over 100,000 column inches through mid-1952.[71] There can be little question that the efforts of the press had of themselves an impact on government action and, beyond this, gave force to citizen complaints. Indeed, the latter may have been the more important. One study of the California situation found that in "the 1950s, when smog was Southern California's most salient public issue, newspapers and radio stations stimulated and focused public outrage with the inadequate control agencies and the opposition of the auto manufacturers."[72] When there were no crises to report, the press shifted to giving warnings of their coming. Thus, a 1955 headline read "Smog Termed Near Disaster Stage in L.A."; one a year later announced that "Deadly Smog Could Kill Thousands, Cause Panic."[73] But the interest of the press waned with time, perhaps for the same reasons discussed earlier in the context of citizen action. "With the lack of shocking new medical discoveries about smog's poisonous effects and the increasing age of the issue, pollution has become far less newsworthy," the study referred to above concludes.[74]

The second way crises might play a part, independent of directly stimulating government response or citizen demand for it, has to do with crusaders, those few zealous individuals whose dedication to solving a problem grows not from an episodic crisis but a long commitment—people like Ralph Nader, Rachel Carson, and (at an earlier time) Upton Sinclair. Crises are hardly crucial to the personal efforts of such people, but they may be essential to their getting any attention and support. When a crisis or something resembling one can be found, crusaders try to capitalize on it, for the crisis makes the crusader's claims seem more urgent and credible. Thus a study of the process of making federal air pollution policy discusses one United States Public Health Service staff member's "exploitation of crisis for purposes of policy change." The staff member was a "grand strategist who had a plan to take advantage of every episode." His efforts to manipulate the situation in times of crisis "confirmed the importance of disaster on air pollution policy as well as policy pertaining to mining safety, food and drug regulation, pesticides, etc." The study states that the "Air Pollution Division [of the Public Health Service] used air pollution episodes in a conscious effort to focus public attention on the problem and 'to get things done.'"[m] If there are no crises on which to capitalize, efforts will be made to fabricate them—whether by the bottles of "smog" marked "poison" and passed around by California's congressional delegation in a successful campaign to defeat federal preemption of California motor vehicle standards, or by the tactics of "environmental activists" designed "to get people to really think about how close we are to disaster."[n]

The burden of the discussion of the causes and functions of crises has been this: Surely crises do not always play a part in environmental policymaking, but there is good evidence that they often do. We lack a sufficient theory of government intervention to engage in anything beyond speculation on this point,[75] but the speculation seems plausible enough. The core structure of environmental problems suggests that crises are likely to be virtually a necessary part of environmental policymaking in two senses. They are likely to be necessary first in the sense that the effects of growth, of common property, and of collective goods, coupled with a process of policy-by-least-steps, lead to a compounding problem but little reaction

[m]Dyck, "Evolution of Federal Policy," pp. 3-5. The account illustrates the ambiguous relationship that might exist between a crisis and government response on its own initiative.

[n]Quoted in P. Janssen, "The Age of Ecology," in *Ecotactics: The Sierra Club Handbook for Environment Activities,* ed. J. Mitchell and C. Stallings (New York: Pocket Books, 1970), pp. 53, 59. Thus, one sees claims today that the "energy crisis" is a fabrication of the oil companies designed to make a convincing case for easing production constraints enacted during the "environmental crisis." See P. Weinberg, "Contrived Crisis: An Environmental Lawyer's View of the Supposed Fuel Shortage," *Buffalo Law Review* 23(1974):435.

until a significant episode; "general interests are asleep, to be awakened only by . . . catastrophe."[76] The catastrophe, the crisis, identifies new problems—it produces knowledge. It also reinforces or substantiates knowledge by transforming an intellectual grasp of the situation into an acute realization that yes, matters really are serious. At times a crisis (sometimes even awareness of a serious problem that has not yet reached crisis proportions) will of itself give rise to government reaction without prodding from outside. But often more is needed, and here crises are necessary again to making pollution policy in that they facilitate effective prodding. We saw a number of ways in which they might do so. One would expect them to work together and reinforce each other: crises attract the interest of government, they help citizens to organize, they add weight to the claims of crusaders, they generate press coverage, which in turn helps citizens organize, or gives publicity to crusaders, or in any event helps stimulate the government, and so forth, on and on—at least for a time.[77]

Thus, while crises are to some extent the result of the troublesome characteristics of environmental problems and policy, they are also one helpful means of overcoming them, and of prodding policy along. This is not to say, however, that crisis planning is altogether free of troublesome consequences of its own.

The Fixations of Pollution Policy

Inertia, uncertainty, and crisis planning all contributed to three facets of pollution policy in the period before 1970. First, they had much to do with the preoccupation with obvious sources and emissions that characterized early control efforts, a preoccupation that tended to right itself through a process discussed in the next section. The second and third facets—indeed, fixations—of the period never did right themselves; they have largely persisted into present policy. One of these is the technological fixation, the other the fixation with regulation as opposed to other means of legal intervention. The two go very much together; here they are discussed in turn. (The focus of the discussion is on California policy, but the reader will recognize that the remarks apply as well to developments at the federal level.)

It is easy to understand how tempting technological solutions—technical "fixes" that called for gadgets and mechanical innovations rather than "change in the individual's social attitudes"[78]—must have appeared to policymakers when Southern California developed its first serious pollution problems in the early 1940s. There was first of all a fundamental

American bias toward such fixes, a "preoccupation with operating technique (the 'practical')" and a "relative impatience with understanding (the 'theoretical')." Given the forces that shaped the bias in the past ("constant pressure to improvise, contrive, and manage"),[79] one would expect it to appear when a crisis demanded action. Technological solutions look quick, easy, and predictable. "Their feasibility is relatively easy to assess, and their success relatively easy to achieve. . . . Social engineering," on the other hand—"changing people's habits or motivation," asking "people to behave more 'reasonably' "—is difficult, time-consuming, and uncertain in the extreme.[80] Technology had solved the pollution problem in other areas (primarily the East Coast, which experienced serious conditions at an earlier time and managed them through technological controls); by employing it, policymakers could "make opportunistic use of the means closest at hand."[81]

Thus the demands of crisis probably helped to shape a preference for technological solutions. The forces of inertia, the very considerations that had contributed to a situation of crisis in the first place, must also have played a part. Not only are technological fixes quicker, easier, and more predictable than "social engineering," they are also less disruptive of the status quo. This, at least, is the general opinion:

> The technological alternatives are in general more straightforward, more amenable to direct application, and less likely to induce major social and economic side-effects. The nontechnological alternatives [such as "modifying use preferences"] are more difficult to effect and involve social dislocation and possibly unpleasant side-effects.[82]

Consistent with the demands of crisis and the forces of inertia, policymakers in the early stages of the pollution problem had their minds open to two methods of control. They could "attack the pollution blanket directly, and seek its dispersion . . . using natural or mechanical means"; they could "attack by controlling the individual sources."[83] The first method consisted of "Rube Goldberg" devices, the bizarre mechanical solutions that aimed to bring the conduct of nature, rather than man, into line. These were especially attractive because they avoided any disruption of activities "so intimately associated with a healthy industrial and economic development."[84] The second method, while it could be taken to include nontechnological controls (for example, simply requiring sources to reduce or eliminate operations), was never viewed this way. Thus, in the 1940s "no one knew that [vehicle] exhaust gases were one of our biggest pollution problems," but it would have mattered little in any event since "no acceptable device was available which we could have compelled motorists to install."[85] It apparently occurred to no one to modify "people's habits or motivation," to somehow reduce the amount of driving pending a suitable control device. The faith of the times was in technology.

That faith was expressed from the outset, and it persisted throughout the years. It was reflected in the call for a "Manhattan Project" to "attack . . . the smog problem" (a project never undertaken); in the confident statement that the problem would "be solved by research and engineering ingenuity"; in the central credo that the "success of a smog control program will be geared to the success of science in discovering what contaminants cause smog and in developing technical devices for eliminating them."[86] Control would reach to the limits of "technical feasibility," but it would reach no further.

Our discussion has already suggested that the technological fixation was understandable in light of the forces at work. It may also have been sensible—at least initially. In the early stages of the problem, existing technology may have been sufficient to control the stationary sources considered at the time, and probably correctly, to be the prime causes of the pollution problem. Moreover, measuring the degree of control by the limits of technical feasibility was about the most that policymakers could do. Some guide, some measure of how far to control, was necessary, but there was great uncertainty concerning the effects of air pollution on health and other factors. In the absence of anything better, and inasmuch as minds had fixed on technological solutions, it was only natural—indeed, it was quite logical—to ask only for all that technology could accomplish. There was nowhere else to turn for a standard. Thus, an Assembly committee noted in 1952 the paucity of information that "would be helpful in . . . establishing quantitative standards. . . . In the absence of this precise information," it said, "the degree of improvement which is required of the various industries must be based upon what the industries *can* do, rather than upon a scientific determination of the significance of each particular waste with respect to the public health."[o]

[o]California Assembly Committee on Air and Water Pollution, *Final Summary Report* (n.d., c. 1952), p. 43 (emphasis in original). The refrain of doing the best that technology can accomplish in the absence of any other reasonable guide to control has been a continuing one. Some public health specialists urge "*control of emissions to the greatest extent feasible, employing the maximum technological capability*. . . . Determining feasibility and technological capability, although difficult, is vastly easier than determining the level in the atmosphere at which a pollutant may be safe or unsafe." E. Cassell, "The Health Effects of Air Pollution and Their Implications for Control," *Law and Contemporary Problems* 33(1968):

There were shortcomings in this approach, to be sure. Most essentially, the approach overlooked that technological feasibility is a relevant but not an exclusive consideration. What is technologically feasible might be economically infeasible; the best technology might be "too good," too costly, in terms of the benefits it would yield. Conversely, feasible technology might not be good enough; nontechnological steps might result in gains that would outweigh any accompanying disruption or inconvenience. The technological fixation ignored these points.

The first worry (that what was technically feasible might be economically infeasible), if it was ever a concern at all, was avoided by cautious administration. Thus the Los Angeles APCD's policy of fair-but-strict, of taking into account the "practical needs of the industry," ensured that technical feasibility would never be carried to the point of hardship. In fact, enforcement tended to be based only on what the least capable polluters could accomplish. There were problems with such a program, but they were probably inescapable given the desire to avoid the disruptive effect of closing down businesses. The second concern (that the best technology might not be good enough) was handled less adroitly. There was, as we saw, a commitment not to go beyond technology; "action of an enforcement nature" had to await "development of technical remedies."[87] Sensible or not in its conception, such an approach could at least have been pursued in constructive fashion. Policy could have been structured to induce as effectively as possible the production of new and better technical controls—through government research and development programs, through subsidies to the private sector, through demanding emission standards that pushed sources to the limits of their ingenuity in developing controls.

There was some of this, but most of all there was a naive faith that polluting industries, for some unannounced reason (their sense of social responsibility, perhaps), would themselves develop the necessary technology. Programs of direct research and development by government were

---

197, 215 (emphasis in original). See also J. Esposito, *Vanishing Air* (New York: Grossman Publishers, 1970), p. 306, calling for "legislation . . . founded on the principle of reducing atmospheric contamination to the greatest extent technologically possible." The Federal Water Pollution Control Act Amendments of 1972 mimic this approach to some extent in the law's concepts of "best practicable control technology currently available" and "best available technology economically achievable." Pub. L. No. 92-500, 86 Stat. 816, codified at 33 U.S.C. §§ 1251 et seq. For critical discussion, in addition to that in the text above, see A. Kneese and C. Schultze, *Pollution, Prices, and Public Policy* (Washington, D.C.: The Brookings Institution, 1975), pp. 60-64; B. Ackerman et al., *The Uncertain Search for Environmental Quality* (New York: Free Press, 1974), pp. 319-330.

quite few and unenthusiastic, started relatively late, and aimed primarily at basic research. They were concerned more with cause and effect than with control. Government subsidies for research and development were of the same tenor. When subsidies to encourage development of control devices were discussed, they reflected a marked naiveté about the incentives of the intended recipients of the funds to put the money to good use. At one time there was under consideration "a bill to grant funds to refineries and the automobile industry to encourage research to develop smog control devices."[88] Pollution sources, of course, had little incentive to develop and employ improved (and more costly) pollution control techniques, even if government subsidized the *research* expenses. In *using* the controls, they would simply be incurring costs for themselves while conferring benefits upon the public at large. If they needed better technologies to meet stringent legal requirements, matters would be different. Since the law required control only to the extent of technical feasibility, however, and since feasibility was left largely in the hands of pollution sources, there was little if any inducement for the development of more effective controls.

Despite this unhappy combination of influences, policy did little more than insist for a considerable time that "responsibility . . . lies with the refineries and the automobile industry to develop the ways and means" of control.[89] In 1953 a member of the Los Angeles County Board of Supervisors wrote to an auto company that he "realize[d] that this problem has to be solved at the engineering level by the automobile industry, rather than by legislation."[90] A spokesman for the Los Angeles APCD said two years later that "of most importance in the entire program" of exhaust controls was "the acceptance of responsibility for the development of effective and workable automobile exhaust control devices by the automotive industry itself."[91] Two years later the APCD said that "the hope for future improvements rests squarely with the automobile industry in Detroit."[92] In 1958 the Los Angeles County counsel repeated the same view:

> One would think that the automobile industry would be gravely concerned, and launch an all-out crash program to tame the lung-searing, eye-smarting monster which its ingenuity has unwittingly created. The pitiful truth is that months and now years have passed since that industry was confronted with the indictment, and to date our automotive experts have received no substantial evidence that any such all-out program has been activated.

Despite this "pitiful truth," the county counsel persisted in his "firm opinion that the primary responsibility in controlling air contaminants is squarely upon the shoulders of those who produce it."[93]

There was, of course, little if any reason at all to suppose "that the automobile industry would be gravely concerned." Morally justifiable or not, its concern was to avoid increasing the price of vehicles through modifications that yielded no individual benefit to consumers. It had no incentive to develop control devices, for it had no reason to believe that vehicle purchasers would wish to buy them. It had no wish to see such control required, for this would increase vehicle prices and might thus reduce sales and revenue—a matter about which the industry was most sensitive. The industry did not say as much, of course. A policy of "technical feasibility," of "responsibility with the industry," was compatible with its best interests, and so the industry courted that policy. It spoke, with remarkable inaccuracy and probably a large measure of downright dishonesty, of its "competitive incentive" to find solutions to the pollution problem; it felt "a responsibility . . . to do everything we can to produce automobiles that are going to give you as pure an exhaust as we know how to make."[94]

Pollution policy virtually asked for delay, and it got it.[p] There is even evidence that what may have been a chief instrument of delay, the "cooperative" research and development agreement among the auto companies, came about partly at the urgings of policymakers. The head of the Los Angeles APCD felt it "reasonable for the auto industry to undertake a large portion of this research" on development of control devices. Earlier he had spoken of "securing the full participation of the automobile industry" in the "development of effective automobile exhaust controls through research and developmental activity." Much later, in 1964, he saw the shortcomings in such an approach, recognizing that "control of air pollution does not make cars easier to sell, it does not make them cheaper to produce, and it does not reduce comebacks on the warranty. To people interested in profits, expenditures for the development and production of exhaust control are liabilities."[95] Others may have seen the problem earlier, but little was done about it.[q] Through "technical feasibility," through the two-device requirement, through steps that frustrated

[p]In the meanwhile, shortsighted policy did its best to aggravate the problem. For example, the proposals to increase the number of freeways in order to attain traffic patterns that would result in fewer emissions per vehicle mile (see footnote d, p. 96) overlooked the fact that freeways would also increase traffic volume and encourage expansion of the metropolitan area, leading to longer average trips. In short, there could well be a net increase in the total amount of emissions. It is possible, however, that the connection between more freeways and more traffic was not understood at the time.

[q]As early as 1952, an Assembly committee recognized the central problem (though its comments also reflected the technological fixation): "Abatement and prevention of pollution is [*sic*] very costly. Even though benefits may exceed the

independent entrepreneurs and rewarded the auto industry for hiding its control techniques until the last possible moment, pollution policy encouraged a retarded development of the very technology upon which it had placed exclusive reliance. Even when, in 1959, knowledge had developed to the point that health-related air quality standards could be established, it was not these but technical feasibility that dictated the reach of control.[r] Even in 1968, after a large body of experience had taught wariness of the motivations of the auto industry, legislation required controls that extended only to the limits of feasibility—as measured by technology. True, in the Pure Air Act steps were taken to ensure independent judgments of the feasible and to set standards that would push technological progress. But this came late in the game, after years of delay, and still left policy dependent upon a technology that was becoming more and more heavily taxed with each new level of control.

The technological fix had become a persistent fixation; there was little consideration given to whether other means of control might help accomplish improvements impossible or too costly to realize through technical

---

costs, those who must expend the money are seldom directly benefited and hence tend to delay remedial measures. Successful control of pollution therefore lies in requiring general application of engineering remedies under effective governmental direction." California Assembly Committee on Air and Water Pollution, *Final Summary Report,* p. 7. Yet as late as 1955, and after the unhappy experiences in the Bay Area, the Los Angeles APCD was calling for "air-consciousness" and "voluntary cooperation"; industries were being "urged to establish voluntary air pollution abatement programs." See Los Angeles County Air Pollution Control District and American Broadcasting Company (Station KABC), "The Smog Story" (Transcript of weekly radio series, Los Angeles, Calif., 1955), pp. 58, 61; S. Griswold, "The Smog Problem in Los Angeles County," in *Proceedings of Southern California Conference on Elimination of Air Pollution* (Los Angeles, Calif., 10 Nov. 1955), pp. 2, 16.

[r]It would not, of course, have been any more sensible to protect health at any cost than to employ all feasible technology at any cost—a point overlooked by some. See the discussion at pp. 122-124. Protection of health, like technological feasibility, is a relevant but not an exclusive consideration, a point suggested in chapter 1. A central issue in air pollution control is to determine as well as one can the worthwhile degree of improvement considering all effects (not simply health effects), all available measures (not simply technological ones), and all the costs and benefits of avoiding the former and employing the latter. The closest that policy came to such an approach was the mention by the 1959 standard-setters of the need to face up to a difficult but inescapable calculation of costs and benefits—a need the standard-setters were largely forced to ignore due to legislative constraints. See the discussion at pp. 122-123. The standard-setters, as directed by the legislature, determined the standards necessary to protect health (at least where they found information sufficient to permit judgment on the matter). But as policy developed in 1960 and after, these were pursued only to the point of technical feasibility.

devices alone. This is not to say such means had not been proposed. In the mid-1950s, when the automobile was coming to be recognized as an important pollution source but when there were also no technical devices available to control vehicle exhaust, suggestions for nontechnical measures came forth—for example, car pools and increased use of rapid transit. More significantly, there was the 1959 proposal for a "smog tax." This proposal suggested a break with the prevailing tradition of regulation. As pointed out in chapter 1 (in the discussion of "Means"), regulation proceeds by mandating certain steps (installation of control equipment, cutback of operations, changes in production processes, attainment of a given emission standard) backed up by sanctions designed to induce compliance. The smog tax, as a kind of pricing system, would have gone forth quite differently. It would have levied a charge on a vehicle's pollution and left to its owner the decision whether to pay that charge and behave as always, reduce driving, tune the vehicle to reduce its emissions, install any available control devices, or mix all of these in some fashion.

Pricing systems have a number of advantages. They take into account, as regulation does not, that pollution control imposes different costs on different sources and individuals, and that measures worthwhile for some may not be for others. Thus they leave the decision about control to the source, making every effort to set the price such that sufficient control will occur in the aggregate to achieve whatever quality level policymakers have in mind. This promotes efficiency. Those who can control at little cost control most, those who can only control at high cost control least, or not at all. In the aggregate, the desired level of quality will be reached at the smallest possible control cost.[8] As to those who do control in order to avoid

[8]See generally W. Baumol, "On Taxation and the Control of Externalities," *American Economic Review* 62(1972):307. Earlier discussion mentioned that the Los Angeles APCD's enforcement policy tended (at least in its early phases) to be based only on what the least capable polluters could accomplish. One problem with such an approach, of course, is that more control could in principle have been achieved without putting sources out of business—the sort of disruption the APCD wished to avoid—by requiring *each* source to control to the extent of *its* capabilities. An Assembly committee appears to have criticized APCD policy on just this ground—the failure to set tailor-made standards for each source, based on what each could "feasibly" accomplish. See California Assembly Committee on Air and Water Pollution, *Final Summary Report,* p. 9. The criticism is probably misplaced. While the tailor-made approach bears the appearance of efficiency on its surface, the administrative costs of implementing it would probably have been so high as to more than consume any gains. Thus uniform standards were set, and—to avoid disruption—they were based only on what the least capable sources could accomplish. A pricing system could have achieved the goal of varying standards the committee had in mind; a uniform emission fee, for example, would

all or some of the pollution charge or tax, pricing can result in a whole range of subtle adjustments that accord with individual preferences and ingenuity, adjustments so diverse and unpredictable that it would be impossible to determine, mandate, and police them effectively through regulation.

Pricing also has advantages concerning technical fixes. First, it is not dependent upon their existence. As the authors of the 1959 smog tax proposal pointed out, "effective smog control need not necessarily await the perfection of special anti-smog devices." The tax would induce nontechnological steps toward reducing emissions. But second, it would also give powerful incentives to the auto industry and independent manufacturers to develop devices (or alternative, nonconventional engine technologies) the total cost of which would be lower than driving without the devices (and thus paying the tax). This was a point made in connection with the smog tax proposals of 1969. Third, pricing might be considered to have an element of fairness not present in the technological-regulatory approach. It would not require, as the latter does, a vehicle owner who drives very little to install the same, probably costly, technical controls required of the traveling salesman who logs one hundred thousand miles annually.

It is not surprising that there were no coherent proposals for pricing systems until 1959. The same forces of crisis, inertia, and uncertainty that nurtured the technological fixation encouraged the regulatory fixation as well. The sudden appearance of severe pollution problems demanded quick and confident action aimed at "wrongdoers," not talk of sophisticated measures to induce subtle alterations in the total pattern of social behavior. Pricing systems, as economists acknowledge, are ill suited to crisis situations, for it is difficult to predict accurately how, and in what time period, polluters will adjust to them. "Direct controls" (regulation), on the other hand, "may be able to *guarantee* substantial reductions" in pollution and "may be necessary to avoid a real catastrophe."[96] In addition, regulation indicated "dramatic action, which conveyed the appearance of decisive resolution of problems." It was most consistent with "more obvious, close-to-hand, shortly-accomplished action as compared with multi-factored, long-term planning and organization."[97] Pricing systems would have induced the very nontechnological responses, the changes in

---

have resulted in degrees of source control that varied with the abilities of each source. As suggested later in this section, however, the information needed to implement such a system probably did not exist in the early 1950s. And as discussed later in this chapter, such a system might not have been worthwhile at that time even if the necessary information had existed.

behavior, that it was in the interests of inertia to avoid. If not they, but technical devices, were the solution, why bother with economic incentives when the direct approach would be simply to mandate the use of technology? Conversely, if alterations in basic social conduct were not to be pursued, what need was there for a system (pricing) a basic aim of which would be to secure such change? Policymakers behaved accordingly. The decision for technology was a decision for regulation, and vice-versa; the two quite naturally went together.

In any event, there was probably insufficient information at the outset to employ a pollution pricing technique. The scientific principles behind such a system had not yet been well worked out by economists, and surely the approach was at best very poorly understood by policymakers and the public at large. Knowledge about the pollution problem itself was also no doubt insufficient to form the basis for a "smog tax." Any pricing system technique is dependent upon information concerning the causes and effects of pollution, yet there was little such information available even by the mid-1950s. Finally, and this is an observation that must await clarification later in this chapter, pricing may not have been worthwhile in the early stages of the past even had the requisite information existed. Technological controls may have held sufficient promise at that time, and the pollution problem in general may not have been sufficiently serious to justify a system that might entail higher net costs than the admittedly less efficient, but also administratively less expensive, approach of uniform regulations.[t]

By 1959 the situation may well have been different. The pollution problem had grown in breadth and intensity, and (as to the automobile in particular) there appeared to be a shortage of suitable technology. Moreover, there was the problem, now well recognized, that growth would require higher and higher degrees of control—90 percent, 95 percent, and more.[98] But still there was a blithe faith that technology would meet the new challenges: "There really is no fundamental limit to the cleaning

[t]Two other considerations, in addition to those discussed, might have contributed to a disposition in favor of regulatory and technological measures and against pricing. First, the *apparent* cost to the citizen of some sort of smog tax might have been higher than the *apparent* cost of buying a vehicle the purchase price of which included some undisclosed amount for pollution control devices. (Recall that the public seemed less inclined to reject time-of-transfer than time-of-registration device-installation requirements, probably because the former had lower apparent costs.) Second, revenues raised through a smog tax could be spent by government in geographical areas other than the areas in which the revenues were raised; there could be a redistribution away from a locality—one that would not occur with regulatory and technological measures.

efficiencies attainable. Singly and in combination, collectors are capable of up to 99.99 percent removal."[99] There seemed little regard for the fact that such control, without a significant technological breakthrough, would be exceedingly expensive;[u] that to require it would, accordingly, meet strong opposition; that, for this reason, perhaps it was time to consider nontechnological measures to be implemented through nonregulatory means. There was little apparent attention to the point that the longer such approaches were delayed, the more drastic they would have to be and thus the more difficult they would be to realize. In short, policy asked for a situation that was only likely to get worse.

And get worse it did. By 1969 the pollution problem was regarded as sufficiently serious, and the shortcomings of sole reliance on technological and regulatory solutions sufficiently plain, that there were proposals for dramatic improvements to be achieved in part through nontechnological controls on individual behavior. In 1970 such proposals took the shape of federal law. The debacle that followed owed a good deal to the fixations of past policy.

Pollution Policy and the Production of Knowledge

There was considerable mystery as to the nature and causes of the smog conditions that first became chronic in Southern California in the 1940s. Given this, it is understandable that public resources would be allocated to study the troublesome new problem. Such systematic production of knowledge was important inasmuch as private study would likely be insufficient of itself, largely for reasons sketched earlier in this chapter. Insights into the problem that might help bring about its cure would be collective goods nearly as much as cure itself, and this suggests they would be underproduced in the absence of government support.

There were private studies, of course—by the Stanford Research Institute, by the automotive industry, by the Air Pollution Foundation, by independent scientists. SRI's work was backed by a well-funded trade association, the Western Oil and Gas Association. The association was, no doubt, very much interested in a private benefit (evidence that its members played a relatively small role in the pollution problem). This, and the fact that it represented a tightly organized group, would tend to meliorate typical collective goods effects. Yet there was a problem on the other side, namely the danger that the findings and disclosures of a group such as SRI would be biased in favor of the interests on whose behalf it

[u]See the discussion at pp. 24–27.

worked. There was little good reason to suppose that SRI would suggest refineries were in fact the major problem and that action should be taken against them immediately. And actually, of course, SRI's research led it to the conclusion that many sources were responsible, that refineries did not appear to be the major cause, and that more work was necessary before taking action. Whether it was bias or good science that yielded this conclusion is beside the point; what mattered was that while groups like the Western Oil and Gas Association could be expected to support research, it would be unreasonable to place exclusive reliance on such efforts. All of the same can be said with respect to the research program of the Automobile Manufacturers Association, where again the evidence tends to support a suspicious attitude.

The Air Pollution Foundation was a different story. Here was a nongovernmental research group that tried very hard to maintain independence and that advanced knowledge a good deal. APF was supported by a number of contributors, including a wide array of industrial interests (which perhaps gave funds as evidence of a conscientious desire to realize a solution to the problem). APF also enjoyed some public support, both directly and by way of the tax advantages given to not-for-profit organizations.

Whether because policymakers understood the need or not, there were in addition to the private (and partially public) research programs noted above a number of direct public efforts to understand the pollution problem and what to do about it. To be sure, some quarters (such as the Los Angeles APCD) seemed to find the research role distasteful for a time, and others (such as the Beckman Committee) appeared to downplay the importance of understanding health effects. The fact remains, however, that government research became more and more a commonplace after 1940, with even the APCD taking a quite active role (by employing Haagen-Smit as a consultant beginning in 1949–1950). There was little study, public or private, prior to 1940, especially at the state and local levels. Perhaps this was because solutions to the relatively minor problems of the early years were regarded as so obvious as not to require study; perhaps it was because the conditions themselves were not regarded as sufficiently serious to justify expenditures on research.

Concerning the state and local levels, the 1943 episodes in Los Angeles stimulated a series of studies and investigations into the causes of the mysterious new pollution problem, and research efforts were stepped up after the 1946 episodes. By 1949 the problem appeared less tractable than had at first been believed, and the Los Angeles APCD started its program. In 1954, again largely in response to severe pollution conditions, Governor

Knight began his "all-out smog war," which consisted primarily of a multifaceted research program. A year later, legislation provided for ongoing studies by the Department of Public Health, and from that time on the state was quite heavily involved in research. The story was much the same at the federal level. The national government had entered into pollution studies relatively early (in 1912), but these were sparse and narrow efforts. Extensive research did not begin until the late 1940s, and from then until 1963 virtually the sole federal role had to do with advancing understanding, as opposed to undertaking control. As to each level of government, these measures were consistent with the stimulating effect of crises, and with the least-step approach to making pollution policy.

The efforts described above were necessary and useful. They reflected policy consciously working to produce systematic knowledge. What is more interesting, however, is the manner in which policy unconsciously produced *unsystematic* knowledge. It did so by a process of gradually exposing, layer by layer, inappropriate or insufficient responses to the pollution problem, at each stage arriving at a better understanding of what to do next. By the very steps it took, policy tended to ensure the existence of conditions that would themselves suggest the shortcomings of past steps and the directions for future ones. This process of "exfoliation"[v] was a product of all the considerations examined earlier in this chapter—inertia, uncertainty, crisis, fixations, and the whole pattern of policy-by-least-steps down the path of least resistance itself. It was a process that helped to right, though very slowly, the very problems those considerations created. The knowledge it produced often worked in conjunction with study; at times it was the primary stimulus to systematic research. But on other occasions it yielded important information on its own. We must not exaggerate the influence of exfoliation as a way policy produced knowledge, but we should not ignore it either.

Consider a few examples of the process both to show it at work in various contexts and to suggest some of the factors that shaped it from time to time. We said above that on occasion exfoliation produced information on its own. The best example, of course, has to do with the great pollution episodes of the fall of 1943. The initial reaction to those events was to lay

[v]We avoid the familiar phrase "trial and error" because it connotes an open acknowledgment of not knowing what to do and a consciously systematic experimental approach to finding out. Prior to 1970, policymakers generally believed they knew what to do (even if it was just to "find out" through *research*), and in any event almost never regarded their actions (other than those aimed at research) as experiments designed to produce knowledge. Yet those actions, by stripping away layers of misconceptions, uncertainties, and so forth—hence the label "exfoliation"—produced more knowledge than did research, or at least as much.

blame entirely on a single synthetic rubber plant. While this might seem foolish today, it was not entirely irrational at the time. There was then an almost total absence of knowledge regarding the "peculiar nature" of the pollution conditions that had so suddenly appeared; they presented a "completely new problem."[100] At the same time, the rubber plant did (by virtue of heavy visible emissions) appear a plausible enough cause, and in any event citizens—who were up in arms over the problem—regarded it as *the* source. Given ignorance, the suggestions of limited evidence, a situation of crisis, and public pressure, what could one expect of officials other than to attack this "obvious" problem? Just such an approach, it should be recalled, had seemed to work in the past. In the years before 1940, policy had relied upon events to describe problems, and then it reacted accordingly. As outbreaks suggested the importance of particular sources—orchard heaters, for example—they were controlled; more generally, the focus of control efforts was on obvious sources and their visible emissions. The same approach did not work with respect to the rubber factory, of course; it was controlled and the smog conditions persisted.[101] Yet the foray on the plant, which ended up being little more than a convenient scapegoat,[w] did quickly produce a useful piece of information that might otherwise have taken considerable time to acquire: the problem was more complicated than had at first appeared, and other sources shared the blame.

The attack against the rubber plant was not the only response to the 1943 crisis. Officials also undertook a program of studies that built with time and eventually provided important information both as to the makeup of the pollution problem and the relative importance of various sources. Exfoliation proved to play a strong supporting role for those studies, however, both by encouraging their direction and corroborating their implications. After failing to abate pollution by controlling a single factory, officials turned their efforts to other stationary sources. Control focused first on visible smoke and sulfur emissions. These had been a problem earlier in the East, and it was presumed they were the problem in the West.[x] Never mind that East Coast conditions arose largely from the burn-

[w]As we shall see very shortly, public pressure to attack scapegoats was to occur again as a part of the exfoliation process; it has also influenced pollution policy in other parts of the country. See, e.g., L. Goldner, "Air-Pollution Control in the Metropolitan Boston Area: A Case Study in Public Policy Formation," in *The Economics of Air Pollution,* ed. H. Wolozin (New York: W.W. Norton and Co., 1966), pp. 127, 144, 150. The Goldner study generally reveals a story quite similar to our own on a number of points.

[x]Thus the Los Angeles County counsel, speaking with reference to the local ordinances of 1932–1945, pointed out that they were designed to "regulate both visible

ing of soft coal, and that Los Angeles industries had not "burned soft coal for fifty years" (they employed primarily natural gas and fuel oil).[102] The fact remained that the smoke emissions were obviously there. "Regardless of their total effects on the entire smog blanket, the public itself knew of the dirt, the smudge, eye-irritation, and even damage to crops and ornamental vegetation that those open sores of pollution were creating in localized areas. And so to those sources the control program directed its first attack."[103] The attack was fruitless, of course, at least in eliminating the new smog problem. "Black smoke and soot never were the real problem in this area. Removing them *makes it more apparent* that there is a serious pollution menace from invisible . . . vapors, sulphur dioxide, and other pollutants in the atmosphere."[104]

Policy, by attacking the obvious, had stripped away the rubber plant as sole cause, and next stripped away smoke emissions generally. Subsequently, it pursued sulfur, the next most obvious problem. Measurements in Los Angeles shortly after passage of the 1947 act had revealed high concentrations of this pollutant. "It had been a very bad actor during eastern smog"; "it was only logical that [it] should be suspect in the industrial economy that had burgeoned in Southern California."[105] But the logic did not work. The Los Angeles APCD began a relentless attack on sulfur oxide emissions. The agency's efforts were so successful that by the early 1950s it could be concluded that "the amount of sulphur dioxide is only a fraction of that which occurs in other cities which do not suffer from smog. . . . It would appear that the only conclusion which can be drawn is that sulphur dioxide, as a pollutant in the general atmosphere, is one of the least important contaminants in the Los Angeles smog."[106]

It was through exfoliation, not study, that policy produced this important conclusion, and with several significant consequences. First, "the time finally came . . . when some people realized that we really did have a tough problem on our hands—a problem different from any place else in the country. They decided we had better do some good, solid research to

smoke and fumes and are similar to those in eastern cities." H. Kennedy, *The History, Legal and Administrative Aspects of Air Pollution Control in Los Angeles County* (Los Angeles County Board of Supervisors, 1954), p. 7. See also A. Will, "Abatement Progress under Existing Air Pollution Laws" (Address to the Statewide Conference on Air Pollution Legislation, California Chamber of Commerce, San Francisco, Calif., 21 Feb. 1955), p. 10 (after passage of the 1947 act, Los Angeles County "officially launched its abatement program against obvious sources of smoke and fumes"); Editorial, "Smog Problem Put under Microscope," *Los Angeles Mirror,* 16 July 1954 ("Smog control here began with the assumption that air pollution in Los Angeles was similar to that of other cities where it was due largely to sulphur and smoke").

find out just what was going on and what really were the objectionable components of the Los Angeles smog."[107] Most notably, the Los Angeles APCD employed Haagen-Smit, and his research quickly indicated the importance of a photochemical reaction involving hydrocarbons, nitrogen oxides, and sunlight. Although Haagen-Smit's early work in fact did little more than confirm findings that had been reached as early as 1947, there was progress nevertheless. The APCD had ignored those earlier findings, but its own policies had now created conditions that made such an attitude impossible—they had eliminated other explanations. So the APCD broadened its horizons to include hydrocarbons.

This was the second significant consequence, and it provided the final link in the process by which exfoliation helped ultimately to establish beyond doubt that the motor vehicle was the major source of pollution in the Los Angeles area. Consistent with Haagen-Smit's findings, the APCD began to direct attention to hydrocarbon emissions, but from stationary not vehicular sources. This came about despite the fact that Haagen-Smit's work (especially when taken in conjunction with the SRI rebuttals) implicitly pointed to vehicles as an important contributor to the pollution problem. There are a number of considerations that might explain the APCD's continued primary concern with stationary sources. Perhaps it regarded its legal authority as insufficient to deal with motor vehicles; it is true that the 1947 legislation provided few tools as to the latter, and that there were later some moves by the county for better laws in this regard. Perhaps it regarded stationary sources as by far the major source; this may well have been a sound conclusion earlier, though growth in the number of motor vehicles after 1945 suggested it was a suspect judgment by 1950 (see table 7 in chapter 7). Or perhaps (and this may be the most plausible conjecture) the APCD had simply become comfortable in dealing with stationary sources and had little taste for taking on the public and its driving habits.[y] When it had the temerity to suggest that public activities like driving and backyard burning might require control, citizens had "screamed," pointing to industrial sources (the scapegoats) as the cause. Thus the "seeming reluctance" of Los Angeles officials to go after the auto may have been because it "did not present an inviting field for police-type enforcement activity."[108]

[y]And at least felt it knew how to handle stationary sources, whether or not they were the primary problem. See L. Tribe, *Channeling Technology through Law* (Chicago: Bracton Press, Ltd., 1973), p. 95: "The famous story of the poor soul who looked for his lost key not where he had dropped it but beneath a lamppost dozens of yards away 'because that's where the light was' illustrates a nearly universal tendency among human beings to work with the tools they think they know how to use best—even if those tools do not happen to fit the task at hand."

In any event, the reasons matter not so much as the fact. The APCD along with state and local policymakers generally were unable to muster the will to bring the auto under control. This might have been due to the great uncertainty surrounding the issue. Some scientists seemed convinced of the vehicle's role, but others found only confusion. The auto industry capitalized on the latter, consistently claiming insufficient proof. Yet the passivity of the policy that resulted eventually provided the best proof of all, proof that simply would not yield to scientific attack or industry doubt-mongering. Smog conditions persisted—indeed, they grew worse—yet every conceivable source other than the automobile had been brought under control. The conclusion was obvious. "The persistence of the smog problem in Los Angeles, even after the elimination of major stationary sources of air pollution, left the motor vehicle as the major probable source capable of producing a problem of that magnitude."[109] The conclusion was so obvious that even the industry relented. A former head of the Los Angeles APCD summed matters up in 1964:

> When I started talking to the automobile industry 10 years ago, at first they weren't to blame. Everybody knew that the contaminants were immediately dispersed vertically and horizontally. When we had eliminated every other possible source of air pollution in Los Angeles, particularly during the summer months when we could get natural gas for our steam power plants, they admitted maybe they were to blame.[110]

The auto companies caved in (though not without qualification) by 1960. It had taken fifteen years of exfoliation to demonstrate what science had suspected in 1943-1947, confirmed in 1950, and considered conclusive in 1957.

The exfoliation process provided one means by which policy tended to make its own best case against prevailing conceptions of what policy should be about. And it made cases convincingly, putting matters beyond the reach of bickering scientists and the claims of vested interests that their responsibility had not been established. Like crises, but in much less dramatic fashion, exfoliation helped overcome the burdens of proof and inertia in ways that the finest pieces of research could not. Like crises, it became both the curse and the redeemer of policy that was, sometimes for understandable reasons but often for foolish ones, shortsighted.

The examples of the exfoliation process sketched above largely involved scientific and technical issues. There are, however, other illustrations that relate more to matters of control philosophy and technique. The approaches of voluntarism and technological feasibility had obvious shortcomings, for example, yet it appears policymakers were largely blind to these (or seduced by claims the approaches would work) until events bore them out—until, that is, layers of confusion and misunderstanding (or pretense) were stripped away by the failures of ill-conceived efforts. Only then was there an awakening, for instance, to the fact that the auto industry had little or no incentive to develop and market control technology.

A clearer example in this context probably can be found in the trend toward control at higher levels of government—from local to state, from state to federal. It required no great expertise (whether of political philosophy or otherwise) to realize that local and state boundaries were too confined to make it likely there would be made within them sound judgments regarding those aspects of the pollution problem that reached outside their range. Yet the move to a higher authority needed to counter this shortcoming was a costly and sensitive step that required strong justification. Early on there was the hope (though it was hardly plausible) that pollution would remain a localized phenomenon. It had to be clear this was not the case before state authorities were willing to expend material resources by undertaking control, or political capital by confronting home rule; or before federal authorities were willing to use federal funds and at the risk of interfering with states' rights; or before first local and then state government would tolerate intervention. From the point of view of all interests, a clear showing of need for a higher level of control was necessary. It was provided by a spreading problem that occurred in part by virtue of the very localism it helped to overcome.

This is not to suggest that exfoliation in this case provided a totally valuable service. Control at higher levels meant almost inevitably that there would be the uniformity that tends to go hand-in-hand with centralization, the result being policy that might take insufficient account of conditions varying from county to county, or from state to state. This problem became more serious as federalization marched forward. The states (at least if California is typical) appeared quite sensitive to local desires and situations, and balanced them with the need for a final central voice in making policy. Thus California exempted rural counties from some vehicular controls, and provided (at least in principle) for varying air quality standards.[z] Even if there were to be substantial uniformity, this would not be so serious at the state level inasmuch as anything smaller than an entire state might well be of suboptimal size for purposes of making air pollution policy.

[z]The ARB had authority to establish air quality standards that varied basin by basin, but it appears the general practice was to establish uniform standards. See California Air Resources Board, *1969 Annual Report* (1970), pp. 29-30.

The same can hardly be said of the entire United States, yet there was the danger that the federal government—once it intervened by way of control—would be less sensitive to variations in local wants and conditions, if only because Congress is more distant from them and less subject to local pressures. For a time this danger was avoided; states, for example, were permitted to set their own air quality standards subject to federal approval. But with the federal preemption of new vehicle standards (with the possibility of waivers for California), and especially with the federal air quality standards in the Clean Air Amendments of 1970, the danger of national uniformity materialized in most concrete terms. Because these measures represent central parts of present pollution policy, their appraisal is saved for the next chapter. Suffice it to say here that there are serious questions about whether they are justifiable in terms either of efficiency or fairness—about whether, that is, they are sensible at all.

The process of exfoliation, then, did not always lead to unmistakable improvements in policy. Generally, however, it helped, perhaps more than any other factors, certainly in conjunction with them. The policy process could fall into error—the product sometimes of uncertainty, sometimes of plain foolishness, sometimes of political considerations—but it had the capacity to benefit from mistakes.

The Notion of Lag

We should not conclude a review of pollution control efforts prior to 1970 without a few words on the notion of "lag"—of presumably unwarranted delay between a problem's recognition and its resolution. There has been a good deal of frustration, and it continues, with the slow response of policy to a situation that began in the early 1940s and persists today. As one observer put the point, "It is one of the great ironies of the twentieth century that we should be threatened by a cloud of pollution created by our own technological progress."[111]

Of course, it is not ironic at all, but exactly what one would expect. Both technical and institutional considerations ensure that a problem like pollution will be resolved only gradually, if ever at all.[aa] Take the years in question in this study. On the technical side, there was at the outset so

[aa]For one brief summary (in addition to that we will present) of the considerations that produced delay in California policy, see A. Carlin and G. Kocher, *Environmental Problems: Their Causes, Cures, and Evolution, Using Southern California Smog as an Example* (Rand Corp., Santa Monica, Calif., Report Number R-640-CC/RC, 1971), pp. 63-93.

much uncertainty about what to do that one could hardly expect the correct thing (judging by hindsight) to be done. Research and the process of exfoliation were necessary (they still are) to clarify the issues and suggest ways to deal with them, and all of this took considerable time. In the early days there were no satisfactory instruments even to measure and study the problem, much less bring it under immediate and total control.[112] When this deficiency was remedied and studies undertaken, findings still had to be interpreted and their ambiguities dealt with and studied further. After understanding (or the illusion of it) came response, but here too technical factors caused delay. As to many sources, automotive in particular, no known methods of control existed.[113] Even once conceived, they could not be made immediately available; the Air Pollution Foundation estimated that an average of seven years was required to go from laboratory development to commercial development.[114] Once available, devices could not be immediately installed; we saw the estimates of the years required to put used-vehicle controls into effect in California.[bb] And this entire process had to begin more or less afresh for each new pollution problem that appeared, for each untried means of control, for each new legal requirement of that control.

Turning to institutional considerations, the very structure of pollution problems promised something less than rapid resolution; there were strong incentives to pollute and weak ones at best to achieve control. Policy, of course, aggravated these unhappy facts. By its inertia, its allocation of the burden of uncertainty, its early reliance on voluntarism, its approach of technological feasibility, it invited delay in many ways. But perhaps it is too easy to condemn, in the context of what we know now, each aspect of policy made in the past. There were reasons—not always good ones, but reasons nevertheless—for what came about. Granted that more research earlier would have been useful,[115] but consider how nice it would have been had research proved unnecessary after all. Policy simply waited to be convinced of a need, and then the pace of research quickened. Granted, too, that technological feasibility meant "vehicle manufacturers had nothing to lose and some costs to save by waiting until certification was imminent and then developing and using their own less expensive, factory installed controls."[116] Yet to go beyond technological feasibility appeared, under all the circumstances, to be unwarranted, and might in any event have only brought more delay as powerful interests tried to fight off stringent requirements—a danger that critics of the technological-feasibility standard themselves recognize[117] (and one that may be said to have materialized after 1970). Beyond these points, the manufacturers were making

[bb]See the discussion in footnote m, p. 148.

promising noises about their earnest efforts. When they were clearly found to be doing nothing more than that, policy tightened.

As two final examples, it is true enough that such market surrogates as emission fees might have been useful in the past, or again that higher levels of government were slow to intervene.[118] But as to the first, the knowledge needed to put pricing systems into effect did not exist for a considerable time, and even if it had the situation may nevertheless have been such that pricing would be inappropriate. As to the second example, legitimately heavy interests were at stake, and thus the move to a higher level had to await clear justification.

One should hardly hasten to bless thirty years of policy-by-least-steps, but those years should not be too roundly condemned either. While policy reflected many errors, its development was not totally irrational. In general, its approach was to protect and maintain the prevailing situation so far as possible, and not without some reason. Unjustified change can be as costly as delay in undertaking needed alterations, an observation that finds some support in developments since 1970. Complaints of lag either overlook the reasons for delay or assume they must always be bad ones. Moreover, they imply the existence of an idealized solution that must be both clear and satisfactory to all, when in fact there is none. If policy in the past had a central fault, it was in proceeding by "present-minded pragmatism rather than long-term rational planning."[119] But this observation only brings us back to where we started, for pragmatism too had its sense. In any event, the irony is not in delay. Given the structure of the problem and the manner in which it was handled, the irony, if any, lies in the fact that policy moved so far and quickly as it had by the close of the 1960s.

THE YEARS SINCE 1970

Comparing the main currents of present policy with those of the years up to 1970, we can see both similarities and differences, continuity and change. For example, the steady march in the direction of federalization, a trend clearly evident in the earlier years, continues into the present. Yet now there is a difference in the kind as well as the degree of federal control. Whereas past policy reflected a sort of "cooperative federalism" consisting in some national but also considerable state authority, that of the present underscores "federal" and, as one might expect, is distinctly uncooperative. Pollution policy is national policy, and the states are little more than reluctant minions mandated to do the dirty work—to implement federal directives often distasteful at the local level. (The fate of the federal mandate, at least as to air pollution, is in some jeopardy as of this writing—

thanks to the federal court decisions discussed in chapter 13.) This is the case not simply with respect to air pollution policy but as to water quality control as well; the Federal Water Pollution Control Act Amendments of 1972 have resulted in demanding new uniform federal standards.[120] Beyond this, and as a sign of the times, there has been consideration of "measures that would increase Federal preemption of tighter state standards" (not surprisingly, the states and conservationists are opposed to such preemption, while industry tends to favor it).[121] The fact that federal policy of today is simply the culmination of a slow but steady trend that began years ago should not obscure the essential difference between old policy and new, between federalism and federalization.

Mention of slow-but-steady signals another point of similarity and difference. Policy in the past proceeded by a process of least steps down the path of least resistance; this is the way it went "forward" (in the direction of stricter control). The policy conceived in the Clean Air Amendments of 1970 went "forward" too, but hardly by least steps, hardly by a tentative reaching out of one federal foot in a halting search for the route that offered least opposition. Rather, there was a dramatic plunge forward, a devil-may-care foray into uncharted and presumably hostile territory. The sovereign states so honored (less and less with time, but still honored) in the past were suddenly stripped of their royal robes; they became commoners subject to enlistment in the war to save the environment. The forces of inertia that had in the past protected the status quo were overcome. In decreeing its new pollution policy, Congress conceded that the stringent federal requirements might well "impose severe hardship on municipalities and States,"[122] but this was considered the necessary and worthwhile price to be paid for environmental improvement. The heavy burden of uncertainty before allocated to those who sought greater control was thrown upon those who resisted it.[cc] Doubts were resolved in favor of health and environment, and states, industries, and the auto companies had to establish that they could not accomplish what Congress asked.

Why did it all happen? We can at least offer a few speculations. The process of exfoliation was at work: events themselves revealed that states were defaulting on their responsibilities (at least to the congressional mind,

[cc]A general trend in this direction appears to be developing. There have been efforts, thus far unsuccessful, to enact environmental legislation providing that once plaintiffs could show that a defendant's polluting activities created a reasonably possible risk to public health, the burden would be on the defendant "to prove that no threat to public health exists, that the risk of any such threat is negligible, or that the physical and economic considerations outweigh the public health threats." Bureau of National Affairs, *Environment Reporter, Current Developments,* Vol. 5, No. 30, 22 Nov. 1975, pp. 1179–1180.

which appears not to have entertained the notion that the states and their citizens might not *want* faster progress; but Congress's judgment in this regard was not entirely unjustified, given the biases against effective collective expression of preferences for environmental improvements, biases aggravated by the forces of inertia and by allocation of the burden of uncertainty). Events also suggested that the auto companies were taking advantage of the abundant opportunities presented by the approach of technological feasibility, and that the approach was insufficient to meet the challenge posed by new evidence of a growing problem and one of more severity than had been thought (as shown especially by new findings on health effects). By virtue of recent exposés (the alleged antitrust conspiracy settled by consent decree, the Nader report on air pollution), both the auto companies and Senator Muskie had lost public confidence and political capital, and thus it was only natural for the senator to attack the industry.

And then there was the great environmental crisis; it most of all is a mystery. That, like crises in the past, this one contributed to further intervention seems unquestionable, but there similarity ends. Crises in the past concerned particular pollution episodes and generated, if anything, rather focused response. But that was not the case with the great environmental crisis of 1969–1970. Its roots were not in any specific cause, and it spawned not narrow response but an entire movement (or the movement spawned it; the connection is hardly clear)—a general "environmental consciousness." Some observers appear to attribute the new crisis to the same forces largely behind the old: growth in population, production, consumption, affluence, and pollution.[123] Surely these factors played a part, but more must also have been at work. For this crisis had no episode to trigger it (unless one lays it all on a few oil spills) and, in any event, if growth factors were the cause, why did not the crisis, and the movement, occur sooner—or later? We cannot answer that question; we do not know what "more" was at work. We only know that this new crisis was different. Those of the past were crises in action; they merely signified that government should do more of the same, take that next small step in policy. The great environmental crisis was rather a crisis in thought; it suddenly occurred to many people that perhaps government did not know what it was, or should be, doing. In this sense, the environmental crisis may simply have been part of a larger whole, a general "loss of citizen confidence in governmental institutions . . . generating a crisis of its own."[dd] Whether the environmental crisis arose from this larger sense of unrest, or from efforts of crusaders

[dd]M. Willrich, "The Energy-Environment Conflict: Siting Electric Power Facilities," *Virginia Law Review* 58(1972):257, 319–320. The idea is an interesting one,

finally bearing fruit, or from growth, or the need for a new issue now that the "urban crisis" had quietly died, or from all or none of these is an open question. But what happened happened, and—like the different crises of the past—it stimulated pronounced citizen reaction[ee] and new governmental intervention.

Very new intervention indeed. Yet much of the large new reach of federal policy was, if one pauses to consider the matter, more or less inevitable. If there were to be new federal initiatives, they could not in some respects have taken a shape much different than they did. Previous policy had filled up so much of the area once left to state authority that any further step would occupy virtually all that remained, as of course it did. But there was more to it than this; there was policy that seemed indeed quite new and unexpected, that amounted to discontinuities rather than simply the carrying on of an established trend. There was a sudden and striking willingness to engage inertia and uncertainty and turn them around. And as part of this, there was an apparent weakening of the old faith in technology and regulation, the fixations of policy in earlier years. One need not look too far for evidence of that loss of faith. The constraint of technological feasibility was consciously and explicitly abandoned in the Clean Air Amendments, and there was an awareness in the legislation that nontechnological steps—controls on patterns of behavior, changes in the comfortable status quo—would be needed in some areas, disruptive or not. And, to implement these steps, consideration was to be given to pricing systems.[ff]

The new federal policy was not the only sign of a change of heart about the use of market forces as a mode of government intervention. Increasing concern about the proper use of air, water, and other natural resources generally has been accompanied by increasing interest in emission fees and other pricing or market mechanisms; the correlation is underscored by the fact that writers commenting on the first phenomenon also note the second in the same breath.[124] Some notion of just how revolutionary the change in attitude has been can be gathered from events. The discussion in chapter 11 mentioned a 1969 news article in the *New York*

---

and there is some evidence for it. The so-called environmental crisis (and movement) erupted in the midst of heightened citizen disenchantment with government decision making and persistent insistence by citizens that they be permitted to participate more actively in decision processes—especially those of administrative agencies. See, e.g., Office of Communication of United Church of Christ v. Federal Communications Commission, 359 F.2d 994 (D.C. Cir. 1966); Sierra Club v. Morton, 405 U.S. 727 (1972).

[ee]For a time. See footnote j, p. 273.

[ff]See, e.g., the discussion at pp. 208-209.

Times that told of "objections" to Senator Proxmire's proposal for an effluent tax to control water pollution. The story listed conservationists' complaints about the measure, and implied that the tax would be nothing more than a license to pollute. But not much later, in the midst of the environmental crisis, "several politically important conservation groups, which previously had opposed the charges approach . . . stepped solidly behind the effort to levy pollution charges or taxes." Indeed, some of them joined together to form the Coalition to Tax Pollution.[125] States began to consider effluent charges for water quality control, the Nixon administration proposed a sulfur tax, the Council on Environmental Quality supported emission charges. The trend continues today, regarding both policymakers and conservationists.[126]

There are a number of factors that might have contributed to this new interest in pricing systems. Among conservationists, there was dissatisfaction with prevailing modes of government intervention; the conservationists' "change in position [with regard to pricing systems] result[ed] mostly from their conclusion that the conventional enforcement-subsidy strategy [was] not working."[127] Many economists shared that feeling, favoring "abandoning regulation in favor of the tax because regulation has 'been tried and it has failed.' "[128] A number of government officials were beginning to see "pollution taxes as an idea whose time is coming."[129] Several specific factors that were related to the general sense of dissatisfaction with the old regime must also have played a part. Policymakers still placed heavy (though no longer exclusive) hopes on technology to remedy environmental ills, but they were also becoming aware that the regulatory approach provided very weak incentives for technological advance.[130] Moreover, there was a new congressional commitment to go beyond the reach of technology and to induce changes in patterns of behavior in order to achieve environmental improvement. Pricing systems have considerable advantages with respect to both points—encouraging technological advance; stimulating a myriad of subtle and efficient behavioral adjustments in accord with individual situations and preferences.[gg]

[gg]See the discussion at pp. 35-36. The decision to go beyond technology can be viewed in terms other than simply the failures of the technological fixation. By the end of the 1960s pollution costs were seen as growing in both absolute and relative terms. (That is, there was evidence of a spreading problem and of significant impacts on health and the environment generally—the absolute; and there were heightened demands, probably brought on in large part by increased affluence, for environmental amenities—the relative.) No doubt this growth, which promised to proceed more or less exponentially, helped stimulate the environmental crisis and its movement. But it may also have stimulated the interest in nontechnological solutions. For just as pollution costs were quickly growing, so too were the costs of avoiding

In addition to the foregoing, there had been significant developments in the information base necessary to implement pricing systems. We noted earlier in this chapter that it was understandable there would be little interest in such systems throughout much of the period prior to 1970. The reasons were based in part on the lack of scientific understanding and concrete operational principles needed to put pricing into effect. By the end of the 1960s that dual void in knowledge had been filled considerably; the causes and effects of pollution were far better understood, and an enormous literature and some experience with pricing had developed.[hh] Finally, another change might also have occurred. In the past, pricing may not have been worthwhile even had sufficient information been in hand; now the opposite could well be the case. This observation was mentioned earlier; it is time to expand on it.

The regulatory approach to pollution problems characteristically proceeds by establishing an ambient standard of quality and a set of emission limitations designed to meet that standard. The important point to note here is that in a regulatory program the emission limitations are virtually always *uniform* for all pollutant sources, or at least for all in a particular class of sources. The uniformity is *allocatively* inefficient—it ignores the fact that different pollution sources have different marginal costs of control, that some can control more cheaply than others, and that there would be net savings to society if some plants were required to reduce emissions to a greater and others to a lesser degree.[131] This allocative inefficiency may, however, be counterbalanced by the savings that result from the fact that uniform emission standards, because they do not differentiate among sources, require relatively little information to implement. They may, in other words, be *administratively* efficient.

One could avoid the allocative inefficiency of uniform standards by establishing emission limitations varying with the marginal control costs

---

them through technological fixes (a factor that might have contributed to the sense of crisis—of a problem running amuck). High degrees of technological control had already been imposed by the close of the 1960s, and it was well understood that further such controls would be exceedingly expensive in the absence of some significant technological breakthroughs. Thus, there may have been an awareness that nontechnological solutions, though costly in terms of disruption and inconvenience, might nevertheless be cheaper than further technological controls and also worthwhile in the benefits they could achieve. See generally the discussion at pp. 25–27.

[hh]The first point is obvious. On the second, the body of literature is too large to cite here. For a small sampling, see the references in J. Krier, "The Pollution Problem and Legal Institutions: A Conceptual Overview," *U.C.L.A. Law Review* 18 (1971):429, 467–475.

of each source. But "to do this would require a fantastic amount of information that in practice would be very difficult and expensive to get."[ii] Much time and effort would be required of the regulator to gather the data needed to formulate and implement variable standards—if the data could be gotten at all! There is little reason to suppose that factory managers have much explicit information about marginal control costs. Moreover, they would have a strong incentive to overstate those costs, for the higher the costs, the less stringent the required degree of control. In short, the information costs associated with variable standards would be enormous. Gains in allocative efficiency would be eaten up by the costs of administrative inefficiency incurred in achieving them. Within the regulatory method, then, there tends to be a tension between allocative and administrative efficiency.

Pricing systems might achieve efficient varying emission outputs from different sources while at the same time avoiding the high administrative costs associated with varying emission regulations. A properly set uniform emission fee that charged a given sum for each unit (say a ton) of pollutant emitted, for example, could achieve a collectively established ambient standard at the least total control cost to society. The reasons for this are relatively simple. Different pollution sources have different marginal costs of control. (That is, it costs some sources more to reduce pollution output by one ton than it costs others.) Assuming that each source wishes to minimize its total costs under the emission fee system, each would blend control expenditures (to reduce its emissions and thus its emission fee liability) and emission fee payments (on those units of pollutants emitted after control) in the way it finds cheapest for it. Sources for whom control is relatively inexpensive would control more and pay a smaller fee; sources for whom control is more expensive would likely behave in the contrary fashion. Each source, however, would try to minimize its costs under the constraint of the emission fee. There would be variations in emission outputs among sources without the need to set a separate standard for each source. If the emission fee were correctly set, the aggregate emissions would result in the established ambient air quality level. More important, this would be accomplished at the least total control cost, simply because the total control cost would be the sum of all the least-cost steps taken by the different pollution sources. The result would be that wanted to begin

[ii] J. Dales, *Pollution, Property and Prices* (Toronto: University of Toronto Press, 1968), p. 85. Hence our earlier comments on the criticism that a California Assembly committee leveled at the Los Angeles APCD's uniform emission standards. See footnote s, p. 284.

with—sources with low control costs controlling more, those with high costs, less.[132]

There would be substantial administrative costs associated with choosing and implementing the proper emission fees,[133] but they would probably be less than the costs of devising and employing a system of varying emission regulations equal to a program of emission fees in allocative efficiency. Even if the costs of obtaining information from pollution sources were the same under both approaches, total information costs of the pricing (emission fees) system would appear to be lower. Information under either must flow in two directions, from sources to government and subsequently from government to sources as well. But with varying regulations, a different piece of information must be determined and communicated to each source. With a uniform emission fee, on the other hand, one piece of information does for all. Similarly, if—as will often be the case—adjustments are desired (whether because experience reveals the ambient standard is not being achieved, or because there has been a decision to alter that standard), only one variable, rather than an infinite number of them, need be changed.[jj]

The upshot of these points is a hypothesis like this: The government initially intervened to allocate the air resource because breakdowns in the system of incentives led to undue pollution. While the resource was sufficiently valuable to justify intervention, its value was small enough (the problem was serious, but not *that* serious) that the gains from an efficient allocation by the government would not be worth the information or administrative costs associated with achieving them. So uniform regulations, which, thanks to their crudeness, entail little in the way of such costs, were used. Over time, however, increasing scarcity (and thus value) of the

[jj]Emission fees, like regulatory standards, would have to be adjusted to take account of increases in the number of sources; they would also have to be adjusted to keep up with increases in general price levels. A system of pollution permits could avoid the need for such adjustments, as well as several other shortcomings of emission fees (the possibility that sources with strong market power could simply pay the fees and pass them on to consumers; the possibility that emission fees could produce widely wrong results where right results are very important, as with certain pollutants that are hazardous in even small amounts). Under a system of pollution permits, "the total amount of permissible pollution is determined by the government and an equivalent number of rights to pollute are then auctioned off to the highest bidders." M. Roberts and R. Stewart, "Book Review," *Harvard Law Review* 88(1975):1644, 1653-1656. See generally W. Montgomery, "Markets in Licenses and Efficient Pollution Control Programs," *Journal of Economic Theory* 5(1972): 395; Dales, *Pollution, Property and Prices.* We consider in chapter 15 the extent to which the EPA might usefully have employed an analog of the permit system as an alternative means to implement its gasoline rationing plan.

resource in both absolute and relative terms made efficiency gains more worthwhile, while associated administrative costs remained more or less fixed. When value increased sufficiently, one might then expect to see some shift toward pricing systems in order to realize those efficiency gains, because the higher administrative costs associated with such systems would now be worthwhile and also lower than the administrative costs of a regulatory system equal in allocative efficiency. Perhaps this is one underlying reason why, early in the 1970s, one heard not only that the supply of environmental amenities "has fallen far short of the rising effective demand . . . and the supply of certain critical goods, such as pure air and water, has virtually vanished"[134] but observed as well a new interest in pricing. Such speculation is underscored by the fact that mention of increasing concern about environmental resources at times went hand-in-hand with comment upon the new interest.

This section began by noting that in comparing past and present, one sees both similarities and dissimilarities, continuity and rather abrupt change. The discussion thus far might suggest, as to each way of putting it, more of the latter than the former. There seems to be much since 1970 that is significantly new. On close scrutiny, however, that impression largely vanishes. Technological constraints continue to exert a tight hold on policy, as revealed for example by the easing of emission standards for new vehicles. Problems of uncertainty continue to bewilder policymakers and to retard control, as in the case of sulfates. The forces of inertia have proved difficult to overcome, policymakers learning that it is one thing to talk about imposing inconvenience they consider worthwhile and another to accomplish it. Thus, as was the case with the pink-slip controversy ten years earlier, the refusal of Californians to tolerate used-car controls required other than at transfer resulted in retreat as to nitrogen oxide devices; the same intolerance for disruption made a laughingstock of the EPA's proposals for transportation controls. And pricing systems, though much discussed, have yet to be implemented. Indeed, some policy advisors have recently refused even to consider pricing, saying it "raises more problems than could be solved in the next century."[kk]

[kk]Quoted in Bureau of National Affairs, *Environment Reporter, Current Developments,* Vol. 6, No. 3, 16 May 1975, p. 137. There is considerable room for doubt that the United States will realize implementation of pricing systems in the foreseeable future, despite their promised efficiencies and despite the work of their supporters. (But see the discussion of an excess emission fee for new motor vehicles in footnote a, pp. 308–309.) More than any other means of intervention, pricing techniques confront strong resistance, in large part because we have become so accustomed to adapting technology to the dysfunctions in our behavior (which pricing systems aim to correct) rather than the other way around; that

Some change, then, is more apparent than real, and in several respects present policy bears especially heavy resemblance to that in the past. As before, it is proceeding as much by the painful and time-consuming process of exfoliation as by action based on reflection and modest insight. Granted that in the past that process was a product of least steps, whereas

---

comfortable habit (if not addiction) will not be easily changed. Congress's unrealistic air quality standards, and the EPA's gross mishandling of gasoline rationing as a means to meet those standards, helped in this regard not at all. Beyond this, industrial opposition to such measures as emission fees has been stiff and is likely to remain so. See, e.g., J. Burby, "White House Plans Push for Sulfur Tax Despite Strong Industry Opposition," *National Journal,* 28 Oct. 1972, p. 1663. While pricing systems tend to minimize the total control costs for society, they increase costs for the target sources. Under regulation, sources expend for control only; with pricing systems they pay for control *and* for amounts of pollution uncontrolled. There is no question that the total costs to sources are higher than they would be with equivalent regulatory controls. See J. Krier and W. Montgomery, "Resource Allocation, Information Cost and the Form of Government Intervention," *Natural Resources Journal* 13(1973):89, 103-104. The fee is not an expense to society (it is merely a transfer payment from one sector to another), but it is to sources, and thus one can anticipate strong opposition. There may be other barriers to pricing. Some economists suggest that "one reason for the congressional propensity to rely upon regulation as the solution to complex social problems is the fact that most congressmen are lawyers. . . . Legal training necessarily, and quite properly, concentrates on the specification of rights and duties in law or in regulations. . . . The economic approach stresses not rights and duties but incentives. . . . Usually the legal and economic approaches are complementary; each has its proper sphere as a means of governing social activity. But sometimes they are competitive. . . . The fact that the incentive approach is often given such short shrift in favor of the regulatory approach is hardly unexpected in view of the predominance of lawyers in Congress." Kneese and Schultze, *Pollution, Prices, and Public Policy,* pp. 116-117. These comments, of course, ignore the question *why* legal training has concentrated its energies as it has. (Here, by the way, important change is underway. Legal education is very quickly making more and more use of economic insights, no doubt giving greater attention to economics than economists do to law—the last point being one which might also have to do with the failure of lawmakers to be convinced by economists' arguments in behalf of pricing.) Perhaps economists have not shouldered their responsibility to design practicable systems. Or perhaps, and this is probably more the case, pricing systems simply were not the best approach until very recently (if now), for reasons sketched above. In any event, many legally trained people, some congressmen, and even a few well-known "industry" lawyers (like Lloyd Cutler) advocate more extensive reliance on market mechanisms. See, e.g., Burby, "White House Plans Push," p. 1666. For further discussion of barriers to pricing systems, see W. Baxter, *People or Penguins—The Case for Optimal Pollution* (New York: Columbia University Press, 1974), pp. 78-83 (offering a number of speculations, including the one "that most legislators have not carefully thought through environmental problems in any very analytical way"—with which one can only agree).

in the case of present policy it started with large ones. Granted too that past policy (generally) enjoyed small movements forward, while now it suffers withdrawal—in new-vehicle emission standards, in transportation controls, in proposed amendments to soften the heady demands of the 1970 federal legislation. But the differences in stride and direction should not obscure the essential similarities in process. Policy continues to be informed by the unsystematic production of knowledge, as to both technical matters (for example, the sulfate problem that results from use of oxidizing catalysts) and institutional ones (for example, the timing and implementation of transportation controls).

Given the continued presence of the exfoliation process, it should not be surprising that lag (to the extent that notion has any content) is still a characteristic feature in the making of pollution policy. Despite stern demands and grand pronouncements, progress has been slow in coming. Federal air quality standards, for example, especially those relating to pollution caused primarily by motor vehicles, have yet to be achieved in many of the major metropolitan areas of the United States; the original controls on new vehicles have yet to be fully realized; most transportation controls have yet to be implemented.[ll] There has been as much talk as action and accomplishment, and for reasons similar to those in the past—uncertainty, inertia, foolishness, and so forth. In the years of past policy, lag—for all its problems—may at times have performed the service of guarding against too abrupt and costly change. Perhaps it is doing so again.[mm]

The failings of present policy to realize announced ends are due in part to the legacy of earlier efforts. Pollution policy in the years up to 1970 nourished a situation that could only grow worse in every relevant respect—by its fixations, its shortsightedness, its inability to impose restraints that would simply become more necessary, and in larger measure, if environmental quality were to be improved or even maintained as time went on. Some shortcomings of past policy were understandable, some not, but all had their effect. Yet present policy exhibits some shortcomings of its own, as to both ends announced and means resorted to, and any lag in accomplishments since 1970 must be attributed in part to these. It is time to consider a few of them.

[ll]See, e.g., U.S. Council on Environmental Quality, *Environmental Quality—1976* (7th Annual Report, 1976), pp. 1-5, 213. Recent legislative proposals to extend deadlines for implementing transportation controls and for meeting air quality standards and new-vehicle emission standards are discussed in the next chapter.

[mm]That the central themes of the policy process appear to be continuing ones should be implicit in our discussion of some 1976 developments in the next chapter.

15

SOME COMMENTS ON PRESENT POLICY

The focus of this chapter is on some fundamental shortcomings in *federal* air pollution policy since 1970—shortcomings that relate in particular to controlling the motor vehicle's contribution to air pollution problems. The discussion will consider not only the design and administration of the Clean Air Amendments of 1970, the embodiment of present federal policy, but will look briefly as well at a bill that would have altered that policy in the years to come. The bill was not enacted (thanks perhaps to a filibuster in the waning days of the Ninety-fourth Congress in fall 1976), but it came awfully close, and its provisions or something very much like them may well be law by the time these pages are read, or soon thereafter.[a] We

[a]The bill referred to was a compromise measure recommended by a conference committee after the Senate and House had each passed bills differing from each other in a number of respects. See H.R. Rep. No. 94-1742, 94th Cong., 2d Sess. (1976) (hereafter cited as *Conference Report*). On the Senate and House bills, see S. 3219, 94th Cong., 2d Sess. (1976), and Senate Committee on Public Works, *Clean Air Amendments of 1976,* S. Rep. No. 94-717, 94th Cong., 2d Sess. (1976) (hereafter cited as *Senate Report*); H.R. 10498, 94th Cong., 2d Sess. (1976), and House Committee on Interstate and Foreign Commerce, *Clean Air Act Amendments of 1976,* H.R. Rep. No. 94-1175, 94th Cong., 2d Sess. (1976) (hereafter cited as *House Report*). For discussion of the earlier (1975) background of the proposed amendments, see pp. 238–240. The conference committee's bill died in late 1976

should, accordingly, give some attention to the measure—not in detail, because details are likely to be changed, but at least in general terms.

Our focus on the federal program should not be taken to suggest that it is to blame for all the setbacks and difficulties that have plagued pollution policy since 1970. State and local governments have had opportunities to be forceful and creative, and they have undoubtedly been guilty at times of default. We saw instances of this in the case of California (for example, the state's clumsy handling of retrofit controls for nitrogen oxide emissions, and its stubborn refusal to take a constructive part in the planning of transportation controls), and a study of other areas would surely reveal similar instances. Nevertheless, many of the troubles in California and other regions have quite clearly been related to problems in federal policy. And this should hardly be surprising: by virtue of the Clean Air Amendments of 1970, after all, present policy *is* federal policy. If this is not entirely the case in general, it is virtually so with particular respect to

when the Senate adjourned, after five hours of filibustering by Utah's two senators, without voting on the measure. Senate staff reportedly said that the Senate might not take early action on air pollution legislation during the Ninety-fifth Congress, turning its attention instead to water pollution. See *Environment and the Economy,* Oct. 1976, p. 3. Both the Senate and the House did, however, begin consideration of the issue early in 1977, an important initial question being whether Congress should try once again to enact a comprehensive air pollution bill similar to the unsuccessful 1976 proposal, or rather should concentrate first on a measure to relax the standards for new motor vehicles that the automobile manufacturers must otherwise meet in 1978 and subsequent years. The auto companies, claiming that the latter standards cannot be met, consider quick relaxation essential to avoiding "the possibility of a shutdown if the 1978 rules remain unchanged." Bureau of National Affairs, *Environment Reporter, Current Developments,* Vol. 7, No. 39, 28 Jan. 1977, pp. 1444-1445. See also ibid., No. 40, 4 Feb. 1977, pp. 1480-1481; No. 42, 18 Feb. 1977, pp. 1589-1590; No. 43, 25 Feb. 1977, pp. 1633-1635.

An alternative approach to the problem of the 1978 standards for new motor vehicles has been suggested: an excess emission fee that would permit the sale of vehicles meeting only the 1977 standards, provided the manufacturers paid a pollution charge calculated to take the economic advantage out of noncompliance with the 1978 standards. Proponents of the approach point out that it would keep the auto industry in business but at the same time give the companies an incentive to produce, and consumers an incentive to purchase, cleaner cars. See ibid., No. 37, 14 Jan. 1977, pp. 1350-1351. See also ibid., No. 42, 18 Feb. 1977, p. 1592.

The excess emission fee approach may have been sparked by a somewhat similar technique recently adopted by Connecticut as an enforcement measure to be used in the case of noncomplying stationary sources. See ibid., Vol. 5, No. 40, 31 Jan. 1975, p. 1512; Connecticut Enforcement Project, *Economic Law Enforcement,* Vol. 1 (1975). It should be understood, of course, that neither the Connecticut program nor the excess emission fee proposal represents a pure (even a relatively pure) pricing system, though each tries to capitalize on some of the advantages of such systems.

problems related to motor vehicle emissions. In mandating ends (uniform minimum air quality standards) and reserving the final voice on the means to meet them (controls on new and used vehicles), Congress reduced the realm of ultimate state and local authority to relative insignificance. Perhaps the federal government does not have a monopoly, but it surely has an awfully large share of the market. If it is not solely responsible for present troubles in policy, it has at least adopted a position such that only it can do much about those troubles. If a more constructive approach is needed, it will have to come at federal initiative. Accordingly, we focus on the federal program.

There has recently been a good deal of discussion about that program, especially as it relates to control of emissions from new vehicles—studies of the costs and benefits of the stringent emission standards, of the wisdom of preemption, of "two-car" strategies, of the 1970 legislation's general approach to technology, of relationships between the timing of new-vehicle controls and the deadlines for achieving air quality standards.[1] Thus far, this book has not had much to say about such matters, and—except for a few comments in this chapter—it will not. The studies mentioned above adequately cover the ground. In any event, there is a more basic issue that deserves discussion it has not received.

The issue is uniform standards (more particularly, uniform standards set without regard to cost considerations), which in our judgment represent a central shortcoming in present federal pollution policy.[b] Uniformity abounds in the Clean Air Amendments of 1970—as to air quality standards, emission standards for new vehicles (with California the lone exception), and emission standards for hazardous pollutants and major stationary sources. Most of the discussion that follows will focus on the uniform air quality standards, primarily because they are the most important aspect of current federal policy: the standards form, in one way or another, the foundation for all the other features of the federal program that have proven so troublesome. It was to achieve the uniform air quality standards in the worst areas of the nation, areas like Los Angeles, that Congress dictated the stringent federal controls on emissions from new

[b]Not simply air pollution policy. See, e.g., the Federal Water Pollution Control Act Amendments of 1972, Pub. L. No. 92-500, 86 Stat. 816, codified at 33 U.S.C. §§ 1251 et seq. One excellent study of water pollution policy notes that this legislation "traded an overly fractionalized system," the federal legislation on water quality that had existed prior to 1972, "for an overly centralized one." B. Ackerman et al., *The Uncertain Search for Environmental Quality* (New York: Free Press, 1974), p. 326. As our remarks in this chapter should make clear, exactly the same can be said of the transformation in federal air pollution policy that occurred in 1970.

vehicles.[c] (Uniformity in another guise, federal preemption of state authority over new-vehicle emissions, made these controls the norm in all areas but California, which Congress has permitted to set more strict standards under certain conditions. Other states must live with the federal standards whether or not they need them or like them.) It was to achieve the uniform air quality standards that Congress required consideration of transportation controls, and it was this requirement that forced the EPA to promulgate such controls—against considerable public resistance—for Los Angeles and many other major urban areas. In particular, the air quality standards led directly to the drastic federal gasoline rationing plan for Los Angeles. As we saw repeatedly in chapter 13, rationing and the other transportation controls became the source of frustration, dissatisfaction, disruption, delay, and ultimately litigation in California's pollution control efforts in the years following the Clean Air Amendments. This has been the case in other areas as well.[d]

Their central place in present federal policy is one reason to focus on the uniform air quality standards, but there are also other reasons. The standards illustrate best the general problems of uniformity and the sorts of thinking that lie behind those problems.[e] Discussing them permits us

[c]See the discussion at pp. 206–207.

[d]On litigation in other areas, see, e.g., the discussion at p. 233. As to problems with transportation controls generally, see, e.g., U.S. Council on Environmental Quality, *Environmental Quality—1976* (7th Annual Report, 1976), p. 4: "TCPs [transportation control plans] to achieve reductions in automobile-related pollutants have been controversial from the beginning. The purpose of these plans is to discourage the use of automobiles in polluted areas and to reduce automobile emissions. . . . Some TCPs stimulated local opposition because they could impose substantial costs on the community and on its residents. There has been substantial pressure to postpone the implementation of TCPs and to restrict in general the conditions under which these plans could be adopted." As will be seen, the 1976 bill discussed earlier would have extended the deadline for transportation controls and would also have limited the sorts of controls that could be required. Latest reports indicate that from twenty-nine to thirty-one areas need transportation controls in order to achieve federal air quality standards, and it is likely more areas will be added to the list in the future. See *Senate Report,* p. 29; *House Report,* p. 190. Given the resistance to transportation controls, the EPA "largely abandoned" them in early 1977—without any apparent authority to do so. See G. Hill, "E.P.A. Sources Say Agency Ended Efforts to Cut Auto Smog in Cities," *New York Times,* 14 Jan. 1977. Earlier, in October 1976, the agency had officially revoked its gasoline rationing requirements. See 41 *Federal Register* 45565 (1976) and the discussion in footnote aa, pp. 336–337.

[e]The bill almost passed by Congress in 1976 would not have undone the problems in the case of air pollution, though we shall see that it did reflect some congressional awareness of them; the bill would have opted in essence to put off resolving most of the problems for ten more years.

to reveal fundamental and widely held misconceptions in how to think about making responsible (fair and efficient) pollution policy; thus the discussion has at least implicit bearing on a range of concerns wider than air quality alone. But air quality is our explicit concern, and after arguing that uniform standards are both wasteful and inequitable, we will suggest a more constructive approach. It would rely on "management standards" designed to achieve significant, time-phased air quality improvements. The standards, however, would not be uniform; they would vary with conditions and circumstances in different parts of the country.

Management standards could avoid much of the inefficiency and unfairness that trouble the present federal program. In sketching how the approach might be implemented, we can address two matters of particular importance. One has to do with transportation controls, especially gasoline rationing. A relatively modest and carefully designed rationing program could be one useful means to the end of improved air quality, yet Congress (owing in some part, we think, to faults in the EPA's proposals) appears ready to abandon the technique. The second matter has to do with uniform emission standards for new vehicles. In preempting new-vehicle standards, Congress has created difficulties not unlike those that arise in the case of uniform air quality standards; we will suggest that the issue might deserve rethinking.

SOME PROBLEMS WITH UNIFORM STANDARDS

Recall that in the Clean Air Amendments Congress required the EPA to promulgate two air quality standards for each pollutant: a primary standard designed to protect public health; a secondary standard to safeguard "public welfare"—aesthetic values, plant and animal life, materials, and so forth.[2] Primary standards were to be set so as to allow an adequate margin of safety. As established, they have been conservatively based on worst-case assumptions, the idea being to protect the most susceptible part of the population in the most polluted areas of the country from adverse effects, leaving considerable room for error. Each state was to submit an implementation plan to achieve the primary standards within three years of EPA approval of the plan, unless specified exceptions were granted (in any event, the primary standards were to be achieved by 1977). Secondary standards, generally more stringent than the primary, must be achieved within a reasonable time. Both sorts of standards represent the minimum requirements for the nation; states may set more stringent standards if they wish.

The discussion in chapter 1 sketched a way to think about the ends and means of making pollution policy. Working with the obvious propositions that both polluting the air resource and keeping it clean are costly, it argued that a rational approach is to try to manage the resource in a way that minimizes the sum of the costs of pollution and its control, but subject to the constraint that the ends and means of policy (in our context, air quality standards and the methods to achieve them) accord with notions of justice. The aim, in short, should be to make policy that satisfies as best we can in an uncertain and imperfect world the dual concerns of fairness and efficiency. There is, of course, little evidence that the makers of pollution policy have consciously approached their tasks in these systematic terms,[f] nor is there much indication that policy nevertheless has managed to evolve as though they had. The themes of the policy process might suggest a groping search for fair and efficient resolutions of pollution problems, but they hardly testify that we have found them. Our failure to do so stands out especially in the present requirement of uniform air quality standards, and in the rationales offered to justify them.

The Failure to Minimize Costs

Let us first consider the efficiency of those standards. To justify uniform standards in terms of cost minimization, one would have to assume that the costs of a given level of pollution and a given level of control are the same across the nation. That assumption, however, is manifestly inaccurate. For example, aesthetic costs and materials losses will be functions of the varying resource endowments, degrees of development, and human attitudes that exist in different regions. Even health costs, which were of the greatest concern to Congress in passing the 1970 legislation, vary from place to place. Since such costs represent the aggregate of individual health effects, and since population varies significantly by region, so too will total health costs. If one believes that per capita and not aggregate health costs should be the relevant measure, efficiency considerations would still suggest some variation in air quality levels. This is because the costs of pollution control will also vary, depending upon population, density and nature of development, and meteorological and topographical conditions in any particular region. In short, since the costs of pollution

[f]The sole exception concerns the views expressed by California's standard-setters in 1959-1960—views that amounted to an academic exercise, given the legislative mandate to establish standards that would protect public health and welfare, period. See the discussion at pp. 122-126.

and the costs of control vary across the country, it is difficult to see how a uniform standard can begin to take the varying costs into account. The standard that minimizes total costs for a region in Iowa is hardly likely to do so for all the regions of California or New York or Colorado as well. To require adherence to the same stringent standard everywhere will in many areas result in the imposition of control costs that are much larger than the pollution costs avoided.[3]

At this level of analysis, then, uniform standards appear to labor under a strong presumption of inefficiency. At a somewhat higher level, however, there are several arguments that might be made to rationalize uniformity. However, only the last of these carries any weight at all, and not much at that.

The first argument is that uniform standards are not inefficient because the law permits each state to set more stringent standards (for the whole state or any region within it). Thus, states where the uniform standards are less strict than dictated by efficient resource allocation can solve the problem by tightening up. The problem with this argument is that those states where the standards are too strict are not permitted, under the law, to relax them. There is a suspicious asymmetry lurking in this tighter-but-not-looser approach of the federal law. The approach recognizes that states have a legitimate interest in setting tighter standards, say, to protect health at the cost of growth; but it denies that states have an equally legitimate interest in setting looser standards, say, to promote growth at the cost of health. One rationale for this approach might be that a relaxed standard in one state could adversely affect a neighboring state by increasing the latter's pollution levels or attracting away its industry,[g] while a strict standard in one state could have no such adverse impact on its neighbors. The difficulty is that the second premise of this response is not so clearly correct. A strict standard in one state could interfere with the interest of a neighbor—for example, by injuring the neighbor's economic well-being. This result is likely where a particular state has features attractive to industry (a harbor, a transportation hub) which are not shared by a neighboring state. Industrial location in the first state might well yield benefits to the neighbor such as growth within its borders of firms manufacturing supplies for industries in the first state, as well as increased employment for its citizens (who might work for the local firms or cross state lines to work for the industries in the first state). Stringent air quality

[g]On the latter point, it might be suggested that air quality improvements can only be achieved through federal standards because no state will establish stringent requirements on its own initiative—out of the fear that if it did so, it would drive industry to other areas. This *is* an argument for federal standard-setting; it *is not* an argument for *uniform* standards. See the discussion in footnote u, p. 332.

standards in the first state could cause enterprises to look elsewhere for the features they want, and the neighboring state—not suitable for location by these industries—could suffer losses in well-being as a result.[h]

Another rationale for the tighter-but-not-looser approach brings us to the second argument in favor of uniform standards. This is the assertion that "protection of health is a national priority."[4] Note first that this statement hardly justifies uniform *secondary* standards, which are not based on a concern for public health. But beyond this, the assertion would not appear to support uniform primary standards either. During congressional debate on the Clean Air Amendments, representatives of the government often tried to justify *uniform* standards with arguments that only justified *federal* standards. The "national priority" argument is one example: given that protecting the health of its citizens is a matter of heavy concern to the national government, it hardly follows that programs to protect public health should ignore the costs of doing so. Yet uniform standards appear to do just that, for they do not take account of the fact that costs of adverse health effects and costs of avoiding them vary from place to place, nor do they consider the relevance of costs not related to health effects. The objective of Congress was to "protect the health of persons regardless of where such persons reside."[5] If this implies uniform standards, and to the congressional mind it apparently does, then it seems unacceptable in efficiency terms. Whether it is made acceptable by reference to considerations of fairness is an issue we consider shortly.

As a further example of attempts to support uniform standards with arguments that only support federal ones, consider the Nixon administration's position that a chief advantage of "uniform nationwide air quality standards" is that they provide "an opportunity to take into account factors that transcend the boundaries of any single state. States cannot be expected to evaluate the total environmental impact of air pollutants, or take it into account in standard setting."[6] This line of reasoning might suggest that air quality standards should be set by the federal government because it is in the best position to assess the interdependent needs of the entire nation, and also perhaps best able to avoid parochialism, delay, and pressure from special interests. But the same reasoning would apply if the federal government set standards varying from area to area. (Moreover, such standards would permit the federal government to take into account the many factors that do *not* transcend relevant boundaries—

[h]For a more straightforward example of the same point, suppose emissions from sources in state *A* interfered with attainment by state *B* of its strict standards, though the emissions did not result in either state *A*'s or state *B*'s air quality being in excess of the federal standards. The Clean Air Amendments would appear to require that state *A* cut back on emissions. See 42 U.S.C. § 1857c-5(a)(2)(E) (1970).

something uniform standards make largely impossible.) The logic, in short, supports federal standards but *not* uniform ones.

This observation introduces the final argument offered to rationalize the efficiency of uniform standards. It is the argument with the most substance, although just how much turns out to be an empirical question. There is some evidence indicating that the argument ultimately amounts to little. In essence, the argument concedes that the cost-minimization test is appropriate, but goes on to point out that in its application, costs other than those thus far considered must be taken into account. These are the information (administrative) costs associated with varying standards—extra costs occasioned by the need to gather and assess specific information for *each* area, extra costs that are avoided by uniform standards. As one expert has noted, "It is possible, of course, for the Federal Government to establish standards according to regional or local situations. This, however, would be a large task and would require studies and hearings in local areas."[7] Government proponents of uniform standards might have had in mind these costs associated with varying standards during debates on the Clean Air Amendments. The Nixon administration offered as a "principal advantage" of uniform standards the fact that "the process of putting [such] air quality standards into effect would be accelerated, primarily because there would no longer be any time consumed in reviewing and approving air quality standards for each air quality region."[8]

The issue, then, is whether the efficiency benefits of varying standards—standards that took at least some account of differences among pertinent factors from area to area—would be greater than the administrative costs associated with setting them. This is quite obviously an empirical question, yet it is also one on which there appears to be little hard data (perhaps because policymakers have failed to grasp the relevance of the inquiry). Still, some reasonable guesses can be made. First, the EPA already needs and acquires so much information about each area in order to assess implementation plans that it is doubtful much more would be required to tailor appropriate standards for each, or at least for categories of areas that share common characteristics (an approach that we pursue later). Second, even if this is not the case, it hardly follows that additional administrative expenses associated with varying standards would not be worth incurring. Achieving some of the present uniform air quality standards in some areas (at least in other than the very long run) would appear to be so costly that it seems almost impossible to conclude that there would not be net savings in a somewhat more tailored approach.

It is true that one cost-benefit study bearing on this issue, though only remotely, finds "no substantial basis" for changing the present standards. Its judgment that the present standards are worth their costs is rather suspect, however, inasmuch as it looks at total costs and benefits for the nation as a whole rather than at marginal costs and benefits on an area-by-area basis; moreover, the range of its calculations is so wide as to justify virtually any conclusion (annual benefits of the standards are estimated to be from $2.5 to $10 billion, annual costs from $5 to $8 billion). Another study pertaining to the nation generally concludes that some of the present standards are unattainable in many regions, at least within anything like the time periods specified by law; we take this to mean that attaining them would not be worthwhile, the costs of attaining the unattainable being exceedingly high. Yet a third study, focused on the Los Angeles area in particular, quite clearly suggests that attaining some of the standards would result in large net losses.[9] It is not difficult to see why. We mentioned in chapter 13 a 1973 estimate that the number of days in excess of the federal standard for oxidant in the Los Angeles area could be reduced from the then-prevailing 250 to about 40 by, among other measures, an approximate 33 percent reduction in gasoline consumption.[i] To achieve the federal requirement of no more than one day per year in excess of the standard, on the other hand, would entail a reduction of at least 90 percent, and probably more. In other words, the price for about 40 more days of improved air quality, after accomplishing 210 days already under the more modest proposal, would be an almost threefold increase in the controls on gasoline consumption! As another measure of the costs of meeting the federal oxidant standard in Los Angeles, consider one (admittedly pessimistic) estimate that attainment would require "not only the maximal fixed-source and retrofit controls and the biggest bus system allowed, but also . . . a $1.28-per-mile surcharge to produce an almost 95 percent VMT [vehicle-miles traveled] reduction. Nearly 50 percent of the uncontrolled trips are foregone, with an associated social-cost proxy of nearly $5 billion."[10]

One needs no fine calculus to conclude that attaining the federal standard for oxidant in Los Angeles would not be worthwhile in anything but a quite long term. The information sketched above amply supports such a conclusion; so do the events related in chapter 13. The EPA does not believe the standard is sensible, the state does not, local officials do not, and residents do not. And Los Angeles is hardly unique; as we saw above, other areas share the same or related problems.[11] At least in instances like these, some of the present uniform standards are simply unduly costly.

[i]See the discussion at p. 220.

A more tailored approach of varying standards could reduce these unjustified costs substantially—surely far more than enough to absorb whatever additional administrative expense might be involved. That, at least, seems an awfully safe guess.[j]

Some Objections to the Foregoing (Including Objections Based on Fairness)

The uniform air quality standards of present federal policy fare poorly in terms of cost minimization, even when the administrative expenses of alternative approaches are taken into account. There are, however, a number of objections typically raised against application of the cost-minimization test to pollution problems, and these must be confronted before any firm conclusions about the uniform standards can be reached. The following discussion considers the three most common objections, addressing them in an order inverse to their persuasive force.

The first objection asserts in essence that when it comes to health and well-being, costs should not be a concern.[k] It should be clear that the

[j]Especially so when one considers that the administrative expenses accompanying varying standards might actually be *lower* than those that have attended attempts to implement the uniform standards in areas where they do not fit. As we have seen, the uniform standards have led to high administrative costs in the forms of hassle, frustration, bickering, delay, and litigation. More acceptable standards would presumably avoid such problems.

We might note here that even if the administrative expenses of setting and implementing varying standards *would* presently be so high as to offset the efficiency advantages such standards would otherwise yield, this situation would probably change with time—and not simply because the EPA would in the future have more information in hand. The air resource is likely to be more valuable in future years than it is now, as the result of increasing demand brought on by population and economic growth. As a resource increases in value, so too do the benefits of allocating it efficiently among competing uses. Since the administrative costs associated with varying standards should remain more or less fixed over time, at some point they are likely to be surpassed by the efficiency gains such standards would realize. (We think that point has already been reached.) This line of reasoning is similar to that applied in the last chapter in an attempt to explain recent interest in pricing systems. See the discussion at pp. 302-305.

[k]See, e.g., the remarks of Louis Fuller, at one time head of the Los Angeles APCD, in "The Advocates" (Transcript of television presentation on KCET-TV, Los Angeles, Calif., and WBGH-TV, Boston, Mass., 5 Oct. 1969), pp. 9-10: "I don't equate cost with the preservation of the health and well-being of the people at any time. I don't care what the cost is. . . . No cost concerns me. As far as health and well-being is [*sic*] concerned, no cost."

objection is nonsensical. Behind it is the assumption that there is some absolute of "good health," and the further assumption that this absolute has infinite value, since it should be chosen no matter what cost that choice implies. Yet neither assumption seems reasonable. There is good health and there is better health. People regularly choose at some point not to opt for better health because they consider that the resulting benefits will not be worth the costs of attaining them. People take jobs that involve a high (and unhealthy) degree of pressure or some measure of danger because the jobs pay more than admittedly more relaxed or safer alternatives; they purchase less than the best possible medical care because the best costs too much. Or, in the particular context of the air pollution problem, there are people—hundreds of thousands of them—who live in the smoggy San Fernando Valley area of Los Angeles because housing in parts of it is moderately priced. To say these people behave in this fashion because they have limited resources and cannot afford any other course of action only underscores the central point: Society too has limited resources, and it cannot afford to spend more on cleaner air than cleaner air is worth. At some point there simply arises a better bargain for the dollar than obtaining still cleaner, more healthful air. This is especially clear when we remember the obvious (but often overlooked) fact that air pollution is hardly the only unhappy phenomenon that inflicts adverse effects on health. The more single-mindedly society spends to protect health from the effects of pollution, the less it can exploit opportunities to respond to other health problems, problems whose resolution might well yield larger benefits. In short, the heaviest cost of protecting too zealously against pollution effects could be in health costs themselves—costs imposed by other ills that we have wasted the opportunity to attack. The first objection ignores all of these points.

There is, of course, an understandable appeal to the objection—especially for policymakers. The message is comforting, and it makes for good press and public relations. People do not enjoy being reminded of the trade-offs in life, notwithstanding that the trade-offs exist nevertheless. Especially during times of crises (and these, real or imagined, have had great influence on the making of pollution policy), it would be odd indeed to hear the officials in charge talk of what they can and cannot afford to do about the problem. But these observations, if they do anything, only suggest what might lie behind the first objection; they do not justify it. The natural appeal of the objection only makes it a greater obstacle to careful thinking.

It should be clear that none of the foregoing endorses doing nothing about very pressing problems; to do nothing in such situations would be too costly. Rather the point is that lines must be drawn. It is, for example, one thing to tolerate 250 days annually of poor air quality in the Los Angeles area; it is quite another to suggest that there should be virtually no such days even though the net costs of achieving this result would be enormous. A middle ground, based among other things on a rough assessment of the relative advantages and disadvantages of alternative policies, would be better in minimizing so far as possible the sum troubles of living in a less than perfect world. Some people, to be sure, might suffer as a consequence—that is harsh reality—but surely not in the numbers or to the degree that would result from approaching that reality as though it did not exist.[l]

The thrust of the second objection to the cost-minimization test has to do with the uncertainties that pervade pollution problems. Those making this objection argue that cost minimization is acceptable in principle only where all relevant costs can be calculated, and that in the case of pollution such a calculation is impossible.[m] The weakness of this argument is that

[l]Given the uncertainties discussed in chapter 1, it is difficult (and perhaps impossible) to judge the health consequences of requiring, say, compliance with the federal ambient concentration standard on photochemical oxidant (.08 parts per million) for only 320 days per year, rather than insisting (as present law does) that the standard be met on all but one day annually. The concentration level itself already has a safety factor built in. At the low concentration level in question, the health effects of more days of exposure per year appear to be indeterminable at present; they might turn not only on the number of days' exposure annually, but also on the length of exposure per day, the degree to which the standard is exceeded in any given day, and on whether exposure is for a long or short term (measured, for example, in years). For a description and discussion of the uncertainties on oxidant, see Report by the Coordinating Committee on Air Quality Studies, National Academy of Sciences and National Academy of Engineering, 93d Cong., 2d Sess., *Air Quality and Automobile Emission Control, Vol. 2, Health Effects of Air Pollutants* 317-382 (Comm. Print 1974). Among other things, the report concludes (at pp. 321-322) that the "technical data base for the oxidant standard was inadequate at the time the standard was set (1971) and remains inadequate, considering the implications for public health and the economic impact. . . . There may be adverse health effects at some concentrations below the lowest for which there is positive evidence of adverse health effects in man. Despite the fact that knowledge of health effects at the very low concentrations is inadequate for scientific purposes, the risks should not be ignored in protecting the health of the public. Thus the concepts of threshold and margin of safety may not be appropriate for assessing the health risks. . . . Adverse health effects from short-term exposure to photochemical oxidants and ozone at the standard concentrations have not been observed in man. There are human and animal data that suggest that adverse health effects might be expected at concentrations near the standard, especially under conditions of long-term exposure or in the presence of copollutants."

[m]See, e.g., " 'Pollution Fee' Idea Rapped," Letter to the Editor from S. Schwartz,

the calculation is unavoidable. In the abstract it probably is impracticable, perhaps impossible, to attach concrete dollar values to intangibles such as the aesthetic benefits of clean air. One must also agree that concrete cost figures for the health effects of various levels of air quality cannot be determined with much precision. However, these concessions have to do with valuations in the abstract. In the practical context of determining air quality standards, on the other hand, implicit valuations are necessarily made. The choice of a given standard, made through a collective judgment, always involves some implicit conception and balancing of costs and benefits.

In choosing among various standards and their accompanying control costs and effects on the level of pollution, policymakers necessarily decide what costs they consider worthwhile to achieve what benefits. In rejecting a slack standard that would involve low control costs but serious health effects, for example, they reveal an implicit valuation of those adverse health effects that will be avoided by a more stringent, but also more costly-to-achieve, standard. As we said, this sort of implicit valuation is both common and necessary; indeed, it is unavoidable. The fact that we cannot make concrete valuations does not negate the further fact that we must make choices. In choosing, we make a value judgment about the merits of the alternatives, and this judgment—an implicit calculation of the incalculable—is as unavoidable as the choice.[n]

This brings us to the third and final objection to the cost-minimization

---

Los Angeles Times, 20 Dec. 1969: "It is necessary to quantify all damages—including aesthetic, sensory irritation, illness, and death—in dollars. The attitude that all values regarding human life can be translated to economic terms is not widely shared, even by economists."

[n]One could, on just such an argument as this, suggest that uniform standards are rational because Congress implicitly concluded that the benefits of achieving them in the areas where the control costs of doing so would be *highest* would nevertheless be worth those costs. (This would imply that other areas should have standards even more demanding than the national uniform standards; in this connection, recall that the 1970 legislation permitted the states to set more stringent air quality standards if they wished.) The difficulty is that the congressional attitude appeared to be that the standards should be achieved without regard to cost—the questionable kind of reasoning dealt with above. See Senate Committee on Public Works, *National Air Quality Standards Act of 1970,* S. Rep. No. 91-1196, 91st Cong., 2d Sess. 2-3, 24 (1970). Thus any implicit valuation is suspect. In any event, one is free to disagree with the conclusions of Congress. In our judgment the situation in Los Angeles and a number of other places suggests that achieving some of the uniform standards in some areas clearly would not be worth the costs of doing so. As we shall see, Congress is coming to agree that this is at least the case for the next decade.

test. It says in essence that the test is concerned only with efficiency and ignores considerations of fairness, which are more important. Of the three objections, this one has the most substance and it is the most difficult with which to deal. Although the cost-minimization test itself is not explicitly concerned with fairness considerations, employing the test hardly means that such considerations are or should be ignored. As the discussion in chapter 1 suggested, cost-minimization refers to desirable air quality goals *from the standpoint of efficiency*; once those goals are determined, they could still be rejected as unfair.

It appears that unfairness in this context could bear two rather different meanings. The first meaning must be that it is unjust for a society to live in less than a pure and wonderfully healthy environment. This line of reasoning, however, comes close to suffering the faults of the first objection. It probably is "unfair," in some loose sense, that the air is not perfectly clean, or even as clean as physically possible. It is a sad thing; it is a sorry state of affairs; it is just plain wrong. But it is also plain hard fact, even if society were to run itself solely in terms of fairness, that using limited resources to make the air or other environmental media cleaner and cleaner would mean diverting those resources from other worthwhile enterprises that should *in terms of fairness* also be pursued. Choices among opportunities must always be made, whatever the rhetoric of resolution. It seems likely that the most principled choices will be made when the full dimensions of both considerations—fairness and efficiency—are kept in mind, and when it is remembered that a step toward a result considered more fair in the context of one problem area, if accompanied by positive economic costs, may mean unfairness in another.

The second meaning of unfairness is more pointed. We take this meaning to be that injustice lies in the distribution of different levels of air quality among people living in different areas of the country, especially if the distribution appears to discriminate against socially or economically disadvantaged groups. The principle of cost minimization does quite clearly imply cleaner air in some places than in others, and there is of course good reason to suppose that in some, perhaps many—but by no means all—cases the poor and otherwise disadvantaged will live in the less desirable locales. (Although hardly by themselves. Air quality levels, local nuisances aside, tend to be more or less the same over large areas inhabited by rich and poor alike. For example, areas to the east of Los Angeles have relatively high pollution levels, but they also consist largely of suburbs ranging from middle class to affluent.) The conclusion is that such inequality among income groups or geographic locations is unfair, and the prescription is to make the air equally healthy for all persons everywhere as a matter of distributive justice; in other words, to "protect the health of persons regardless of where such persons reside."[12]

This position, because it ignores the costs that accompany it, suffers most of the problems discussed above. But, more to the point, it is extremely problematic even in its own terms—in terms, that is, of distributive justice. Not only does the position pay no heed to the costs of uniformly stringent standards, and thus to the fact that these will be exorbitantly high in some areas, but it also overlooks that the costs will largely be borne by—distributed to—the residents of such areas. They will have to pay more for automobiles to the extent that additional control devices are mandated; more for gasoline to the extent that government policy limits the supply; more for consumer goods made in the area by firms facing higher production expenses by virtue of pollution abatement requirements. They will lose employment opportunities to the extent that new firms cannot come into the area; or existing ones cannot expand, must cut back, or are forced out of business by the high costs of control. They will be inconvenienced, and not merely a little, to the extent that transportation controls and other measures disrupt the ordinary flow of everyday affairs. If the total of these burdens is high, and we saw that in many instances it would be,[o] residents of the areas in question might well be worse off despite better air. Such outcomes are rather difficult to justify in terms of distributive justice, especially when we consider that their impact will be felt disproportionately by the economically and socially disadvantaged.[p] And such outcomes, of course, are precisely what a policy of cost-minimizing varying standards would aim to avoid.

[o]The EPA has conceded that a program to achieve some of the federal standards in the Los Angeles area would "lead to significant economic disruption and will certainly result in a major transformation in the life style of residents of the South Coast Air Basin." There would be "a great economic and social impact" and "serious dislocation" to the local economy. 38 *Federal Register* 2195, 2197 (1973). In a later suit against the agency, the court noted that "all parties agree that implementation of a program to achieve the standards [in particular, gasoline reductions] would produce extreme social and economic disruption." City of Santa Rosa v. Environmental Protection Agency, 534 F.2d 150, 155 (9th Cir. 1976). See also the discussion at p. 317. And as we said above, the Los Angeles area is not unique in these respects.

[p]See, e.g., U.S. Council on Environmental Quality, *Environmental Quality—1973* (4th Annual Report, 1973), p. 107: "The aggregate impact of government financing [for pollution control] is predominately progressive—i.e., the wealthy pay proportionately more than the poor. The aggregate impact of private financing is somewhat regressive—the poor pay a higher proportion of their income than

Conclusions and Some Comments on Recent Congressional Action

The foregoing remarks should make clear that the present federal program, insisting as it does upon attainment of uniformly stringent air quality standards to the same degree and by the same final date in all parts of the country, is undesirable in terms of both efficiency and fairness. And Congress, it appears, is beginning to agree with that conclusion, though it nevertheless persists in thinking that uniformity should remain as the ultimate end of pollution policy.

The conference committee bill recommended (but not enacted) during the closing days of the Ninety-fourth Congress would have permitted extensions of up to ten years in the original 1977 deadline for attainment of federal primary ambient air quality standards in all areas where attainment by that deadline would require transportation controls having "serious adverse social or economic effects." More particularly, the bill provided that the governor of a state in question could, after mid-1976, apply to the EPA for a first extension of up to five years (until May 31, 1982). Under the terms of the bill, extensions could be granted only if the state would begin to implement controls on stationary sources of vehicle-related

---

the wealthy. Combining public and private financing shows that in total the net incidence of all incremental expenditures is slightly regressive." On the same note, one study prepared for the EPA concluded that the distributional impacts of state transportation control plans, not only in Los Angeles but generally, would be significantly regressive. TRW Transportation and Environmental Operations, *Socio-Economic Impacts of the Proposed State Transportation Control Plans: An Overview* (prepared for the Environmental Protection Agency) (1973), pp. 202-203 especially. The regressive distributional impact is particularly disturbing given that low-income persons probably prefer improvements in air quality *less* than increased or maintained economic well-being. See the discussion at p. 30, and U.S. Council on Environmental Quality, *Environmental Quality—1973,* pp. 81-82. Thus, it is not simply that present policy may force the poor to pay relatively more, but—worse—to do so for what they value relatively less. L. Tribe, "Legal Frameworks for the Assessment and Control of Technology," *Minerva* 9(1971):243, 253, quite properly makes the general point that an inefficient allocation of a good like improved air quality could be justified in that it achieved an indirect income redistribution that would be politically impossible to achieve directly. We are sympathetic to the observation, but in the present context it carries little weight, simply because those for whom the indirect redistribution is intended would likely lose more than they would gain. In the view of the Council on Environmental Quality, analysis does *not* suggest "that our environmental policies should be directed toward creating either income or geographical equity. Quite to the contrary, environmental programs are an unlikely and probably inefficient mechanism for pursuing such goals." U.S. Council on Environmental Quality, *Environmental Quality—1973,* p. 111.

pollutants (for example, hydrocarbons and nitrogen oxides) by mid-1977; would begin to implement all reasonably available transportation controls by the same date; and had agreed to complete by the end of 1978 a detailed planning study of alternative control measures. (Planning studies were to examine the measures needed to meet the federal standards; describe the projects and timetables to put the measures into effect; and identify and analyze the social, economic, and environmental effects of the measures.) States would also have been required to submit transportation control plans by the end of 1978. Among other things, the plans were to demonstrate attainment of the primary standards as quickly as practicable but no later than the May 31, 1982, deadline, or demonstrate that attainment of the standards by the 1982 deadline would be impossible despite implementation of all reasonably available controls. In the latter event, the bill provided for another round of applications to be submitted after mid-1981. The process just described would then be repeated: an additional extension of no more than five years could be granted subject to the same terms as outlined above. All areas of the country would have been required to meet the federal standards by mid-1987.[q]

There is no great mystery about what moved members of Congress to recommend the foregoing provisions. The House and Senate committees alike recognized what reflection in 1970 would have suggested, and what the process of exfoliation since that year had in any event made abundantly clear—that a large number of areas could only meet the federal standards by 1977 through the use of transportation controls, and that many of those controls would simply be unreasonable.[13] In the view of the House committee, "changes and restrictions in transportation systems may

[q]See *Conference Report,* pp. 49-52. The terms of the conference committee bill discussed above were based, with but a few changes, on the original Senate bill. See ibid., pp. 104, 106; *Senate Report,* pp. 28-33, 160-163. The House bill differed from that of the Senate in a number of respects, a chief one being that the two extensions provided for by the House bill would have run to January 1, 1980, and January 1, 1985. See *Conference Report,* pp. 105-106; *House Report,* pp. 11-12, 189-192, 337-340.

The discussion of the conference committee bill in the text omits a number of details. The most important of these will be raised in context later in this chapter, but one bears mention here: the bill provided that if the EPA found a state application inadequate, the agency was to notify the state of deficiencies and specify a date for a revised application. If the EPA denied an extension application, or if a state that did not meet the federal standards failed to submit an application or a revised application, the agency (after local consultation and the opportunity for public hearings) was to promulgate a transportation control plan subject to the same terms and requirements governing plans submitted by states. *Conference Report,* p. 50.

impose severe hardships on municipalities if imposed too quickly"; "implementation of many of the [transportation control] measures is impracticable within the time frame permitted under the current Act. Some of the measures may never be practicable."[14] The Senate committee held similar views.[15] Both committees seemed aware that the Clean Air Amendments had gone awry in ignoring costs, though they articulated this in different ways. The House committee wished to "insure the protection of public health and the environment," but "at the same time take into account the energy and economic needs of the Nation." The Senate committee was more to the point, noting that many of the transportation controls promulgated by the EPA in order to satisfy the 1970 legislation had "imposed vast economic and social costs, for relatively small improvements in the quality of the environment."[16]

Given these remarks, it is disappointing that the House and Senate bills, and the conference committee bill that grew out of them, neglected to address the central shortcoming of the Clean Air Amendments—the requirement of uniform standards. It was that requirement, after all, that had motivated the extension proposals in the first place. Those proposals reflected a sensitivity to considerations of fairness and efficiency that appears to have been lacking when Congress enacted the Clean Air Amendments; they showed an awareness that the air quality problem is a complicated and deep-seated one calling at the least for a moderately complex resolution that takes varying circumstances into account. Yet, for all of this, the proposals urged only that compliance with the uniform standards be postponed for ten more years (until 1987 under the terms of the conference committee bill); by then, at the latest, all areas of the country would have been required to conform.

The implications of such an approach should by now be clear. Were it to become law, doomsday would not be repealed but only delayed. Every region would once again have to achieve air quality meeting the federal concentration standards all but one day per year[r]—notwithstanding that this would "impose severe hardships" or require "measures that may never be practicable" (in the words of the House committee); notwithstanding that it would mean "vast economic and social costs, for relatively small improvements" (in the words of the Senate committee); notwithstanding,

[r]As mentioned earlier, most of the federal air quality standards are expressed in terms of ambient concentrations not to be exceeded more than one day per year. None of the 1976 proposals appears to have contemplated a change in this approach, though several of them called for a "review" of the standards. See *Conference Report,* p. 187.

in short, that it would not be worthwhile in some areas and would be impossible in others.

It is true, of course, that the proposals discussed above would have granted time to the states, but then so too did the 1970 legislation—seven years, in fact. That span has quite clearly been shown to be an inadequate one in which to achieve the unreasonable, and ten years more would surely prove no better. Within those years, states would be expected to allocate scarce resources to achieving ends that were not worthwhile; as doomsday approached, they could be expected to devote them to fighting unreasonable demands—especially after having learned that opposition can succeed. The same sorts of waste, debacles, and delays that burdened the Clean Air Amendments would likely arise again. (There could then be, of course, yet another extension, but this is a questionable strategy by which to make policy—it encourages gamesmanship, damages congressional credibility, and results in unnecessary expense.)[17] Congress should learn more than it has from the lessons of theory and experience that have grown out of present policy. It should concede that the nation is not uniform and never will be, and that no legislation can or should try to make matters otherwise. To ignore such obvious facts is only to ask for inefficiency and inequity. A more responsible and realistic approach should be pursued.

TOWARD MANAGEMENT STANDARDS

A responsible and realistic approach to the air pollution problem would take into account, as well as possible, all the advantages and disadvantages (not simply those pertaining to health) of various levels of air quality improvement in various areas. It would acknowledge that while the calculation of costs and benefits is difficult, it is also unavoidable; that one is likely to get further and waste less by making reasoned guesses[18] than by surrendering to an irrational pursuit of uniformity in a nonuniform world. Such an approach, though rightly paying heed to the fact that state policymakers probably have insufficient regard for the impact of their policy on neighbors, would give sufficient attention to factors that might not transcend state lines—local conditions, local preferences, and so forth. It would recognize that while the expense of tailor-made standards for each region could possibly be so high as not to be worthwhile, breaking regions into even fairly rough categories might yield net gains.[s] And perhaps most

[s]For example, two obvious polar categories would be, first, a class consisting of those regions that can reasonably be expected to meet, say, existing standards by

important, it would recognize that it will take time to overcome the results of three decades of halting policy, and that the time necessary will vary from area to area.

Even these brief comments suggest that we should probably discard three approaches to the air quality problem that might otherwise appear attractive. The first of these is the approach used from 1967 to 1970, under which the states set their own standards in light of federal criteria. Such an approach would likely be seen as too great a step backward; accordingly, it would prove so politically unpalatable that to propose it would hardly be realistic. Political constraints and commitments are as real as any others, and they must be given consideration. In any event, the 1967–1970 approach gave too little regard to the problem of conflicts of interests among neighboring states, and it gave too much leverage in each state to those favoring little pollution control. By leaving standard-setting to the states, the Air Quality Act of 1967 pitted the burdens of inertia and uncertainty against environmental groups, and these—as we have seen—are poorly equipped to combat highly organized, well-financed industrial associations.

There is a second approach that could take these problems (of parochialism, of leverage) into account. It would provide for uniform federal standards set more or less as they are now, it would provide that states could (as they may now) set more stringent standards, but it would further provide that states could also set *less* stringent ones for particular regions on a demonstration that the federal standards would be too costly to achieve and could be exceeded without undue effects on other states. Before concluding that under such a system the federal standards would serve no useful function, one should consider whether the approach might not sufficiently guard against narrow state policy and polluter influence. States would have the obligation to establish the justifications for less stringent standards; polluters would have the substantial burden of overcoming the inertia of the legislative process. Federal standards would have to be affirmatively rejected by state legislatures, something legislative

---

1977 through present stationary source measures plus federal new vehicle controls; and, second, a class consisting of those that require very stringent transportation controls to meet the federal standards by that date. We build on this approach below. It will become clear that many, but not all, of our suggestions are similar to proposals in the 1976 conference bill; we will note important points of congruity and incongruity. The two most significant of the latter are: (1) we would abandon uniform standards (the bill, as has been seen, would retain them); (2) we would retain the possibility of reasonable gasoline rationing measures as a means to air quality improvements (Congress, as will be seen, is probably ready to abandon them).

bodies might be reluctant to do lightly in an era of increasing environmental consciousness unless justification were clear and strong. If a state could not muster the will and the facts to meet the burden, federal standards would prevail.

Perhaps this approach also looks too much like that of old,[t] or perhaps federal policymakers more knowledgeable than we might conclude that it would put less balance into decision making than our discussion suggests. A third alternative would be for the federal government to set standards as it does now, but standards that were made to order for each region in light of its wants and circumstances. The difficulty here, as we have already mentioned, is that while the approach could yield quite satisfactory standards in each case, its implementation might possibly be so expensive as to more than consume the gains of relatively efficient standards.

While none of the foregoing is perfectly satisfactory, it does appear that each has advantages, whether by way of avoiding costly uniformity, guarding against parochialism, or recognizing relevant local circumstances. The question quite naturally arises whether there might be an approach that capitalizes on these advantages while avoiding their pitfalls and those of present (and newly proposed) policy. We believe there is such an approach, which we call management standards. Let us first sketch some of its elements (with no notion that we are addressing all details, our purpose being to stimulate further thinking, not to present a final plan). Then we can conclude with a few comments bearing on implementation.

A Sketch of the Management Standards Approach

The concept of management standards is based on technical, economic, and social feasibility. It envisions a series of time-phased steps in each of which there must be achieved substantial percentage reductions in the number of days per year in excess of federally specified uniform ambient *concentration* standards. Ultimately, this number would be reduced to the point where further reductions in the area in question would not be worth the costs of attaining them. Management standards thus aim at long-term (but not uniform) improvements, but they insist in the meantime upon short-term accomplishments that exhaust all feasible controls and that enhance air quality relative to what it was before. The central purpose of the management standards approach is to take into account,

[t]Though one should recognize its similarity to the EPA's approach to the nondegradation problem. See the discussion in footnote l, p. 238.

in a manageable way, the varying problems and conditions that exist in different areas. The approach would preserve desirable features of the 1970 legislation; it would also (as would the 1976 proposals discussed earlier) require certain planning steps that have in fact been pursued in efforts to implement the Clean Air Amendments even though not formal parts of the law; but, unlike the 1970 law and the 1976 proposals, it would abandon uniform air quality standards as presently prescribed.

Our focus in the following is on primary air quality standards, those intended to protect public health. It should be noted, however, that focus on secondary standards would present us with an easier task, for at least two reasons. First, most of the arguments made in behalf of uniform standards evince a concern with protecting *health* on a nationwide basis. The direct concern of secondary standards is not with air pollution effects on health but with those on such economic factors as materials, plants, and livestock. This being so, the central arguments offered in support of uniformity are far less applicable. Further, inasmuch as secondary standards are largely concerned with economic factors, they pose issues of economic trade-offs in clear fashion—and thus can be forcefully said to call for a nonuniform, region-by-region, benefit-cost approach.

Second, present policy on secondary standards can in fact be seen as one of nonuniformity for other than the long term. While secondary standards are expressed in terms of uniform concentrations, they are to be achieved within a reasonable time. Presumably the reasonable time will vary by region, depending upon regional conditions. Since nonuniformity results either from varying ambient concentration standards or from varying deadlines for achieving a uniform concentration standard, present policy on secondary standards would appear to call for nonuniformity at least until the last region reaches the reasonable time deadline. It is largely that policy, already a part of federal law, which the management standards approach attempts to mimic. What, after all, is more reasonable than the reasonable?

There are, of course, a number of ways that one could carry out a program of management standards. We propose a series of steps designed to mesh quite closely with present law so as to facilitate as much as possible a change in approach. But it should be kept in mind that our proposal is only one example; there may prove to be better alternatives.

We would begin with a requirement, much like that of present law, that the EPA promulgate uniform primary ambient standards. Unlike present law, however, the standards would be expressed only in terms of concentration levels; the practice of providing that these levels are not to be exceeded more than one day per year would be abandoned. This would be done simply because getting down to that one day (as opposed to five or ten or twenty days) would not be worthwhile (or possible) in many instances, and to proceed as though this is not the case asks only for trouble and waste. The management standards approach would permit regional variations in the number of days in excess of standards, and in the schedule of required improvements, as will be illustrated below.

For each pollutant the EPA would be required to develop criteria that would be applied to determine whether any particular region fell into one or the other of two classes: Class A Regions, consisting of those areas in which it is feasible, in the judgment of the agency, to achieve no, or virtually no, days in excess of (for example) the present primary ambient concentration standards by 1977 in accord with present law; Class B Regions, consisting of those regions in which the foregoing is not feasible in the judgment of the agency. Criteria on feasibility would deal with constraints imposed by technological, economic, administrative, political, and other social considerations—considerations that would bear on the issue of what schedule of compliance would just approach (but not exceed) that point where any more demanding schedule would not be worthwhile in light of its costs and consequences. It is fruitless to say that such considerations are intractable, for confronting them is unavoidable. Moreover, they are the very sorts of considerations that the EPA has taken into account, as instructed by Congress, in passing on state requests for extensions; and the very sorts of considerations that the EPA would have to take into account under the terms of the 1976 proposals. Thus federal policymakers can hardly claim that the considerations are beyond human capacities.

As the next step, the EPA would be required to promulgate, within a specified period of time, management standards for each pollutant, expressed in terms of the *minimum* percentage reductions (in the number of days in excess of the prescribed ambient concentration standard) required by given dates for Class B Regions. The percentage reductions required of each region would depend upon the average number of days annually each region's ambient air quality, based on the best data available, exceeded the ambient standard. As an example, for oxidant the EPA might promulgate the reductions and dates of attainment illustrated in table 10.

It is important to note two features of the hypothetical management standards schedule illustrated in table 10. First, as already mentioned, the schedule sets forth the *minimum* improvements generally required of any region. To apply the schedule across the board would be little less arbitrary than the present approach of uniform standards. Because the schedule would set forth the generally applicable minimum requirements,

TABLE 10
ILLUSTRATIVE MANAGEMENT STANDARDS FOR CLASS B REGIONS

Days per year standard is exceeded during base period (e.g., 1972-1976)	10	30	50	100	200	365
Allowable days per year in excess of standard in 5 years from base period (e.g., by 1981)	8	23	39	74	138	219
Allowable days per year in excess of standard in 7 years from base period (e.g., by 1983)	6	17	28	53	92	137
Allowable days per year in excess of standard in 9 years from base period (e.g., by 1985)	4	11	18	33	52	55

it should be based on worst, not best or average, areas. Then the expectation would be that most areas could achieve more than the minimum. In the discussion below we comment on the implementation of this expectation, as well as on those peculiar local instances where even achieving the minimum requirements might be infeasible in the near term. Second, the schedule illustrated operates such that the worse the quality of a region's air (in terms of number of days in excess of the standard annually), the more the region would generally be required to improve both in absolute and relative terms. In other words, the more serious a region's problem, the more resources the region must devote to it if necessary to comply with the management schedule.[u]

[u]The management standards approach, because it insists on less than perfectly tailored standards, and because it requires little if any more information to implement than does the present program of uniform standards, would not in our judgment entail administrative costs so large as to outweigh its advantages. The skeptic might agree, but go on to argue that the political implications of the approach—of defining regional boundary lines, of permitting more or less economic development in some areas than in others—are so troublesome as to suggest that anything other than the uniform standards of present policy would be infeasible. But such an argument would ignore the fact that uniform standards have precisely the same troublesome implications. They define the entire nation as, in essence, one region within which some areas (the more polluted ones) face greater constraints on economic activity than others (the less polluted ones). As we have seen, political problems developed as a result. It is difficult to believe that the management standards approach could give rise to greater political difficulties than have been experienced already in the short, troubled life of uniform standards. Indeed, one would expect fewer difficulties, simply because management standards can find rational justifications that appear to elude present policy.

Some Comments on Implementation

The process of implementing the management standards approach could obviously follow many patterns. Let us simply outline one possible sequence here and, in the course of doing so, comment briefly on a few points that deserve emphasis.

Implementation might begin by requiring state governments (working in conjunction with local agencies) to prepare for each of their regions a preliminary plan, using public hearings and other procedures to encourage local participation in the process.[v] Preliminary plans would be submitted to the EPA, which would review them subject to something like the following guidelines. First, each region would initially be presumed, as to each pollutant, to fall into Class A. In drafting its preliminary plan for each region, a state could accept this presumption, in which case the plan would be required to set forth in detail the measures to be employed to realize the federal standards by (in our example) 1977. Alternatively, the preliminary plan could attempt to overcome the working presumption by showing the measures necessary to achieve an ambient standard and the considerations that would make that achievement infeasible. As suggested above, feasibility would be judged with reference to such factors (none of them conclusive) as availability of necessary technology within a given time period, direct economic costs, administrative shortcomings, possibility of significant social dislocation, induced unemployment, and so forth.

As can be seen, the effect of the working presumption built into the implementation process would be to require for each region a rough Class A plan; states, in short, would necessarily have to think about ways to attain the standards before insisting they could not be attained. To overcome the working presumption, they would have to submit several items of information: the best data available concerning the average number of days annually on which a particular region's ambient air quality exceeded the federal standard in question; a description of the measures necessary

[v]The 1976 proposals aimed to encourage local participation in the planning process not only by requiring it but also by funding local planning costs. See *Conference Report,* pp. 49-51; *Senate Report,* pp. 29-30. The latter noted: "To date, a major problem has been a deficiency in local involvement in transportation control planning. To correct this, the bill requires that locally elected officials participate in the development of transportation control plans to obtain the post-1977 extensions. This recognizes that transportation control planning is a local political process affecting the daily lives and transportation patterns of local voters. To augment the ability of local officials to undertake this work, the bill provides 2 years of 100 percent Federal grants for planning transportation controls."

to comply with the management standard applicable to the region; a detailed discussion of the considerations that make infeasible compliance with any timetable more demanding than that proposed in the preliminary plan (which would be at least the minimum timetable designated for the particular region's circumstances by the management standards schedule).[w]

Suppose a state claimed that it could not meet even that minimum timetable? Such claims would be quite foreseeable, and they might at times be justified. For some pollutants in some regions, technological and economic and social considerations do suggest that *any* percentage improvement in air quality is impossible and that perhaps even some deterioration may have to be tolerated in the short term in order to realize other, more important ends. This hardly means, however, that the only feasible response is to do nothing; there are measures to cope with such short-term problems. For example, Los Angeles happens to have a rather difficult sulfur dioxide situation. Nevertheless, some utilities in the area have been able to plan ways to manage their supplies of high-sulfur and low-sulfur oil so as to minimize the adverse environmental impact of the former. With the aid of a network of ground monitoring stations downwind of a given power plant, utilities can take advantage of favorable meteorological conditions (high inversion layers, strong winds) to burn high-sulfur fuel, and switch to low-sulfur fuel when conditions are less favorable. The state should have an affirmative obligation to explore measures such as these and discuss them in its preliminary plan; perhaps of more importance, it should also be required to discuss the measures needed to achieve at least the minimum requirements of the management standards schedule, and show—just as it was required to do with respect to Class A standards—why these would not be feasible.

In reviewing preliminary plans, the EPA would consider whether adequate data had been submitted in their support, and whether justifiable conclusions had been drawn from the data. It is important that good data be required on the technologies specified in the plan to reduce pollution from various sources, including timetables on the availability of these technologies and the rates of their introduction. Where nontechnological, social, and economic control measures are to be employed, again the times and rates of introduction should be stated. (The EPA should be prepared to provide technical assistance with respect to gathering and employing data for the plans.)[x]

[w]The presumption, by placing the burdens of inertia and uncertainty on those opposed to strict air quality standards, would help correct the systematic biases against conservation that tend to inhere in environmental policymaking. Compare the similar provisions of the 1976 conference committee bill discussed at pp. 324–325.

[x]The 1976 conference committee bill would have required the EPA to prepare

Our mention of nontechnological, social, and economic control measures should make clear that management standards would not place exclusive reliance on technical fixes. Quite to the contrary, there will be many instances where technology will be unavailable or too expensive to realize improvements, and yet where controls such as reductions in the amount of driving will be well worthwhile. It is important, however, to introduce nontechnological controls gradually and sensibly. Since there has been little experience with them, the strong possibility of serious error in policy design dictates a program of caution. Moreover, the forces of inertia must be dealt with in a sensitive fashion. We have seen the difficulties encountered in the course of efforts—separated by a decade—to impose even mild inconvenience on an area's citizens (the pink-slip and nitrogen oxide controversies). The residents of Southern California appeared ready to accept the costs of used-vehicle controls if imposed in a manner (at the time of transfer of a vehicle) that did not require anything of them while they owned their vehicles. They accepted the expense of such controls if added on as a small (and, to some extent, hidden) part of the price of a large purchase, but rebelled when required to purchase controls outright. Whether this behavior was rational or not (time-of-transfer requirements, as opposed to registration or license plate schedules, probably save vehicle owners little if anything in the way of direct costs, but they almost certainly save time and inconvenience), it suggests an unhappy but important lesson to the makers of environmental policy: the more that controls disrupt comfortable habits of citizen behavior, the less acceptable they will be, and they may be unacceptable even if they appear to policymakers not to be very disruptive at all.[y] If that lesson is ignored in the future, the likely result will be further resistance and further frustrated policy.

---

and make available guidelines and information regarding transportation controls. See *Conference Report,* pp. 4-6.

[y]This is one temptation of the "wholesale" approach to control—the approach that aims at manufacturers or used-car dealers, for example, rather than at vehicle operators (the "retail" approach). See, e.g., S. Brubaker, *To Live on Earth* (Baltimore: Johns Hopkins Press, 1972), p. 184 (noting, with considerable understatement, that "the consumer is difficult to regulate"). See also U.S. Department of Commerce, *The Automobile and Air Pollution,* Part 2 (1967), p. 111. It is also the case, of course, that the wholesale approach, because it has fewer targets, brings with it lower administrative costs than does retail control. See, e.g., E. Mills, "Economic Incentives in Air Pollution Control," in *The Economics of Air Pollution,* ed. H. Wolozin (New York: W. W. Norton and Co., 1966), pp. 40, 49. The weakness of the wholesale approach is that it focuses its efforts on an organized group able to offer effective opposition and, beyond this, is simply not available as a means to some sorts of control (reductions in driving, for example). The difficulties

Given the foregoing, there is no surprise in the failure of federal demands—expressed especially through gasoline rationing—that the residents of Southern California and other areas reverse their life styles virtually overnight. Thirty years of preaching a faith in technology, with habits developing accordingly, cannot be overcome by one ill-reasoned sermon that the old-time religion is no longer good enough. Conversion will only be achieved through modest and carefully designed efforts that explore a number of avenues and that have as their aim not only instructions for policymakers about what works best, but lessons for citizens that they can adjust to new ways after all. If we are to escape the technological fixation other than in rhetoric, we must not ask too much of any nontechnological measure, and we must construct each of the measures sensibly (proceeding first by pilot projects). The federal gasoline rationing proposals for the Los Angeles region failed miserably in these respects. In part this is the fault of Congress in insisting upon standards that *no* reasonable controls (or combination of them) could achieve in a number of areas of the country. But there is also fault with the EPA in putting forth a miserable series of proposals—miserable not for the degree of rationing required (as to that, Congress is responsible), but for the means by which to accomplish it. The agency by its proposals performed both a small service and a large disservice. It made quite apparent the costs attached to what Congress asked, reminding citizens that when politicians voiced the pleasing and powerful rhetoric of wonderfully healthful air everywhere, such an outcome would mean a high price to be paid by the public in one way or another. There is some evidence that the EPA did this intentionally.[z] Whether or not this is the case, we should be grateful for the service. We should also, however, be annoyed that the agency unnecessarily planted in the public mind visions of rationing so senselessly designed as to undermine seriously the already shaky chances that transportation controls have in the foreseeable future—whether to reduce pollution, conserve energy, or both.[aa] Because such controls could be worthwhile

---

of the retail approach are its high administrative costs and the ability of a large and diffuse public to resist control not so much by organized effort as by individual stubbornness that is hard to combat primarily for the very reasons that the targets of control are diffuse and large in number. And it is also the case, as we have seen, that if even a small percentage of the controlled public is sufficiently annoyed to take political action, the absolute number of complaints can appear great enough to move policymakers.

[z]See the discussion at p. 222.

[aa]Thus the House committee would have authorized "the Administrator [of the EPA] to delete gas rationing from a State plan." *House Report,* p. 192. The Senate committee concluded that "gasoline rationing is not a reasonable procedure, and

elements of future policy, we want to comment on a few fundamental errors in the EPA's ideas.

Some reduction in driving in Los Angeles and other areas is clearly necessary and would no doubt be worthwhile to achieve significant improvements in air quality, and some sort of gasoline rationing, properly implemented, could be a fair and efficient means for achieving the reduction. Rationing would make motor vehicle driving a more scarce and more valuable commodity than it is now. If efficient use of this scarce commodity is to be realized, a system is needed to allocate the gasoline—which is to say the driving—to those who value it most highly. The most familiar system is the price system. Caution is necessary, however, lest allocation through the price system discriminate inequitably against the poor. The shortcomings of the EPA proposals all centered about the inefficient and inequitable allocation systems it considered. The agency's proposals properly recognized that essentially two rationing techniques exist.[bb] One would employ coupons, such as during World War II. A coupon must be surrendered for each gallon of gasoline purchased, and government, by limiting the total number of coupons, controls gasoline sold and thus miles driven. Coupons could be initially sold by the government or they could be distributed without charge—so many, say, to each registered vehicle. But the most important ingredient of a coupon system if it is to work fairly and efficiently is this: the coupons *must* be transferable. Only in this way can the limited supply of gasoline find its way to those who value it most highly. With unrestrained transferability, those who valued additional gasoline more than the going price of coupons would purchase the coupons. Those who sold their coupons would obviously value the money more than the right to the gasoline, otherwise they would have held out for a higher price. If thought necessary to achieve equity, coupons could be distributed progressively, such that the poorer the owner of a vehicle, the more coupons that person would receive. In this manner, the system could yield what would seem to be a desirable form of wealth redistribution.[cc]

---

cannot be expected to become a reasonable strategy," then went on to contradict itself, "recogniz[ing] that any measure, pushed to the extreme, is unreasonable, and almost any measure, when implemented with moderation, can be judged reasonable." *Senate Report,* pp. 30-31. This congressional reaction to gasoline rationing prompted the EPA to revoke the rationing requirement in October 1976. See 41 *Federal Register* 45565 (1976) and footnote d, p. 311. A moderate rationing program, we believe, would clearly be one reasonable means to air quality improvements in many instances.

[bb]See the discussion at p. 222.

[cc]Of course, this redistribution would not be without cost. Low-income persons would most likely sell some of their coupons to others, and the process of transfer

These observations suggest that transferability is wise; experience with rationing in other contexts shows that it is inevitable. Where legal transfer is prohibited, black markets in coupons spring up. But since these markets represent an illegal, risky business, an extra premium is charged by the broker. The result is higher prices and transfers of wealth to dishonest entrepreneurs. Yet, despite all the considerations favoring a policy of transferability, the EPA hardly seems committed to such an approach, for it called for comments on the issue. This is cause for some concern. Transferability should be a foregone conclusion. If the EPA disagrees, then it apparently understands very little about the valuable social functions served by pricing mechanisms. That this might well be the case becomes more plausible when one evaluates the agency's views on the second rationing technique.

The second technique is based on the fact the coupons are not really necessary, since there already exists an extensive retail gasoline market effectively employing the price system. Thus it is possible to achieve the wanted effects of rationing simply by limiting the total supply of gasoline sold. The limitation on supply would drive the retail price up to an equilibrium point and efficiently allocate the available gasoline. Oddly enough, however, the EPA proposed to ignore the convenience of this existing price system. Instead, it would have employed "price controls at the retail level (to prevent windfall profits)" and would have sold the available gasoline "on the basis of first-come-first-served." These measures represent the very worst of the EPA's ideas.[dd] First-come-first-served is what economists call a queue. One need only recall recent experience with gasoline shortages to appreciate that queues are notoriously inefficient, simply because waiting in line is hardly socially productive.[19] The inefficiency is especially striking in the case of air pollution, for first-come-first-served would mean long trains of automobiles idling their engines or constantly starting and stopping them, both of which increase emissions.[20]

The EPA was forced to resort to first-come-first-served because it would employ price controls, and only free-floating prices would eliminate the

---

would not be frictionless: costs would arise in establishing a market in the coupons. Strict efficiency considerations dictate against giving coupons to people we know would subsequently transfer them. Equity considerations, however, might well suggest that these costs would be warranted in light of the redistribution of income achieved.

[dd]It is quite possible, of course, that the EPA consciously put forth ill-conceived measures in order to maximize local resistance. See the discussion at p. 222. Or perhaps the agency sensed a "crisis" that demanded sacrifice on the part of everyone.

need for queues. The EPA wanted price controls in order to prevent windfall profits. The same objective could have been accomplished, however, with an excess profits tax, while retaining the advantages of the price system. The difficulties are that the construction of such a tax is not easy, and it is doubtful in any event that the EPA has authority to impose one. But the agency could have proceeded in another fashion that would not only prevent windfall profits but avoid another problem with its proposal as well. Recall the EPA's comment that the technique presently being discussed would "be enforced at the manufacturer's level." What the agency meant was that the amount of gasoline that refineries would be allowed to sell would be limited. But how would the EPA decide which refinery could sell how much? One way to deal with that question would be to *auction* selling rights to the refineries, with one right for each gallon to be sold. Competition among refineries for these rights would give the rights a positive price roughly equal to the windfall profits. The auction would automatically decide which refinery could sell how much; it would prevent windfall profits; and in making price controls unnecessary, it would avoid the need for queues at the retail level.[21]

It might appear at first glance that queues, as compared with an auction system with free-floating prices, would be more desirable from the standpoint of low-income persons, because the poor would suffer less hardship by confronting long lines than by confronting high prices. It is not clear, however, that this follows. Standing in line might mean losing income or recreational opportunities, and to make a judgment on the trade-off we would have to know what the price increase would be compared with the length of the queue (and the value of the opportunities lost as a result of its length) in any particular case. On the other hand, it is likely that the cost of standing in line is lower for the poor than for the more well-to-do, whose time is generally compensated at a higher rate, and who tend to place a relatively high value on leisure. If so, it seems clear that with a queue system the poor could improve their position by standing in line for the wealthy at a mutually agreeable price. Nevertheless, queues are burdened with so many inefficiencies that they should be avoided. If considerations of equity weigh heavily, then a system of progressively distributed, transferable coupons should be used to allocate gasoline.

So much for rationing or other transportation controls. Policy should insist that consideration of such measures be reflected in implementation plans, and the federal government should be prepared to assist in their design and to present good (not horrendous) examples. No one measure should attempt to accomplish too much, and each should be phased in gradually (perhaps beginning, as we suggested, with pilot projects). With this approach, the likelihood of serious resistance is minimized.[ee]

Once preliminary implementation plans prepared as outlined above were submitted to the EPA, the agency—based on the state submittals and its own independent study—would be required either to accept a preliminary plan as final or prepare a proposed revised plan for each region in response. In any event, the plan coming from the federal government would be either a Class A plan, a Class B plan, or a plan fixed at any point in the spectrum between A and B.[ff] For example, an EPA counterproposal might reflect the conclusion that a region cannot meet Class A requirements but can do better than the minimum applicable Class B requirements. In the event of EPA counterproposals, these would be published and local hearings held. Thereafter, EPA and state (and local) representatives would negotiate the final plan. "Final" plans would, of course, be final only for a time; replanning would obviously be in order as conditions, the available technology, and other relevant variables changed. Thus there should be provisions for periodic review by the EPA of its classifications and the management standards requirements. There should also be provisions, in conjunction with or independent of the foregoing, for repetition of the entire implementation process outlined above—a requirement, say, that each state start anew ever five years.

The subject of negotiation deserves a few more words. The Clean Air Amendments of 1970 took far too rigid and polar an approach to "cooperative" federalism, with the result that there has been very little cooperation. Under the legislation, the federal government has dictated standards and left implementation in the first instance to the states and in the last to the federal government, with too little room for interaction in between. The decision to proceed in this fashion ignored two points. First, standards themselves must take into account matters of implementation.

[ee]It is worth noting in this regard that proposals for more modest reductions in the amount of driving (more modest in both amount of reduction and time of achievement) put forth by groups in the Los Angeles area were received quite favorably by the public—at least judging from the press. See J. Dreyfuss, "Caltech Tells Plan for Breathable Air by 1976," *Los Angeles Times,* 4 July 1971; idem, "Smog Scientists Urge Taxing Car Users on Mileage," *Los Angeles Times,* 18 Oct. 1972; Editorial, "The Environment Fight," *Los Angeles Times,* 11 July 1971. See also the discussion at p. 220.

[ff]The exceptions would be those instances where the EPA concludes that particular areas cannot reasonably be expected to meet even the minimum applicable Class B requirements for a time. See the discussion at p. 334. Here the agency would have to specify the measures necessary to achieve all feasible improvements, as well as a timetable of steps designed to accomplish more as soon as possible.

Ends cannot rationally be established without regard to the means available to realize them. What is needed is a dynamic approach to the pollution problem that considers an array of air quality standards and how (and at what cost) they might be achieved, and then arrives at ends and means together. Any attempt to proceed otherwise is likely to ask too much, with the result that the states charged to implement the standards will balk—as of course they have. This introduces the second point. The 1970 legislation overlooked the chief danger of the "you do what we say" approach to pollution policy—that state government, if not sufficiently consulted as to ends and means, might well be reluctant to carry out federal mandates. Reluctance here is due not simply to stubbornness and pious claims of "states' rights" (though there has been some of this) but as well to justified frustration with the federal failure to consider local wants and conditions. Thus ends and means must be established by a process that includes participation by all levels of government.

The little progress made in California under the 1970 legislation has come from cooperation, negotiation, and compromise—the federal government giving in a bit to the state, the state to the federal government, localities to the state, and so forth. When the EPA responded to state resistance with federal insistence, the result was default, delay, and a series of threats that reflected poorly on state and federal policymakers alike. When the EPA encouraged and assisted the creation of local task forces, and paid heed to their recommendations, some positive attitudes began to develop.[gg] There is still, of course, a good deal of friction and difference of opinion. They are likely to continue until federal policy recognizes the need for a more realistic and constructive approach to the pollution problem. Happily, there is a growing awareness of that need. In mid-1974 the administrator of the EPA (no doubt speaking from experience) observed that "success in carrying out the Clean Air Act as well as other legislation depends on the willingness of the Environmental Protection Agency to work with state and local government." In the administrator's view, state and local governments must be "full partners . . . in the formulation of . . . regulations, guidelines, and plans. . . . EPA cannot fulfill its mandate 'by means of edicts issued with a high and heavy hand from the Olympian heights of our ineffable wisdom.' "[22] It appears that the Senate Committee on Public Works has come to agree. One can only hope that the entire Congress will do so as well.[hh]

[gg]See the discussion at p. 229.

[hh]The Senate committee said of its 1976 bill: "It must be emphasized again that this bill does not attempt to define for State and local governments those measures that may be reasonable. That is a responsibility resting with State and local officials,

In the course of rethinking its approach to air quality standards and the means to establish and attain them, Congress might also reconsider its policy of preempting controls on new motor vehicles. The relationship between the federal air quality standards and the federal program on new vehicles was handled inadequately to begin with, and has only become less coordinated by virtue of extensions in the program's deadlines.[23] Preemption forces states other than California to abandon new-vehicle controls stricter than the federal as one technique to achieve air quality gains. This might mean either a slower rate of improvement than could otherwise be achieved, or the need to rely on transportation control measures—both of which a state might regard as less desirable than more stringent and more expensive controls on new vehicles. Yet, thanks to present policy on preemption, all states other than California are foreclosed from opting for those controls, and for reasons that are hardly persuasive.

There were several arguments advanced in support of preemption, and each of them had superficial appeal: preemption would avoid duplication of federal standards; it would protect vehicle manufacturers and owners against multiple standards; it would minimize unnecessary expense to owners; it would avoid "chaos" and "confusion."[24] Yet, given the shape that preemption took, manufacturers were expected to produce two basic sorts of vehicles—one for the California market, one for the rest of the country (and they have done so). In light of this, it is difficult to see how the supposed problems that moved Congress to preemption in the first place could possibly be increased by permitting *any* state, under specified circumstances, to choose either the federal standards or an alternative set of more stringent ones established to meet conditions in California and other areas with similar problems. Such an approach could only mean economies of scale (and thus lower costs) for the alternative type of vehicle; at the same time, the approach would double the flexibility of a too rigid federal policy.[ii]

---

in cooperation with EPA, because maximum flexibility must be preserved." *Senate Report,* p. 31.

The litigation discussed earlier (see the text and footnote g at pp. 232–234) could of course force the national government to adopt a more accommodating approach to state-federal relations. If the broad claims of federal authority involved in that litigation are withdrawn or rejected, the states' leverage in policy formulation will be vastly increased: Congress's only alternative to giving in would be to federalize completely the planning, implementation, and enforcement processes. But, as the *Senate Report,* p. 2, noted, "The Federal Government does not have and will not have the resources required to do an effective job of running the air pollution control programs of the States."

[ii]Whether states should be permitted to establish requirements more stringent than the federal government's but also different from the "California" standards is

In reconsidering preemption, Congress should also examine lems inherent in denying states the freedom to set new-vehicle st *less* stringent than the federal. In some areas, the federal controls a needed to realize even the present uniform air quality standards; in oth they are but one of many feasible means to do so. In such places, peopl are being forced to pay for unnecessary (and unnecessarily expensive) technology. There is, of course, little question that a policy permitting states to establish less stringent standards would result in higher administrative and production costs for vehicle controls generally. Perhaps they would not be too much higher, however, were Congress to permit three sorts of vehicles, and three only—one complying with federal standards, one with more stringent standards (the option mentioned above), and one with standards less demanding than either of these, with the states free to choose, subject to federal oversight and approval.[ii] Even if such an approach (not to mention the alternative of a virtual state license to set unique standards, or none at all) increased the costs of pollution control technology because of the loss of economies of scale in production and administration, the matter would hardly be settled. The question would remain whether those higher costs might nevertheless be more than offset by the savings to motorists no longer required to purchase unnecessary controls. Congress, so far as we can tell, did not consider this issue of net efficiency. Nor did it consider the equity of requiring residents of relatively unpolluted areas to subsidize the purchase price of vehicles in more polluted areas. (Citizens in relatively unpolluted areas of the country have been forced by preemption policy to purchase unnecessary controls in order to reduce the average cost of those controls—hence the subsidy. Congress regarded the higher expenses of a decentralized program as "unnecessary"; it ignored the subsidy that conclusion necessarily implies.)

In adopting its preemption policy, Congress was also concerned, of course, that multiple standards would not only be expensive by way of

---

a more complicated question. Our subsequent discussion suggests some of the issues and leads us to a very tentative conclusion that such a policy might not be undesirable so long as the states in question bore the full costs (administrative and production) of their decisions. Our guess is that most states would find the option we suggest a satisfactory compromise.

[ii] Compare the discussion of "two-car" strategies (low- and medium-pollution vehicles) in F. Grad et al., *The Automobile and the Regulation of Its Impact on the Environment* (Norman: University of Oklahoma Press, 1975), pp. 104–107; U.S. Office of Science and Technology, *Cumulative Regulatory Effects on the Cost of Automotive Transportation* (RECAT) (1972), pp. 30–34; D. Harrison, *Who Pays for Clean Air* (Cambridge, Mass.: Ballinger Publishing Co., 1975), pp. 127–134 (noting the advantages of a two-car strategy in terms of both efficiency and distributive effects).

[illegible] ›nvenience to drivers who cross state lines. [illegible] ısistently left used-vehicle controls to the [illegible] that more new than used vehicles venture [illegible] the case. As to the congressional concern [illegible] lards, there would be none if either the [illegible] l above were adopted as policy; if states [illegible] lards, the EPA could ensure in its review [illegible] rds were sufficiently different from the [illegible] . As to chaos and confusion, there was [illegible] would result by virtue of permitting variations in used-car [illegible]ntrols (and none has)—so again it is difficult to imagine why the new-vehicle situation would be much different. If there was confusion, it was in formulating the arguments for preemption. Pertinent issues exist, but Congress has pretty much managed to avoid them.[kk]

Present pollution policy, like that of the past, reveals a recurring failure to pose relevant questions; perhaps this is one reason why current efforts have continued to stumble along in much the fashion that typified earlier years. There are, to be sure, differences between then and now: policy before 1970 was unduly fractionalized and tended to ask too little; policy since has been unduly centralized and has demanded too much. But extremes in either direction are likely to produce as much failure and frustration as they are success and satisfaction, and the facts bear this out. The dominating federal policy adopted in 1970 promised great improvements

[kk]Matters are, however, looking up, even if only slightly. The Senate committee appears to have been aware of one problem in preemption policy mentioned earlier—that preemption might force states wishing to achieve relatively quick improvements in air quality either to forego that end or rely on transportation controls to achieve it. See the discussion at p. 342. Noting that the Senate bill would have relaxed and extended the federal new-vehicle standards (and this is true of the House and conference committee bills as well; see *Conference Report*, p. 107), the *Senate Report*, p. 29, acknowledged that this "would place even greater burdens on transportation controls." This is one reason why the Senate committee proposed "reasonable modifications . . . in the timetable" for achieving the federal air quality standards. But suppose a state wished more rapid improvements than that timetable would demand, but without the need to rely on transportation controls? Even here the conference committee bill would have given some relief—a slight change in preemption policy permitting states with air quality in excess of federal standards to adopt in 1979 the stricter new-vehicle standards that would otherwise be applicable only in 1980, provided the state could satisfy the EPA that adoption in 1979 would be required to achieve the air quality standards by 1982. *Conference Report*, p. 63. The provision, based on a Senate committee recommendation (*Senate Report*, p. 74), might be taken to suggest that some members of Congress are growing skeptical of the arguments behind preemption. In any event, it reflects a small step in the direction of a more flexible preemption policy.

in air quality—thirty-five years of lethargy were soon to be undone. There have, of course, been some accomplishments,[25] but also much waste, hassle, and delay. Indeed, "data collected by the state Air Resources Board indicates that air quality in the Southcoast [Los Angeles] Air Basin deteriorated in 1974, almost equaling levels not reached since 1971"; the picture in 1976 was equally "gloomy."[26] Might we have gotten more by asking less?

The trends of three decades cannot be reversed at once. Insistent demands for large and almost immediate improvements across the board may be powerful political rhetoric, but in terms of progress they are probably counterproductive.[27] Happily, it appears that Congress might be coming to agree. The 1976 proposals discussed in this chapter suggest that lessons are indeed being learned, and that the federal government may soon be ready to recede to a more moderate position. They also reveal, however, a failure to come finally to grips with the central problems of uniformity. Pollution policy may be about to head in a better direction, but it has still to acknowledge, once and for all, that there is little sense in striving for unreasonable—and in some instances unattainable—ends.

NOTES

Introduction

1. D. Currie, "Motor Vehicle Air Pollution: State Authority and Federal Pre-Emption," *Michigan Law Review* 68(1970):1083.

2. Pub. L. No. 91-604, 84 Stat. 1705, codified at 42 U.S.C. §§ 1857–1858a.

3. U.S. Environmental Protection Agency, 1972: personal communication.

4. S. Griswold, "Air Pollution in California—The Past" (mimeo., n.d.), p. 17.

1. A Framework for the Problem

1. Unless otherwise indicated, the discussion in this section is based on J. Seinfeld, *Air Pollution—Physical and Chemical Fundamentals* (New York: McGraw-Hill Book Co., 1975), chap. 1.

2. U.S. Council on Environmental Quality, *Environmental Quality—1973* (4th Annual Report, 1973), p. 78.

3. J. Trijonis, "An Economic Air Pollution Control Model—Application: Photochemical Smog in Los Angeles County in 1975" (Ph.D. diss., California Institute of Technology, 1972), pp. 5–6.

4. The description is based on D. Lynn, "Air Pollution," in *Environment,* 2d ed., ed. W. Murdoch (Sunderland, Mass.: Sinauer Associates, Inc., 1975), pp. 223, 239–242; Seinfeld, *Air Pollution,* chaps. 7 and 8.

5. L. Lees et al., *Smog: A Report to the People* (Los Angeles: Ward Ritchie Press, 1972), p. 21.

6. H. Wolozin, "The Economics of Air Pollution: Central Problems," *Law and Contemporary Problems* 33(1968):227, 228–233.

7. See R. Heilbroner, "The Arithmetic of Pollution," *Economic Problems Newsletter,* Spring 1970, p. 3.

8. Trijonis, "An Economic Air Pollution Control Model," pp. 180–187.

9. See, e.g., C. Kaysen, "The Computer That Printed Out Wolf," *Foreign Affairs* 50(1972):660; C. Starr and R. Rudman, "Parameters of Technological Growth," *Science* 182(1973):358.

10. K. Arrow, "Classificatory Notes on the Production and Transmission of Technological Knowledge," *American Economic Association Papers and Proceedings* 59(1969):29.

11. H. Demsetz, "Toward a Theory of Property Rights," *American Economic Association Papers and Proceedings* 57(1967):347.

12. See G. Hardin, "The Tragedy of the Commons," *Science* 162(1968):1243.

13. H. Demsetz, "The Private Production of Public Goods," *Journal of Law and Economics* 13(1970):293, 295.

14. See generally M. Olson, *The Logic of Collective Action* (New York: Schocken Books, 1971).

15. N. Jacoby, "The Environmental Crisis," *Center Magazine,* Nov.–Dec. 1970, p. 37.

16. See, e.g., Snyder v. Harris, 394 U.S. 332 (1969); Zahn v. International Paper Co., 414 U.S. 291 (1973); Eisen v. Carlisle and Jacquelin, 417 U.S. 156 (1974); R. Posner, *Economic Analysis of Law* (Boston: Little, Brown and Co., 1972), pp. 349–351.

17. J. Krier, "The Pollution Problem and Legal Institutions: A Conceptual Overview," *U.C.L.A. Law Review* 18(1971):429, 458–459.

18. See, e.g., W. Oates and W. Baumol, "The Instruments for Environmental Policy" (mimeo., 1972); A. Kneese and B. Bower, *Managing Water Quality: Economics, Technology, Institutions* (Baltimore: Johns Hopkins Press, 1968), p. 88; Boomer v. Atlantic Cement Co., 26 N.Y.2d 219, 257 N.E.2d 870, 309 N.Y.S.2d 312 (1970); Diamond v. General Motors Corp., No. 947429 (Los Angeles, Calif. Superior Court, 20 Aug. 1969).

19. For expanded discussion, see Krier, "The Pollution Problem," pp. 459–475.

2. A Brief Prehistory

1. Based on California Department of Public Health, *Clean Air for California* (Initial Report, 1955), p. 18; R. Schiller, "The Los Angeles Smog," *National Municipal Review* 44(1955):558, 560.

2. Based on California Department of Public Health, *Clean Air for California,* pp. 18–19; Schiller, "The Los Angeles Smog," p. 560; Stanford Research Institute, *The Smog Problem in Los Angeles County* (Menlo Park, Calif., Second Interim Report, 1949), pp. 7, 46–51; H. Kennedy, *The History, Legal and Administrative Aspects of Air Pollution Control in Los Angeles County* (Los Angeles County Board of Supervisors, 1954), pp. 1–5; "Weather Explains Smog Attacks," *East Los Angeles Tribune,* 11 July 1957.

3. Kennedy, *Los Angeles History,* p. 4.

4. Ibid., p. 5; "Weather Explains Smog Attacks."

5. J. Edinger, 1975: personal communication.

6. California Assembly Committee on Air and Water Pollution, *Final Summary Report* (n.d., c. 1952), p. 22.

7. "Los Angeles Basin's Smog History Dates Back to 1542," *South Bay Breeze* (Redondo Beach, Calif.), 31 July 1957.

8. Kennedy, *Los Angeles History,* p. 3.

9. See Stanford Research Institute, *The Smog Problem in Los Angeles County* (Menlo Park, Calif., Third Interim Report, 1950), p. 14.

10. Quoted in S. Griswold, "Air Pollution in California—The Past" (mimeo., n.d.), p. 2.

11. A. Carlin and G. Kocher, *Environmental Problems: Their Causes, Cures, and Evolution, Using Southern California Smog as an Example* (Rand Corp., Santa Monica, Calif., Report Number R-640-CC/RC, 1971), p. 50.

12. F. Fredrick and L. Lowry, "Legislative History and Analysis of the Bay Area Air Pollution Control District" (mimeo., n.d.), p. 1.

13. From data compiled by the authors.

14. S. Griswold, Testimony in Hearings before Special Subcommittee on Air and Water Pollution of the Senate Committee on Public Works, 88th Cong., 2d Sess. 37, 39 (1964).

15. J. Davies, *The Politics of Pollution* (New York: Pegasus, 1970), p. 33.

16. J. Bregman and S. Lenormand, *The Pollution Paradox* (New York: Spartan Books, 1966), p. 6.

17. "State Air Pollution Control Legislation," *Boston College Industrial and Commercial Law Review* 9(1968):712, 731.

18. S. Edelman, "Air Pollution Control Legislation," in *Air Pollution,* Vol. 3, ed. A. Stern (New York: Academic Press, 1968), pp. 553, 560.

19. R. Dyck, "Evolution of Federal Air Pollution Control Policy" (Ph.D. diss., University of Pittsburgh, 1971), p. 18.

20. R. Ripley, "Congress and Clean Air: The Issue of Enforcement, 1963," in *Congress and Urban Problems,* ed. F. Cleaveland et al. (Washington, D.C.: The Brookings Institution, 1969), pp. 224, 228.

21. Ibid.; Davies, *The Politics of Pollution,* p. 103; Dyck, "Evolution of Federal Policy," pp. 18–19.

22. Ch. 425, § 13, 30 Stat. 1152, codified at 37 U.S.C. § 407.

23. Davies, *The Politics of Pollution,* p. 38.

24. Ibid., pp. 25, 38.

25. Ibid., p. 37; Dyck, "Evolution of Federal Policy," p. 210.

26. Davies, *The Politics of Pollution,* pp. 108–109.

27. Carlin and Kocher, *Environmental Problems,* p. 50.

28. "State Air Pollution Control Legislation," p. 730; Edelman, "Air Pollution Control Legislation," p. 554.

29. "State Air Pollution Control Legislation," pp. 730–731.

30. Kennedy, *Los Angeles History,* p. 3.

3. A New Problem

1. A. Will, "Abatement Progress under Existing Air Pollution Laws" (Address to the Statewide Conference on Air Pollution Legislation, California Chamber of Commerce, San Francisco, Calif., 21 Feb. 1955), p. 1.

2. R. Dyck, "Evolution of Federal Air Pollution Control Policy" (Ph.D. diss., University of Pittsburgh, 1971), p. 19.

3. "Text of Report and Conclusions of Smog Expert," *Los Angeles Times*, 19 Jan. 1947.

4. E. Ainsworth, "Fight to Banish Smog, Bring Sun Back to City Pressed," *Los Angeles Times*, 13 Oct. 1946.

5. "Bill That Created L.A. County Air Pollution Control District Now Recognized as Model for Entire Nation," *Post Advocate* (Alhambra, Calif.), 9 Aug. 1957.

6. See S. Griswold, "Air Pollution in California—The Past" (mimeo., n.d.), p. 3; T. Roberts, "Motor Vehicular Air Pollution Control in California: A Case Study in Political Unresponsiveness" (Honors thesis, Harvard University, 1969), p. 3.

7. California Assembly Committee on Air and Water Pollution, *Final Summary Report* (n.d., c. 1952), p. 36.

8. Griswold, "Air Pollution in California," p. 3.

9. Ainsworth, "Fight to Banish Smog."

10. E. Ainsworth, "Synthetic Rubber Industry Shows It's Possible to Eliminate Smog," *Los Angeles Times*, 4 Nov. 1946.

11. Griswold, "Air Pollution in California," p. 3.

12. R. Chass and E. Feldman, "Tears for John Doe," *Southern California Law Review* 27(1954): 349, 360-361.

13. Ibid., pp. 360-361. See Los Angeles City Ordinance No. 77,000, § 37.19 (Smoke), as amended, 19 Nov. 1945.

14. Griswold, "Air Pollution in California," p. 4; Chass and Feldman, "Tears for John Doe," p. 361; H. Kennedy, *The History, Legal and Administrative Aspects of Air Pollution Control in Los Angeles County* (Los Angeles County Board of Supervisors, 1954), p. 6. See Los Angeles County Ordinance No. 4460 (n.s.), 20 Feb. 1945 (creating Division of Air Pollution Control and prescribing duties); Los Angeles County Ordinance No. 4547 (n.s.), 18 Sept. 1945 (violation of prescribed Ringelmann Standard constitutes misdemeanor).

15. E. Ainsworth, "Thirteen Suits to Be Filed in Smog Fight," *Los Angeles Times*, 15 Oct. 1946. See also Kennedy, *Los Angeles History*, p. 7.

16. Chass and Feldman, "Tears for John Doe," p. 361; Griswold, "Air Pollution in California," p. 4; F. Fredrick and L. Lowry, "Legislative History and Analysis of the Bay Area Air Pollution Control District" (mimeo., n.d.), p. 1.

17. Ainsworth, "Thirteen Suits to Be Filed."

18. California Civil Procedure Code § 731(a), prior to amendment in 1947.

19. E. Ainsworth, "Divided Opinions on Smog Control Complicate Issue," *Los Angeles Times*, 5 Nov. 1946.

20. "Smog Problem Put Under Microscope," *Los Angeles Mirror*, 16 July 1954.

21. See Kennedy, *Los Angeles History*, pp. 6-7.

22. See Ainsworth, "Fight to Banish Smog"; idem, "Official Steps Under Way Against Smog Summarized," *Los Angeles Times*, 10 Nov. 1946; idem, "Smoke Abatement District Sought for Smog Control," *Los Angeles Times*, 25 Oct. 1946.

23. M. Schneider, "The Nature and Origin of Smog" (mimeo., 1970).

24. Griswold, "Air Pollution in California," p. 5.

25. Kennedy, *Los Angeles History*, p. 7. For an account of the early studies, see C. Senn, "General Atmospheric Pollution," *American Journal of Public Health* 38(1948):962.

26. E. Ainsworth, "St. Louis Has Key to Smog," *Los Angeles Times*, 9 Nov. 1946.

27. See Kennedy, *Los Angeles History,* pp. 5, 15; Ainsworth, "Fight to Banish Smog"; idem, "War on Smog Intensified; New Factors Speed Fight," *Los Angeles Times,* 2 Dec. 1946.

28. Ainsworth, "St. Louis Has Key to Smog."

29. Griswold, "Air Pollution in California," p. 5.

30. Ibid.; Kennedy, *Los Angeles History,* p. 5. See "Text of Report and Conclusions of Smog Expert." (All the quotations in the discussion of the Tucker report are from the last reference unless otherwise noted.)

31. See E. Ainsworth, "Bus Links to Smog Analyzed," *Los Angeles Times,* 3 Dec. 1946; idem, "Diesel Trucks Spout Smog; Officials Move to Control It," *Los Angeles Times,* 6 Dec. 1946.

32. E. Ainsworth, "Report on Smog to Spur Cleanup," *Los Angeles Times,* 21 Jan. 1947.

33. Kennedy, *Los Angeles History,* p. 7.

34. Ibid.

35. E. Ainsworth, "Smoke Abatement District Sought for Smog Control," *Los Angeles Times,* 25 Oct. 1946.

36. See E. Ainsworth, "Legality of Bans on Smog Studied," *Los Angeles Times,* 4 Dec. 1946.

37. Ainsworth, "War on Smog Intensified."

38. California Assembly Bill No. 1, 1947 Regular Session.

39. Kennedy, *Los Angeles History,* p. 10.

40. Ibid., pp. 11-14.

41. Ibid., p. 16, See Ch. 632, § 1, [1947] Cal. Stats. Ex. Sess. 1640, adding to California Health and Safety Code §§ 24198-24341.

4. State Legislation

1. Ch. 632, § 1, [1947] Cal. Stats. Ex. Sess. 1640, adding to California Health and Safety Code §§ 24198-24341. See generally H. Kennedy. *The History, Legal and Administrative Aspects of Air Pollution Control in Los Angeles County* (Los Angeles County Board of Supervisors, 1954), pp. 53-57; H. Kennedy and A. Porter, "Air Pollution: Its Control and Abatement," *Vanderbilt Law Review* 8(1955):854, 869-875.

2. California Health and Safety Code §§ 24199-24212 (1954).

3. California Health and Safety Code §§ 24242-24243 (1954).

4. California Health and Safety Code §§ 24260-24262 (1954).

5. California Health and Safety Code §§ 24263-24264 (1954).

6. California Health and Safety Code § 24265 (1954).

7. Kennedy, *Los Angeles History,* pp. 9, 53; Kennedy and Porter, "Air Pollution," p. 869; California Assembly Committee on Air and Water Pollution, *Final Summary Report* (n.d., c. 1952), p. 9 (hereafter cited as *Final Report*).

8. Kennedy, *Los Angeles History,* p. 53.

9. Ibid., p. 55.

10. Kennedy and Porter, "Air Pollution," p. 871.

11. Ibid.

12. Kennedy, *Los Angeles History,* p. 9.

13. Kennedy and Porter, "Air Pollution," p. 864.

14. People v. Plywood Mfrs. of California, 137 Cal. App. 2d Supp. 859, 291 P.2d 857 (1955), appeal dismissed, 351 U.S. 929, rehearing denied, 351 U.S. 990 (1956).

15. Kennedy, *Los Angeles History,* p. 20.

16. E. Ainsworth, "Smog Controller McCabe Arrives to Assume Duties," *Los Angeles Times,* 2 Oct. 1947.

17. E. Ainsworth, "Public Aid Sought in War on Smog," *Los Angeles Times,* 11 Oct. 1947.

18. E. Ainsworth, "Smogbound Industrial Area Viewed by Dr. McCabe as His D-Day Nears," *Los Angeles Times,* 9 Oct. 1947.

19. E. Ainsworth, "Smog Drive to Get Under Way," *Los Angeles Times,* 28 Sept. 1947.

20. Subcommittee of California Assembly Interim Committee on Governmental Efficiency and Economy, *Study and Analysis of the Facts Pertaining to Air Pollution Control in Los Angeles County* (1953), p. 5 (hereafter cited as *Study and Analysis*).

21. Ibid., pp. 7-11; S. Griswold, "Air Pollution in California—The Past" (mimeo., n.d.), pp. 6-7.

22. *Study and Analysis,* pp. 8, 10.

23. Ibid., pp. 8, 12.

24. Kennedy, *Los Angeles History,* p. 65.

25. Ibid., p. 51.

26. Ibid., pp. 14-15.

27. Griswold, "Air Pollution in California," p. 7.

28. Kennedy, *Los Angeles History,* p. 74.

29. Griswold, "Air Pollution in California," p. 7.

30. Kennedy, *Los Angeles History,* p. 15.

31. Griswold, "Air Pollution in California," p. 8.

32. California Department of Public Health, *Clean Air for California* (Initial Report, 1955), p. 47; *Final Report,* pp. 8, 25.

33. R. Schiller, "The Los Angeles Smog," *National Municipal Review* 44(1955): 562.

34. California Department of Public Health, *Clean Air for California,* p. 47; University of California Bureau of Public Administration, *Air Pollution Control* (prepared for Joint Subcommittee on Air Pollution, California Assembly Committees on Conservation, Planning, Public Works, and Health, 1955), p. 32.

35. Ch. 1185, § 1, [1949] Cal. Stats. Reg. Sess. 2104, adding to California Health and Safety Code §§ 24330-24341.

36. California Department of Public Health, *Clean Air for California,* p. 47.

37. *Final Report,* pp. 8-10, 25.

38. Ibid., p. 9.

39. Ibid.

40. Ibid.

41. Ibid., p. 43 (emphasis added).

42. Ibid., pp. 43, 45.

43. Ibid., p. 43.

44. University of California Bureau of Public Administration, *Air Pollution Control,* p. 45.

45. *Final Report,* p. 43.

46. Kennedy, *Los Angeles History,* p. 73.
47. *Final Report,* pp. 34, 36.
48. Ibid., pp. 49, 50.
49. Ibid., p. 50.
50. Ibid.
51. Ibid., p. 52.
52. Ibid., pp. 52-53.
53. Ibid., p. 50.
54. Ibid., p. 23.
55. *Study and Analysis,* pp. 5, 12, 17.
56. *Final Report,* p. 23.
57. Ibid.
58. Ibid., p. 24.
59. Ibid.
60. Ibid.
61. Ibid., p. 25.
62. Ibid., p. 23.

5. Changes in Direction

1. F. Fredrick and L. Lowry, "Legislative History and Analysis of the Bay Area Air Pollution Control District" (mimeo., n.d.), p. 4, footnote 17.
2. G. Larson, "Reduction at the Source," in *Proceedings of First National Air Pollution Symposium* (Los Angeles, Calif., 1949), p. 77.
3. L. McCabe, "National Trends in Air Pollution," in *ibid.,* pp. 50, 51-52.
4. California Assembly Committee on Air and Water Pollution, *Final Summary Report* (n.d., c. 1952), p. 41 (hereafter cited as *Final Report*).
5. L. McEwen, "Exhaust Gases—Their Relation to Atmospheric Pollution" (mimeo., Los Angeles County Office of Air Pollution Control and U.C.L.A. Engineering Dept., Los Angeles, Calif., 1947), pp. 1, 2, 6, 7.
6. *Final Report,* p. 35.
7. Ibid.
8. Air Pollution Foundation, *Final Report* (Los Angeles, Calif., 1961), p. 7.
9. *Final Report,* p. 41.
10. Ibid., p. 40.
11. Ibid.
12. Ibid., pp. 40-41.

6. Discovering and Documenting the Role of the Automobile

1. Los Angeles County Air Pollution Control District and American Broadcasting Company (Station KABC), "The Smog Story" (Transcript of weekly radio series, 1955), pp. 23-24.
2. Stanford Research Institute, *The Smog Problem in Los Angeles County* (Menlo Park, Calif., Second Interim Report, 1949), pp. 6-10; Los Angeles County Air Pollution Control District, *Annual Report, 1949-1950* (1950), p. 8.
3. A. Haagen-Smit, "Chemistry and Physiology of Los Angeles Smog," *Industrial and Engineering Chemistry* 44(1952):1342.

4. D. Fisher, "Haagen-Smit on Smog," *Westways,* Aug. 1972, p. 36; interviews with A. Haagen-Smit, July 1971; Haagen-Smit, "Chemistry and Physiology."

5. Interviews with A. Haagen-Smit; A. Carlin and G. Kocher, *Environmental Problems: Their Causes, Cures, and Evolution, Using Southern California Smog as an Example* (Rand Corp., Santa Monica, Calif., Report Number R-640-CC/RC, 1971), p. 58.

6. California Assembly Committee on Air and Water Pollution, *Final Summary Report* (n.d., c. 1952), pp. 10, 36, 39–40 (hereafter cited as *Final Report*).

7. Ibid., p. 41; Carlin and Kocher, *Environmental Problems,* p. 58.

8. *Final Report,* pp. 41–42; Carlin and Kocher, *Environmental Problems,* p. 58; A. Haagen-Smit, "The Air Pollution Problem in Los Angeles," *Engineering and Science* 14(1950):7; W. Barton, "Puzzle of Smog Production Solved by Caltech Scientist," *Los Angeles Times,* 20 Nov. 1950.

9. *Final Report,* p. 42; Barton, "Puzzle of Smog."

10. See Carlin and Kocher, *Environmental Problems,* pp. 58–59.

11. Subcommittee of California Assembly Interim Committee on Governmental Efficiency and Economy, *Study and Analysis of the Facts Pertaining to Air Pollution Control in Los Angeles County* (1953), p. 15 (hereafter cited as *Study and Analysis*).

12. R. Harbinson, "Scientists Trying to Develop Plants Immune from Smog," *San Bernardino* [*Calif.*] *Evening Telegram,* 11 Dec. 1952.

13. B. Barger, "Shaw Promises Anti-Smog Bill," *Riverside* [*Calif.*] *Daily Press,* 8 Dec. 1953; *Study and Analysis,* p. 15.

14. Fisher, "Haagen-Smit on Smog," p. 38; R. Schiller, "The Los Angeles Smog," *National Municipal Review* 44(1955):558, 563.

15. Stanford Research Institute, *The Smog Problem in Los Angeles County* (Second Interim Report), pp. 5–8.

16. *Final Report,* p. 10.

17. Stanford Research Institute, *The Smog Problem in Los Angeles County* (Menlo Park, Calif., Third Interim Report, 1950), p. 32.

18. B. Barger, "Smog Probably Result of Sins of Both Industry, Politicians," *Riverside* [*Calif.*] *Daily Press,* 8 Dec. 1953.

19. Fisher, "Haagen-Smit on Smog," p. 38; A. Haagen-Smit, "The Control of Air Pollution in Los Angeles," *Engineering and Science,* Dec. 1954, p. 11.

20. Barger, "Smog Probably Result of Sins"; idem, "Regional Smog Control Need to Be Studied," *Riverside* [*Calif.*] *Daily Press,* 5 Dec. 1953.

21. "Researchers Say Ozone Damaging Part of Smog," *Riverside* [*Calif.*] *Daily Press,* 22 Oct. 1953.

22. B. Barger, "Smog Scientists Join Study Forces," *Riverside* [*Calif.*] *Daily Press,* 22 July 1954.

23. "L.A. Farm Bureau Head Tells of Smog Damages," *Riverside* [*Calif.*] *Daily Press,* Nov. 1953.

24. P. Magill, D. Hutchison, and J. Stormes, "Hydrocarbon Constituents of Automobile Exhaust Gases," in *Proceedings of Second National Air Pollution Symposium* (Los Angeles, Calif., 1952), pp. 71, 82.

25. *Final Report,* pp. 10–11.

26. Stanford Research Institute, *The Smog Problem in Los Angeles County* (Menlo Park, Calif., 1954), p. 53.

27. B. Barger, "Extended Smog Research Near," *Riverside* [*Calif.*] *Daily Press,* 24 Nov. 1953.

28. Air Pollution Foundation, *1955 President's Report* (Los Angeles, Calif., 1955), p. 12.

29. Quoted in B. Zimmerman, "Smog Sessions Bring Discord," *Riverside* [*Calif.*] *Daily Press,* 4 June 1954.

30. Quoted in Fisher, "Haagen-Smit on Smog," p. 38.

31. Air Pollution Foundation, *Final Report* (Los Angeles, Calif., 1961), pp. 7-8.

32. S. Griswold, "Air Pollution in California—The Past" (mimeo., n.d.), pp. 12-13; "Supervisors Vote to Replace Larson as Smog Control Chief," *Los Angeles Times,* 16 Dec. 1953; "L.A. Supervisor Demands Ouster of Smog Chief," *Riverside* [*Calif.*] *Daily Press,* 26 Dec. 1953; "Pure Air Group Backs Larson," *Los Angeles Examiner,* 26 Nov. 1954.

33. Air Pollution Foundation, *Final Report,* p. 20.

34. Quoted in Griswold, "Air Pollution in California," p. 10.

35. Air Pollution Foundation, *Final Report,* p. 7.

36. See Griswold, "Air Pollution in California," pp. 9, 12, 14; Air Pollution Foundation, *1957 Annual Report* (Los Angeles, Calif., 1957), p. 11.

37. Air Pollution Foundation, *1957 Annual Report,* p. 11.

38. Griswold, "Air Pollution in California," p. 12; A. Beckman et al., *Report of Special Committee on Air Pollution to Governor Goodwin Knight's Air Pollution Control Conference* (Los Angeles, Calif., 1953), p. 1.

39. Beckman et al., *Report of Special Committee,* pp. 7-8.

40. Ibid., pp. 13-14.

41. "New Name Picked for Smog Fighters," *Riverside* [*Calif.*] *Daily Press,* 27 Dec. 1954.

42. Air Pollution Foundation, *Final Report,* p. 8.

43. Air Pollution Foundation, *1954 President's Report* (Los Angeles, Calif., 1954), p. 19.

44. Air Pollution Foundation, *Final Report,* pp. 11-12.

45. Ibid.

46. See W. Faith, "Combustion and Smog" (Air Pollution Foundation Report, Los Angeles, Calif., 1954).

47. Air Pollution Foundation, *Final Report,* pp. 12-14.

48. Ibid., pp. 16, 24, 25.

49. California Department of Public Health, *A Progress Report of California's Fight against Air Pollution* (Third Report, 1957), p. 7.

50. Air Pollution Foundation, *1957 Annual Report,* p. 7; idem, *Final Report,* p. 7.

51. U.S. Department of Commerce, *The Automobile and Air Pollution,* Part 2 (1967), p. 26.

52. Letter from C. Chayne, vice pres. engineering, General Motors Corp., to K. Hahn, Los Angeles County Board of Supervisors, 28 Dec. 1953; letter from J. Nance, pres. Packard Motor Car Co., to K. Hahn, 30 Dec. 1953, both in *Smog—A Factual Record of Correspondence between Kenneth Hahn and Automobile Companies,* 6th ed. (Los Angeles, Calif., 1970), pp. 9, 10.

53. Quoted in R. Bedingfield, "A Student of Air Pollution by Automobiles," *New York Times,* 1 Feb. 1970, sec. 3.

54. J. Campbell, "1956—A Year of Decisions," in *Proceedings of Southern*

California Conference on Elimination of Air Pollution (Los Angeles, Calif., 10 Nov. 1955), pp. 112, 115; interview with W. King, 24 Aug. 1971.

55. Griswold, "Air Pollution in California," p. 13.

56. "Auto Engineers Arrive in L.A. to Study Smog," *Riverside* [*Calif.*] *Daily Press,* 27 Jan. 1954.

57. "Automotive Industry Aid Pledged in Fumes Study," *Los Angeles Times,* 3 Feb. 1954.

58. J. Chandler, "Extent to Which Available Control Techniques Have Been Utilized by Industry," in *Proceedings of National Conference on Air Pollution* (Washington, D.C., 18-20 Nov. 1958), pp. 348, 349.

59. H. Williams, "Accomplishments in Air Pollution Control by the Automobile Industry," in ibid., pp. 57, 59; Chandler, "Extent to Which Techniques Utilized," pp. 349-352.

60. J. Esposito, *Vanishing Air* (New York: Grossman Publishers, 1970), p. 41.

61. Williams, "Accomplishments in Control," p. 60.

62. Schiller, "Los Angeles Smog," p. 563.

63. Campbell, "Year of Decisions," p. 117; idem, "Remarks at Press Session," in *Proceedings of Southern California Conference on Elimination of Air Pollution,* pp. 122, 126.

64. See S. Griswold, "Reflections and Projections on Controlling the Motor Vehicle" (Paper presented at Annual Meeting of the Air Pollution Control Association, Houston, Texas, 21-25 June 1964), pp. 5-7.

65. Esposito, *Vanishing Air,* p. 44.

66. Quoted in Bedingfield, "A Student of Air Pollution."

67. Esposito, *Vanishing Air,* p. 37; interview with W. King.

68. Letter from D. Chabek, News Dept., Ford Motor Co., to K. Hahn, 3 March 1953, in *Smog—A Factual Record,* p. 4.

69. Letter from J. Campbell, administrative director, Engineering, General Motors Corp., to K. Hahn, 26 March 1953, in ibid., p. 6.

70. Air Pollution Foundation, *1954 President's Report,* p. 8.

71. L. Hitchcock, "Diagnosis and Prescription," in *Proceedings of Second Southern California Conference on Elimination of Air Pollution* (Los Angeles, Calif., 14 Nov. 1956), pp. 76, 77-78.

72. Air Pollution Foundation, *1956 President's Report* (Los Angeles, Calif., 1956), p. 12.

73. See, e.g., "Smog Fighters Claim Car Fumes Last Unlicked Pollution Trouble Source," *Riverside* [*Calif.*] *Daily Press,* 13 Feb. 1958.

74. Quoted in Esposito, *Vanishing Air,* p. 38.

7. Responses to an Increasing Problem: The Search for Technological Solutions

1. H. McKee, "What's in the Air?" in *Proceedings of National Conference on Air Pollution* (Washington, D.C., 18-20 Nov. 1958), pp. 31, 33, 34.

2. B. Barger, "New York 'Smaze' Smokier, Dirtier than LA Area Smog," *Riverside* [*Calif.*] *Daily Press,* 30 Nov. 1953.

3. Stanford Research Institute, *The Smog Problem in Los Angeles County* (Menlo Park, Calif., 1954), p. 9.

4. "Philly Copies LA," *Riverside* [*Calif.*] *Daily Press,* 31 May 1957.

5. A. Stern, "Changes in Identity and Quantity of Pollutants, Past, Present, and Future," in *Proceedings of National Conference on Air Pollution,* pp. 86, 92-93.

6. T. Patterson, "Middleton Urges Air Control Action to Fight Smog Damage," *Riverside* [*Calif.*] *Daily Press,* 21 Feb. 1959.

7. California Department of Public Health, *A Progress Report of California's Fight Against Air Pollution* (Third Report, 1957), p. 5; J. Lefler, "Smog Said Growing into Major Statewide Problem," *Riverside* [*Calif.*] *Daily Press,* 17 Jan. 1957; California Department of Public Health, *Clean Air for California* (Second Report, 1956), p. 5.

8. California Department of Public Health, *Progress Report,* p. 6.

9. California Assembly Committee on Air and Water Pollution, *Final Summary Report* (n.d., c. 1952), p. 7.

10. California Department of Public Health, *Technical Report: California Standards for Ambient Air Quality and Motor Vehicle Exhaust* (n.d., c. 1960), p. 7.

11. Ibid., p. 94, figure 24.

12. Stanford Research Institute, *The Smog Problem,* p. 8.

13. California Department of Public Health, *Clean Air for California* (Initial Report, 1955), p. 51, table 1.

14. "Press Session," in *Proceedings of Southern California Conference on Elimination of Air Pollution* (Los Angeles, Calif., 10 Nov. 1955), pp. 45, 54-55; L. DuBridge, "Conference Summation," in ibid., pp. 130, 136.

15. L. Hitchcock, "What the Air Pollution Foundation Is Doing," in ibid., pp. 65, 75.

16. A. Carlin and G. Kocher, *Environmental Problems: Their Causes, Cures, and Evolution, Using Southern California Smog as an Example* (Rand Corp., Santa Monica, Calif., Report Number R-640-CC/RC, 1971), p. 100, figure B-1; p. 102, figure B-2; p. 104, figure B-3; p. 105, figure B-4.

17. M. Neiburger, "Weather Modification and Smog," *Science* 126(1957):637, 638; California Department of Public Health, *Clean Air for California* (Initial Report), p. 43.

18. Neiburger, "Weather Modification," pp. 639, 641-642.

19. "Smoke Umbrella Suggested as Possible Smog Answer," *Riverside* [*Calif.*] *Daily Press,* 21 April 1955.

20. Neiburger, "Weather Modification," p. 644.

21. "Press Session," pp. 50, 55.

22. Neiburger, "Weather Modification," p. 641.

23. Los Angeles County Air Pollution Control District and American Broadcasting Company (Station KABC), "The Smog Story" (Transcript of weekly radio series, 1955), p. 21 (hereafter cited as "Smog Story").

24. California Department of Public Health, *Clean Air for California* (Initial Report), p. 43. See also Neiburger, "Weather Modification," p. 641.

25. See Neiburger, "Weather Modification," pp. 638-644.

26. "Knight Opposes Smog Session," *Riverside* [*Calif.*] *Daily Press,* Jan. 1954.

27. Neiburger, "Weather Modification," p. 644.

28. Subcommittee of California Assembly Interim Committee on Governmental Efficiency and Economy, *Study and Analysis of the Facts Pertaining to Air Pollution Control in Los Angeles County* (1953), pp. 6, 12.

29. California Department of Public Health, *Clean Air for California* (Initial Report), p. 43.

30. "Knight Opposes Smog Session."

31. Letter from K. Hahn, Los Angeles County Board of Supervisors, to L. Colbert, pres. Chrysler Corp., 14 Dec. 1953, in *Smog—A Factual Record of Correspondence between Kenneth Hahn and Automobile Companies,* 6th ed. (Los Angeles, Calif., 1970), pp. 7, 8.

32. B. Barger, "Drastic Refinery Controls Urged by Governor's Group," *Riverside* [*Calif.*] *Daily Press,* 7 Dec. 1953, quoting A. Beckman et al., *Report of Special Committee on Air Pollution to Governor Goodwin Knight's Air Pollution Control Conference* (Los Angeles, Calif., 1953), p. 9.

33. H. Kennedy, "Levels of Responsibility for the Administration of Air-Pollution-Control Programs," in *Proceedings of National Conference on Air Pollution,* pp. 389-395.

34. See, e.g., R. Schiller, "The Los Angeles Smog," *National Municipal Review* 44(1955):558, 563; M. White, "5 Years More Smog Feared for L.A. Area," *Los Angeles Examiner,* 17 Nov. 1954.

35. D. Fort et al., *Proposal for a "Smog Tax"* (Rand Corp., Santa Monica, Calif., Report Number P-1621-RC, 1959), pp. 2, 4.

36. Air Pollution Foundation, *Final Report* (Los Angeles, Calif., 1961), p. 14.

37. Air Pollution Foundation, *1955 President's Report* (Los Angeles, Calif., 1955), pp. 21-22.

38. DuBridge, "Conference Summation," pp. 130, 142.

39. Air Pollution Foundation, *1956 President's Report* (Los Angeles, Calif., 1956), p. 26.

40. See Air Pollution Foundation, *1957 Annual Report* (Los Angeles, Calif., 1957), pp. 12-16.

41. California Department of Public Health, *Progress Report,* p. 26.

42. Air Pollution Foundation, *1957 Annual Report,* pp. 17-18.

43. L. Chambers, "Transportation Sources of Air Pollution—Comparison With Other Sources in Los Angeles," in *Proceedings of National Conference on Air Pollution,* pp. 167-170.

44. Air Pollution Foundation, *1958 Annual Report* (Los Angeles, Calif., 1958), pp. 7, 9.

45. Air Pollution Foundation, *Final Report,* pp. 31-32, 36-39, 46.

46. H. Kennedy, "The Legal Aspects of Air Pollution Control with Particular Reference to the County of Los Angeles," *Southern California Law Review* 27(1954):373, 393. A substantial portion of the opinion, which was issued 27 Oct. 1953, is reprinted at p. 394 of this reference.

47. Letter from Hahn to Colbert, p. 8.

48. Kennedy, "Legal Aspects," p. 394 (quoting from Opinion of Counsel).

49. Quoted in "Smog Devices Can Be Ordered on Autos," *Los Angeles Daily News,* Nov. 1953.

50. Air Pollution Foundation, *Final Report,* pp. 20, 22.

51. "Smog Story," p. 27.

52. S. Griswold, "Reflections and Projections on Controlling the Motor Vehicle" (Paper presented at Annual Meeting of the Air Pollution Control Association, Houston, Texas, 21-25 June 1964), p. 2.

53. "Smog Story," pp. 27, 51.

54. Letter from K. Hahn to Henry Ford II, pres. Ford Motor Co., 13 April 1956, in *Smog—A Factual Record,* p. 14.

55. "Panel Discussion," in *Proceedings of Second Southern California Conference on Elimination of Air Pollution* (Los Angeles, Calif., 14 Nov. 1956), pp. 11, 46.

56. Letter from K. Hahn to L. Colbert, 28 March 1958, in *Smog—A Factual Record,* pp. 22, 23.

57. See, e.g., Air Pollution Foundation, *Final Report,* pp. 11-46.

58. M. White, "Knight in All-Out Smog War," *Los Angeles Examiner,* 26 Oct. 1954. See also *Resolution by the California Assembly Public Health Committee and Assembly Conservation, Planning, and Public Works Committee,* 16 March 1954, in University of California Bureau of Public Administration, *Air Pollution Control* (prepared for Joint Subcommittee on Air Pollution, California Assembly Committees on Conservation, Planning, Public Works, and Health, 1955), pp. 56-60.

59. See, e.g., University of California Bureau of Public Administration, *Air Pollution Control,* pp. 44-45; Air Pollution Foundation, *Final Report,* pp. 22, 39, 44.

60. See J. Esposito, *Vanishing Air* (New York: Grossman Publishers, 1970), pp. 26-68; U.S. Department of Commerce, *The Automobile and Air Pollution,* Part 2 (1967), pp. 26-28, 47-48.

61. California Department of Public Health, *Clean Air for California* (Initial Report), p. 42.

62. See the correspondence between representatives of the various motor companies and K. Hahn, in *Smog—A Factual Record,* pp. 15-22.

63. Letter from J. Chandler, engineering staff, Ford Motor Co., to K. Hahn, 9 April 1958, in ibid., pp. 24-25.

64. "Legislature to Hear of Smog Research," *Los Angeles Herald Express,* 14 Feb. 1959.

65. See M. White, "Experts Seek Auto Anti-Smog Device," *Los Angeles Examiner,* July 1957.

66. U.S. Department of Commerce, *The Automobile and Air Pollution,* Part 2, pp. 26-27.

8. Responses to an Increasing Problem: Moves for Federal and State Involvement

1. See, e.g., M. White, "Knight in All-Out Smog War," *Los Angeles Examiner,* 26 Oct. 1954.

2. The discussion is drawn from J. Davies, *The Politics of Pollution* (New York: Pegasus, 1970), pp. 34, 49-51; R. Dyck, "Evolution of Federal Air Pollution Control Policy" (Ph.D. diss., University of Pittsburgh, 1971), pp. 18-41, 160-165, 182-189; J. Fromson, "A History of Federal Air Pollution Control," *Ohio State Law Journal* 30(1969):516; R. Ripley, "Congress and Clean Air: The Issue of Enforcement, 1963," in *Congress and Urban Problems,* ed. F. Cleaveland et al. (Washington, D.C.: The Brookings Institution, 1969), pp. 224-234; J. Sundquist, *Politics and Policy: The Eisenhower, Kennedy, and Johnson Years* (Washington, D.C.: The Brookings Institution, 1968), pp. 331-333.

3. Davies, *The Politics of Pollution,* p. 34; Dyck, "Evolution of Federal Policy," pp. 19-20.

4. Davies, *The Politics of Pollution,* pp. 49-50.

5. Dyck, "Evolution of Federal Policy," pp. 20-23.

6. Ripley, "Congress and Clean Air," p. 228.

7. L. McCabe, "Technical Aspects—The 1950 Assessment," in *Proceedings of National Conference on Air Pollution* (Washington, D.C., 18-20 Nov. 1958), pp. 25, 27.

8. H. Kennedy, "Levels of Responsibility for the Administration of Air-Pollution-Control Programs," in ibid., pp. 396-397.

9. Davies, *The Politics of Pollution,* p. 50; Dyck, "Evolution of Federal Policy," pp. 23-26; Ripley, "Congress and Clean Air," pp. 228-229.

10. S. Griswold, "The Smog Problem in Los Angeles County," in *Proceedings of Southern California Conference on Elimination of Air Pollution* (Los Angeles, Calif., 10 Nov. 1955), pp. 2, 12.

11. Dyck, "Evolution of Federal Policy," pp. 28, 161; McCabe, "Technical Aspects," p. 27. See generally P. McDaniel and A. Kaplinsky, "The Use of the Federal Income Tax System to Combat Air and Water Pollution: A Case Study in Tax Expenditures," *Boston College Industrial and Commercial Law Review* 12(1971):351.

12. Dyck, "Evolution of Federal Policy," pp. 28-29, 161; Davies, *The Politics of Pollution,* p. 50; Ripley, "Congress and Clean Air," pp. 229-230.

13. Editorial, "Uncle Sam Should Get into Smog Fight," *Los Angeles Mirror,* 17 July 1954.

14. Dyck, "Evolution of Federal Policy," pp. 29-33, 161-162, 183-185; Davies, *The Politics of Pollution,* pp. 50-51; Ripley, "Congress and Clean Air," pp. 230-231.

15. Quoted in Sundquist, *Politics and Policy,* pp. 331-332.

16. T. Kuchel, "Public Interest Demands Clean Air," in *Proceedings of National Conference on Air Pollution,* pp. 15, 16-17.

17. Sundquist, *Politics and Policy,* p. 332; Ripley, "Congress and Clean Air," pp. 320-331; Dyck, "Evolution of Federal Policy," pp. 31-32.

18. An Act to Provide Research and Technical Assistance Relating to Air Pollution Control, Ch. 360, 69 Stat. 322, codified at 42 U.S.C. § 1857.

19. Staff of the Senate Committee on Public Works, 88th Cong., 1st Sess., *A Study of Pollution—Air* 24 (Comm. Print 1963).

20. H.R. Rep. No. 968, 84th Cong., 1st Sess. 4 (1955).

21. California Department of Public Health, *Clean Air for California* (Second Report, 1956), p. 22.

22. Staff of the Senate Committee on Public Works, *A Study of Pollution,* pp. 24-30.

23. L. Chambers, "What Has Been Done, Is Being Done, Will Be Done about Smog in Southern California: The Role of the Federal Government," in *Proceedings of Southern California Conference on Elimination of Air Pollution,* pp. 39, 42-43. See also California Department of Public Health, *Clean Air for California* (Second Report), p. 22; idem, *A Progress Report of California's Fight Against Air Pollution* (Third Report, 1954), pp. 18-19.

24. Dyck, "Evolution of Federal Policy," pp. 33-36, 164, 186-187; Ripley, "Congress and Clean Air," pp. 232, 237, 243-244. See *Proceedings of National Conference on Air Pollution.*

25. Dyck, "Evolution of Federal Policy," pp. 33-35, 40-42; Ripley, "Congress and Clean Air," pp. 232, 234.

26. Quoted in Sundquist, *Politics and Policy,* p. 332. See also Ripley, "Congress and Clean Air," p. 238.

27. Sundquist, *Politics and Policy,* pp. 332-333; Dyck, "Evolution of Federal Policy," pp. 37-39, 41, 188-189; Ripley, "Congress and Clean Air," pp. 232-233.

28. California Assembly Committee on Air and Water Pollution, *Final Summary Report* (n.d., c. 1952), p. 49.

29. W. Fredrick, "Biological Aspects of Air Pollution," in *Proceedings of Second National Air Pollution Symposium* (Los Angeles, Calif., 1952), pp. 106, 107.

30. California Assembly Committee on Air and Water Pollution, *Final Summary Report,* p. 52.

31. See, e.g., ibid.

32. J. Haggerty, "City Asks State Control of Smog," *Riverside* [*Calif.*] *Daily Press,* 13 Oct. 1954.

33. "Smog Not Dangerous to Health, Scientists Assure Gov. Knight," *Los Angeles Times,* 17 Oct. 1954.

34. White, "Knight in All-Out Smog War"; California Department of Public Health, *Clean Air for California* (Second Report), p. 5.

35. Interview with L. Breslow, 2 Aug. 1971.

36. California Department of Public Health, *Clean Air for California* (Initial Report, 1955), pp. 8-9, 49.

37. "Smog Aid Sought from U.S., State," *Los Angeles Daily News,* 29 Sept. 1953.

38. B. Barger, "Regional Smog Control Need to Be Studied," *Riverside* [*Calif.*] *Daily Press,* 5 Dec. 1953.

39. J. Haggerty, "Repeated Efforts Fail to Win State Action on Smog," *Riverside* [*Calif.*] *Daily Press,* Oct. 1954; idem, "City Asks State Control of Smog."

40. See University of California Bureau of Public Administration, *Air Pollution Control* (prepared for Joint Subcommittee on Air Pollution, California Assembly Committees on Conservation, Planning, Public Works, and Health, 1955), p. 40.

41. California Department of Public Health, *Clean Air for California* (Initial Report), p. 47.

42. Haggerty, "City Asks State Control of Smog."

43. J. Haggerty, "League Rejects City's Smog Plea," *Riverside* [*Calif.*] *Daily Press,* 18 Oct. 1954.

44. "Smog Aid Sought from U.S., State"; "Smog Problem May Go to Legislature," *Riverside* [*Calif.*] *Daily Press,* 19 Oct. 1953; Barger, "Regional Smog Control Need to Be Studied."

45. Ch. 1312, § 1, [1955] Cal. Stats. Reg. Sess. 2385, adding to California Health and Safety Code §§ 425, 426.

46. J. Maga, "Environmental Sanitation Activities of the State Department of Public Health in Air Pollution," in *Proceedings of Southern California Conference on Elimination of Air Pollution,* pp. 33-38.

47. California Department of Public Health, *Clean Air for California* (Second Report), pp. 5, 23; idem, *Progress Report,* pp. 5, 6, 10, 26, 32.

48. See, e.g., J. Maga, Testimony in Hearings before Special Subcommittee on Air and Water Pollution of the Senate Committee on Public Works, 88th Cong., 2d Sess. 57 (1964).

49. See *Riverside* [*Calif.*] *Daily Press*: "Smog Termed Near Disaster Stage in L.A.," 31 Aug. 1955; "Deadly Smog Could Kill Thousands, Cause Panic," 26 Sept. 1956; B. Boden, "Record Smog Attack Hits," 7 Sept. 1957.

50. "Smog Bill," *Riverside* [*Calif.*] *Daily Press,* 26 Jan. 1957.

51. The description of events in 1958–1959 is taken from T. Roberts, "Motor Vehicular Air Pollution Control in California: A Case Study in Political Unresponsiveness" (Honors thesis, Harvard College, 1969), pp. 28–30; Los Angeles Air Pollution Control District, *Administration and Enforcement of Vehicular Contaminant Control Measures in Los Angeles County* (1963), pp. 1–3; G. Siegel, "The Status of Motor Vehicle Emission Control Public Policy" (mimeo., n.d.), pp. 30–33.

52. Los Angeles Air Pollution Control District, *Administration and Enforcement,* p. 1.

53. Siegel, "Status of Policy," p. 30; Roberts, "Motor Vehicular Air Pollution Control," pp. 28–29.

54. M. Landsberg, "Measure Urged to Force Car Exhaust Curbs," *Riverside* [*Calif.*] *Daily Press,* 10 Feb. 1959.

55. Cf. 42 Opinions of the California Attorney General 47, 48 (1963) (Opinion No. 63-444).

56. Ch. 200, § 1, Ch. 835, § 1, [1959] Cal. Stats. Reg. Sess. 2091, 2885, adding to California Health and Safety Code §§ 426.1, 426.5.

9. Dealing with Uncertainty

1. California Department of Public Health, *Clean Air for California* (Initial Report, 1955), pp. 33–36, 41.

2. See "Panel Discussion," in *Proceedings of Second Southern California Conference on Elimination of Air Pollution* (Los Angeles, Calif., 14 Nov. 1956), pp. 147–148.

3. "Press Session," in ibid., pp. 165, 168, 171.

4. California Department of Public Health, *A Progress Report of California's Fight against Air Pollution* (Third Report, 1957), pp. 10, 27, 31.

5. California Department of Public Health, *Technical Report: California Standards for Ambient Air Quality and Motor Vehicle Exhaust* (n.d., c. 1960), p. 86 (hereafter cited as *Technical Report*).

6. Statement of S. Griswold (mimeo., n.d., c. 1955).

7. California Department of Public Health, *Clean Air for California* (Initial Report), p. 8.

8. "Panel Discussion," in *Proceedings of Second Southern California Conference on Elimination of Air Pollution,* pp. 147–148.

9. California Department of Public Health, *Clean Air for California* (Initial Report), p. 48.

10. California Department of Public Health, *Clean Air for California* (Second Report, 1956), pp. 5–6.

11. California Department of Public Health, *Progress Report,* p. 31.

12. See California Health and Safety Code §§ 426.1, 426.5 (1961).

13. *Technical Report,* p. 3 (emphasis added).

14. Ibid., pp. 13–16, 96.

15. Ibid., pp. 31-32.
16. Ibid., p. 25.
17. Ibid., p. 14.
18. See, e.g., ibid., pp. 37, 86.
19. Ibid., pp. 13, 37.
20. Ibid., p. 78.
21. Ibid., pp. 16, footnotes 1 and 2, 78.
22. J. Maga, Testimony in Hearings before Special Subcommittee on Air and Water Pollution of the Senate Committee on Public Works. 88th Cong., 2d Sess. 61 (1964).
23. See *Technical Report,* pp. 27, 30-33.
24. Ibid., p. 31.
25. Ibid., p. 27.
26. Ibid., pp. 13-14 (emphasis added); Maga, Testimony in Hearings, p. 70.
27. California Health and Safety Code §§ 426.1, 426.5 (1961).
28. See *Technical Report,* p. 3.
29. Ibid., pp. 93, 96. See also J. Maga and G. Hass, "The Development of Motor Vehicle Exhaust Emission Standards in California" (Paper presented at 53d Annual Meeting of the Air Pollution Control Association, Cincinnati, Ohio, 25 May 1960), p. 3.
30. *Technical Report,* p. 16; Maga and Hass, "Development of Standards," p. 3.
31. California Health and Safety Code § 426.5 (1961). See Maga and Hass, "Development of Standards," p. 3.
32. *Technical Report,* p. 107; Maga and Hass, "Development of Standards," p. 10.
33. *Technical Report,* pp. 16, 96, 107; Maga and Hass, "Development of Standards," p. 3.
34. Maga and Hass, "Development of Standards," p. 3; *Technical Report,* p. 96.
35. *Technical Report,* pp. 96, 98; Maga and Hass, "Development of Standards," p. 3.
36. Maga and Hass, "Development of Standards," pp. 5-6; *Technical Report,* pp. 104-106, 110; J. Askew, Testimony in Hearings before Special Subcommittee on Air and Water Pollution of the Senate Committee on Public Works, 88th Cong., 2d Sess. 62, 67 (1964); interview with G. Taylor, 23 June 1971.
37. Maga and Hass, "Development of Standards," pp. 6-8; *Technical Report,* pp. 104-106; Maga, Testimony in Hearings, p. 70.
38. *Technical Report,* pp. 3, 17.
39. Maga and Hass, "Development of Standards," p. 3.
40. *Technical Report,* pp. 96, 107, 110 (emphasis added).
41. Ibid., pp. 16, 78.
42. Ibid., pp. 96, 107.
43. Maga and Hass, "Development of Standards," p. 3.
44. See, e.g., *Technical Report,* pp. 16, 77-78.
45. California Health and Safety Code § 426.5 (1961).
46. Maga and Hass, "Development of Standards," p. 3.
47. *Technical Report,* pp. 15-16, 81-83; Maga and Hass, "Development of Standards," p. 10.
48. Maga and Hass, "Development of Standards," p. 10; *Technical Report,* p. 16.
49. *Technical Report,* pp. 15-16.

50. Ibid., pp. 79, 96; Air Pollution Foundation, *1955 President's Report* (Los Angeles, Calif., 1955), pp. 19-20; California Department of Public Health, *Progress Report,* p. 32.

51. Maga and Hass, "Development of Standards," p. 10.

52. See *Technical Report,* p. 16; Maga and Hass, "Development of Standards," p. 10.

53. A. Haagen-Smit, Testimony in Hearings before Special Subcommittee on Air and Water Pollution of the Senate Committee on Public Works, 88th Cong., 2d Sess. 103, 109 (1964).

10. State Experimentation

1. T. Roberts, "Motor Vehicular Air Pollution Control in California: A Case Study in Political Unresponsiveness" (Honors thesis, Harvard College, 1969), p. 30.

2. Ibid., p. 113.

3. J. Morganthaler, "Anti-Smog Bill Wins Victory," *Riverside [Calif.] Daily Press,* 1 April 1960; Roberts, "Motor Vehicular Air Pollution Control," pp. 30, 49, 53, 113-114.

4. See J. Morganthaler, "Legislature Receives New Plans for Smog Control," *Riverside [Calif.] Daily Press,* 4 March 1960; Los Angeles County Air Pollution Control District, *Administration and Enforcement of Vehicular Contaminant Control Measures in Los Angeles County* (1963), p. 2.

5. See "Tougher Smog Bill Requested," *Riverside [Calif.] Daily Press,* 5 March 1960; "Fresh Attack on Car Exhaust Seen," *Riverside [Calif.] Daily Press,* 7 March 1960.

6. J. Reynolds, "Brown Proposes Car-Smog Action," *Riverside [Calif.] Daily Press,* 23 March 1960.

7. See "Anti-Smog Bill Wins Approval," *Riverside [Calif.] Daily Press,* 25 March 1960; "Auto Smog Board Provided," *Los Angeles Examiner,* 30 March 1960; C. Greenberg, "Assembly OKs Auto Smog Law," *Los Angeles Examiner,* 30 March 1960; Morganthaler, "Anti-Smog Bill Wins Victory"; "Legislature Approves Car Smog-Cutter Bill," *Riverside [Calif.] Daily Press,* 6 May 1960.

8. "Dilworth Slaps Exhaust Action," *Riverside [Calif.] Daily Press,* 6 May 1960. See also "Legislature Approves Car Smog-Cutter Bill."

9. Ch. 23, § 1, [1960] Cal. Stats. 1st Ex. Sess. 346, adding to California Health and Safety Code §§ 24378-24398. See generally W. Brestel, "The California Motor Vehicle Pollution Control Law," *California Law Review* 50(1962):121; D. Haydel, "Regional Control of Air and Water Pollution in the San Francisco Bay Area," *California Law Review* 55(1967):702, 707-711; H. Kennedy and M. Weekes, "Control of Automobile Emissions—California Experience and the Federal Legislation," *Law and Contemporary Problems* 33(1968):297; Q. Hamel, "Air Pollution—Automobile Smog: A Proposed Remedy," *DePaul Law Review* 14(1965):436, 439-440.

10. California Health and Safety Code § 24383 (1960).

11. California Health and Safety Code § 24384 (1960).

12. California Health and Safety Code §§ 24386(4), 24397, 24398 (1960).

13. California Health and Safety Code § 24386.5 (1960).

14. California Health and Safety Code § 24388 (1960).

15. California Health and Safety Code § 24394 (1960).

16. California Health and Safety Code §§ 24390-24393; California Vehicle Code §§ 4000, 4750, 27156 (1960).
17. California Health and Safety Code § 24386(5) (1960).
18. Interview with D. Jensen, 8 Sept. 1971; interview with L. Breslow, 2 Aug. 1971.
19. Interview with W. King, 24 Aug. 1971.
20. Interview with L. Breslow.
21. Ibid.
22. Interview with E. Starkman, 2 Dec. 1971.
23. Interview with J. Havenner, 20 July 1971. See also Roberts, "Motor Vehicular Air Pollution Control," pp. 76-78.
24. Ibid., pp. 49-50.
25. Ibid., p. 85.
26. Interview with D. Jensen.
27. Interview with J. Middleton, 10 Dec. 1971; interview with L. Breslow; interview with D. Jensen.
28. Interview with J. Askew, 26 July 1971; interview with M. Sweeney, 23 July 1971.
29. Interview with R. Kovitz, 15 July 1971.
30. Interview with J. Askew; interview with M. Sweeney.
31. Interview with D. Jensen.
32. Interview with J. Middleton.
33. A. Carlin and G. Kocher, *Environmental Problems: Their Causes, Cures, and Evolution, Using Southern California Smog as an Example* (Rand Corp., Santa Monica, Calif., Report Number R-640-CC/RC, 1971), p. 66.
34. 45 Opinions of the California Attorney General 79, 80 (1965) (Opinion No. 64-304).
35. Kennedy and Weekes, "Control of Emissions," p. 300.
36. M. Brubacher, Testimony before the California Assembly Transportation and Commerce Committee (mimeo., 15 Nov. 1967), p. 1; Roberts, "Motor Vehicular Air Pollution Control," p. 31; B. Dredge, "Smog Device for Cars Given Boost by State," *Los Angeles Times,* 15 July 1961.
37. See "UCR Man Heads Car Smog Fight," *Riverside* [*Calif.*] *Daily Press,* 22 July 1960.
38. See Brubacher, Testimony, p. 1; "UCR Man Heads Car Smog Fight"; "Smog Device for Cars Given Boost by State"; D. Jensen, California Motor Vehicle Pollution Control Board, "Report on Concept of 'Undue Cost,'" in U.S. Department of Health, Education, and Welfare, *Motor Vehicles, Air Pollution, and Health,* H.R. Doc. No. 489, 87th Cong., 2d Sess. 339-342 (1962).
39. "Board Approves Auto Smog Control Device," *Los Angeles Times,* 16 Sept. 1961; Brubacher, Testimony, p. 1.
40. Roberts, "Motor Vehicular Air Pollution Control," p. 64; Carlin and Kocher, *Environmental Problems,* p. 66, footnote 1; U.S. Department of Commerce, *The Automobile and Air Pollution,* Part 2 (1967), pp. 26-27.
41. Letter from J. Gordon, pres. General Motors Corp., to K. Hahn, Los Angeles County Board of Supervisors, 18 Oct. 1960, in *Smog—A Factual Record of Correspondence between Kenneth Hahn and Automobile Companies,* 6th ed. (Los Angeles, Calif., 1970), p. 33; Carlin and Kocher, *Environmental Problems,* p. 66; "Smog Device for Cars Given Boost by State."
42. Letter from J. Chandler, supervisor, Engineering and Research Staff, Ford

Motor Co., to K. Hahn, 25 Oct. 1960, in *Smog—A Factual Record,* p. 34; Roberts, "Motor Vehicular Air Pollution Control," p. 31; U.S. Department of Commerce, *The Automobile and Air Pollution,* Part 2, pp. 26-27.

43. U.S. Department of Commerce, *The Automobile and Air Pollution,* Part 2, p. 28.

44. "Board Approves Auto Smog Control Device."

45. Roberts, "Motor Vehicular Air Pollution Control," p. 86.

46. "Smog Control Device on Used Cars Delayed," *Los Angeles Times,* 20 Sept. 1962; Brubacher, Testimony, p. 1.

47. G. Getze, "Crankcase Controls Given OK," *Los Angeles Times,* 19 Dec. 1962. See also "Smog Control Device on Used Cars Delayed."

48. Getze, "Crankcase Controls Given OK"; *California Motor Vehicle Pollution Control Board Bulletin,* Vol. 1, No. 12, December 1962; G. Siegel, "The Status of Motor Vehicle Emission Control Public Policy" (mimeo., n.d.), pp. 34-35.

49. Ch. 999, § 1, [1963] Cal. Stats. Reg. Sess. 2264, amending California Health and Safety Code §§ 24379, 24386, 24388-24394. See also Getze, "Crankcase Controls Given OK"; "Auto Smog Control Law Takes Effect Sept. 20," *Los Angeles Times,* 18 July 1963; "Timetable Outlined for Key Areas," *Los Angeles Times,* 15 Aug. 1963.

50. Getze, "Crankcase Controls Given OK"; Siegel, "Status of Policy," pp. 34-35.

51. Getze, "Crankcase Controls Given OK."

52. Los Angeles County Air Pollution Control District, *Administration and Enforcement,* p. 3.

53. *California Motor Vehicle Pollution Control Board Bulletin,* Vol. 1, No. 12, December 1962; ibid., Vol. 2, No. 1, January 1963.

54. See Roberts, "Motor Vehicular Air Pollution Control," pp. 33, 77; Siegel, "Status of Policy," p. 35; Ch. 2028, § 1, [1963] Cal. Stats. Reg. Sess. 4169, amending California Vehicle Code § 27156 and adding §§ 4000.1, 28500-28507.

55. Letter from E. Belasco, deputy attorney general, State of California, to D. Jensen, executive director, California Motor Vehicle Pollution Control Board, 14 Oct. 1963.

56. Los Angeles County Air Pollution Control District, *Administration and Enforcement,* pp. 3-4, 7-8; *Los Angeles County Air Pollution Control District Report,* Vol. 11, No. 7, October 1963, pp. 1-2; Siegel, "Status of Policy," pp. 35-36; Roberts, "Motor Vehicular Air Pollution Control," p. 33; interview with D. Jensen.

57. "Timetable Outlined for Key Areas."

58. Ibid.; *California Motor Vehicle Pollution Control Board Bulletin,* Vol. 2, No. 10, October 1963; ibid., No. 12, December 1963.

59. Ibid., Vol. 3, No. 7, July 1964; interview with J. Askew.

60. See "Timetable Outlined for Key Areas"; Dredge, "Smog Device Given Boost"; Roberts, "Motor Vehicular Air Pollution Control," p. 35; A. Wiman, "A Breath of Death" (Transcript of special report on KLAC Radio, Metromedia, Inc., Los Angeles, Calif., 1967), p. 45.

61. Interview with J. Maga, 23 June 1971; interview with J. Askew; interview with R. Kovitz; interview with D. Jensen; interview with E. Grant, 12 July 1971.

62. Roberts, "Motor Vehicular Air Pollution Control," pp. 34-35; Siegel, "Status of Policy," p. 38.

63. Interview with E. Grant.

64. Roberts, "Motor Vehicular Air Pollution Control," p. 35; Siegel, "Status of Policy," p. 38; interview with J. Maga; interview with D. Jensen; interview with J. Havenner.

65. Interview with C. Biddle, 31 Aug. 1971.

66. Interview with L. Bintz, 19 July 1971.

67. Ch. 3, § 1, [1965] Cal. Stats. Reg. Sess. 872, adding to California Vehicle Code § 27156.5, and to California Health and Safety Code § 24393.4.

68. Siegel, "Status of Policy," p. 38; Roberts, "Motor Vehicular Air Pollution Control," p. 35.

69. Ch. 3, § 3, [1965] Cal. Stats. Reg. Sess. 872, 873.

70. Ch. 2031, § 1, [1965] Cal. Stats. Reg. Sess. 4606, adding to and amending portions of California Health and Safety Code and Vehicle Code relating to air pollution. See generally Roberts, "Motor Vehicular Air Pollution Control," p. 35; Siegel, "Status of Policy," pp. 38-40.

71. Interview with J. Maga; interview with J. Askew; interview with E. Grant; interview with R. Kovitz.

72. Wiman, "A Breath of Death," p. 45.

73. *California Motor Vehicle Pollution Control Board Bulletin,* Vol. 4, No. 1, January 1965; "Auto Smog Device Law Supported," *Los Angeles Times,* 21 Jan. 1965; "Delay Urged in Car Smog Law Change," *Los Angeles Times,* 22 Jan. 1965.

74. Interview with J. Middleton; interview with J. Askew; interviews with R. Chass, 28 July, 12 Aug. 1971.

75. Interview with J. Askew.

76. Letter from H. Barr, vice pres., Engineering Staff, General Motors Corp., to K. Hahn, Nov. 15, 1965; in *Smog—A Factual Record,* pp. 47-48.

77. Interview with E. Starkman.

78. Interview with J. Middleton; interview with L. Bintz; interview with J. Havenner; interview with D. Jensen.

79. Ch. 10, §§ 1 and 2, [1965] Cal. Stats. Reg. Sess. 880-881, adding to California Revenue and Taxation Code § 402.7.

80. L. Wagner, "Car Anti-Smog Device Ready," *Los Angeles Times,* 24 Oct. 1960. See also "Pollution Board Views Smog Device for Cars; Cost Varies," *Fresno [Calif.] Bee,* 20 Dec. 1960.

81. Editorial, "State Lagging on Smog Battle," *Los Angeles Mirror,* 24 Oct. 1960.

82. T. Patterson, "Car Smog Law Nears," *Riverside [Calif.] Daily Press,* 12 Jan. 1961; California Motor Vehicle Pollution Control Board, *Chronological History of California Motor Vehicle Pollution Control Board* (1964), p. 1.

83. "Supervisors Rap State Lag on Smog," *Los Angeles Times,* 28 June 1961.

84. Brubacher, Testimony, p. 4; California Motor Vehicle Pollution Control Board, "Memorandum to All Applicants for Certification of Exhaust Control Devices and Interested Parties," in U.S. Department of Health, Education, and Welfare, *Motor Vehicles, Air Pollution, and Health,* pp. 335-337; "Board Approves Auto Smog Control Device."

85. Roberts, "Motor Vehicular Air Pollution Control," p. 31; "Smog Control Device on Used Cars Delayed"; "2 Changes Sought in Smog Law," *Los Angeles Times,* 15 Nov. 1962; *California Motor Vehicle Pollution Control Board Bulletin,* Vol. 1, No. 5, May 1962.

86. "State to Require Smog Devices on Cars in 1963," *Los Angeles Times,* 12 April 1962.

87. Letter from J. Chandler, manager, Engineering and Research Staff, Ford Motor Co., to K. Hahn, 17 July 1962, in *Smog—A Factual Record,* pp. 36-37. See also the letters from General Motors and Ford at ibid., pp. 42-44.

88. Quoted in "Smog Control Device on Used Cars Delayed."

89. "Supervisors Unhappy with Smog Reports," *Los Angeles Times,* 17 Oct. 1962.

90. Getze, "Crankcase Controls Given OK"; "2 Changes Sought in Smog Law."

91. "2 Changes Sought in Smog Law"; Roberts, "Motor Vehicular Air Pollution Control," pp. 31-32.

92. *California Motor Vehicle Pollution Control Board Bulletin,* Vol. 2, No. 4, April 1963.

93. Quoted in "Mayor Calls for Action on Auto Smog Control," *Los Angeles Times,* 30 Dec. 1963.

94. *California Motor Vehicle Pollution Control Board Bulletin,* Vol. 1, No. 9, September 1962.

95. "Board Flays Auto Makers for 'Appalling' Delay on Controls," *Los Angeles Times,* 24 Jan. 1964.

96. Ibid.

97. Ibid.; Siegel, "Status of Policy," p. 36; Roberts, "Motor Vehicular Air Pollution Control," p. 32.

98. Interview with D. Jensen; interview with J. Askew; interview with E. Plesset, 27 Aug. 1971; Wiman, "A Breath of Death," p. 23.

99. Interview with E. Belasco, 27 July 1971; Wiman, "A Breath of Death," pp. 23-24.

100. "Board Flays Auto Makers for 'Appalling' Delay on Controls."

101. Ibid. See also Wiman, "A Breath of Death," p. 22.

102. Wiman, "A Breath of Death," p. 22.

103. Interview with D. Jensen.

104. Statement by a representative of Ford Motor Company quoted in Wiman, "A Breath of Death," p. 42.

105. Roberts, "Motor Vehicular Air Pollution Control," pp. 32, 65; *California Motor Vehicle Pollution Control Board Bulletin,* Vol. 3, No. 3, March 1964.

106. Roberts, "Motor Vehicular Air Pollution Control," pp. 32, 65, 90; *California Motor Vehicle Pollution Control Board Bulletin,* Vol. 3, No. 6, June 1964; "State Approves 4 Devices for Control of Auto Fumes," *Los Angeles Times,* 18 June 1964.

107. "Auto Exhaust Devices Selected for Approval," *Los Angeles Times,* 9 June 1964.

108. Roberts, "Motor Vehicular Air Pollution Control," pp. 32-33, 65, 90-91; Siegel, "Status of Policy," p. 37; H. Jacoby et al., "Federal Policy on Automotive Emissions Control" (Harvard University Environmental Systems Program, 1973), p. 8; *California Motor Vehicle Pollution Control Board Bulletin,* Vol. 3, No. 8, August 1964.

109. "State Asked to OK Smog Device Offer," *Los Angeles Times,* 14 Oct. 1964; *California Motor Vehicle Pollution Control Board Bulletin,* Vol. 3, No. 11, November 1964; ibid., Vol. 4, No. 7, July 1965; Siegel, "Status of Policy," p. 40.

110. Carlin and Kocher, *Environmental Problems,* p. 66.

111. "It Doesn't Show Now But Smog's Decreasing, Pollution Aide Insists," *Los Angeles Times,* 1 April 1964.

112. Editorial, "New Step Toward Smog Control," *Los Angeles Times,* 19 June 1964.

113. Quoted in "Public Urged to Stand Present Smog until '70," *Los Angeles Times,* 14 May 1964. See also Roberts, "Motor Vehicular Air Pollution Control," p. 34; Siegel, "Status of Policy," p. 37.

114. Siegel, "Status of Policy," p. 38; *California Motor Vehicle Pollution Control Board Bulletin,* Vol. 4, No. 6, June 1965.

115. Ibid., Vol. 3, No. 5, May 1964.

116. S. Griswold, "Reflections and Projections on Controlling the Motor Vehicle" (Paper presented at Annual Meeting of the Air Pollution Control Association, Houston, Texas, 21-25 June 1964), pp. 7-8.

117. "Auto Smog Trap Barred by Board," *Los Angeles Times,* 17 Dec. 1964. See also Roberts, "Motor Vehicular Air Pollution Control," p. 34; Siegel, "Status of Policy," p. 38.

118. "County Urged to Quit State Pollution Board," *Los Angeles Times,* 22 Jan. 1965; "Hahn Asks State Inquiry of Vehicle Pollution Board," *Los Angeles Times,* 28 Jan. 1965.

119. See California Health and Safety Code § 24390 (1965).

120. Roberts, "Motor Vehicular Air Pollution Control," p. 35; Siegel, "Status of Policy," p. 41.

121. *California Motor Vehicle Pollution Control Board Bulletin,* Vol. 3, No. 12, December 1964; "Swift Approval of 2nd Smog Device Demanded," *Los Angeles Times,* 14 April 1965.

122. Roberts, "Motor Vehicular Air Pollution Control," p. 92.

123. Ibid., pp. 65, 91.

124. Quoted in ibid., p. 91.

125. Quoted in ibid., p. 92.

126. Ibid., pp. 92-93.

127. "Senate Unit OKs Auto Exhaust Device Bill," *Los Angeles Times,* 13 April 1966. See also Siegel, "Status of Policy," p. 41.

128. Ch. 111, §§ 1 and 2, [1966] Cal. Stats. 1st Ex. Sess. 582-584, amending and adding to California Health and Safety Code §§ 24386, 24386.2; Siegel, "Status of Policy," p. 42; "Senate Unit OKs Auto Exhaust Device Bill."

129. Siegel, "Status of Policy," p. 42; Roberts, "Motor Vehicular Air Pollution Control," pp. 36-37; "The Advocates" (Transcript of television presentation on KCET-TV, Los Angeles, Calif., and WBGH-TV, Boston, Mass., 5 Oct. 1969), p. 81; G. Getze, "Criticism of Smog Devices Mounts," *Los Angeles Times,* 10 May 1967.

130. "State Smog Group Critical of Devices," *Los Angeles Times,* 17 Nov. 1966.

131. "State to Warn Auto Makers on Smog Devices," *Los Angeles Times,* 4 Nov. 1966.

132. "State Smog Group Critical of Devices."

133. Interview with A. Wiman, 5 Aug. 1971; Roberts, "Motor Vehicular Air Pollution Control," p. 97.

134. Roberts, "Motor Vehicular Air Pollution Control," p. 34. See also ibid., p. 96; Wiman, "A Breath of Death," pp. 20, 21.

135. See Wiman, "A Breath of Death," pp. 24-25; interview with R. Barsky, 23 July 1971.

136. Quoted in Wiman, "A Breath of Death," p. 41 (emphasis added).

137. Ibid., p. 20.

138. Interview with J. Havenner.

139. Getze, "Criticism of Smog Devices." See also *California Motor Vehicle Pollution Control Board Bulletin,* Vol. 6, No. 3, March 1967.

140. Wiman, "A Breath of Death," p. 20; Roberts, "Motor Vehicular Air Pollution Control," p. 98.

141. Interview with R. Barsky.

142. Roberts, "Motor Vehicular Air Pollution Control," pp. 69, 85.

143. Interview with J. Havenner.

144. Interview with D. Jensen; interview with E. Plesset; interview with W. King.

145. Roberts, "Motor Vehicular Air Pollution Control," p. 85.

146. J. Maga, H. Wong-Woo, and M. Macon, "A Status Report on Motor Vehicular Pollution in California," *Journal of the Air Pollution Control Association* 17(1967):436.

147. R. Dyck, "Evolution of Federal Air Pollution Control Policy" (Ph.D. diss., University of Pittsburgh, 1971), p. 41.

148. Pub. L. No. 86-493, 74 Stat. 162. See Dyck, "Evolution of Federal Policy," pp. 41–42; R. Ripley, "Congress and Clean Air: The Issue of Enforcement, 1963," in *Congress and Urban Problems,* ed. F. Cleaveland et al. (Washington, D.C.: The Brookings Institution, 1969), pp. 234, 237.

149. Ripley, "Congress and Clean Air," p. 236.

150. Ibid., pp. 234–236; J. Sundquist, *Politics and Policy: The Eisenhower, Kennedy, and Johnson Years* (Washington, D.C.: The Brookings Institution, 1968), p. 351; J. Davies, *The Politics of Pollution* (New York: Pegasus, 1970), pp. 51–52; Dyck, "Evolution of Federal Policy," pp. 42–47. See Pub. L. No. 87-761, 76 Stat. 760, amending 42 U.S.C. §§ 1857b, 1857d.

151. See Dyck, "Evolution of Federal Policy," p. 48; U.S. Department of Health, Education, and Welfare, *Motor Vehicles, Air Pollution, and Health.*

152. Pub. L. No. 87-761, 76 Stat. 760, § 2, amending 42 U.S.C. § 1857b.

153. Dyck, "Evolution of Federal Policy," pp. 48–50, 131–133, 141; Sundquist, *Politics and Policy,* pp. 351–352; Davies, *The Politics of Pollution,* p. 52. For more detailed discussion see Ripley, "Congress and Clean Air," pp. 237–244.

154. Dyck, "Evolution of Federal Policy," pp. 51–58, 133–139; Sundquist, *Politics and Policy,* pp. 352–355; Davies, *The Politics of Pollution,* pp. 52–53. For more detailed discussion see Ripley, "Congress and Clean Air," pp. 244–278.

155. Pub. L. No. 88-206, 77 Stat. 392, codified at 42 U.S.C. §§ 1857-1857l. For general discussions see, e.g., S. Edelman, "Air Pollution Abatement Procedures under the Clean Air Act," *Arizona Law Review* 10(1968):30; "The Federal Air Pollution Program," *Washington University Law Quarterly* 1968(1968):283; J. Fromson, "A History of Federal Air Pollution Control," *Ohio State Law Journal* 30(1969):516, 520–526; L. Green, "State Control of Interstate Air Pollution," *Law and Contemporary Problems* 33(1968):315, 317–319.

156. See Sundquist, *Politics and Policy,* p. 368, footnote 110.

157. Special Subcommittee on Air and Water Pollution of the Senate Committee on Public Works, 88th Cong., 2d Sess., *Steps Toward Clean Air* (1964).

158. Pub. L. No. 87-272, 79 Stat. 992, adding to the Clean Air Act, 42 U.S.C. § 1857f-1. See "The Federal Air Pollution Program," pp. 315–317, 323; Fromson, "History of Federal Control," pp. 526–528; Davies, *The Politics of Pollution,* pp. 53–54; Sundquist, *Politics and Policy,* pp. 367–371; Dyck, "Evolution of Federal Policy," pp. 59–70, 142–144.

159. Kennedy and Weekes, "Control of Automobile Emissions," pp. 309-310; D. Currie, "Motor Vehicle Air Pollution: State Authority and Federal Pre-Emption," *Michigan Law Review* 68(1970):1083, 1087-1089; "The Federal Air Pollution Program," p. 323.

160. Jacoby et al., "Federal Policy on Control," p. 8; M. Greco, "The Clean Air Amendments of 1970: Better Automotive Ideas from Congress," *Environmental Affairs* 1(1971):384, 386-387; "The Federal Air Pollution Program," p. 315.

11. State Reorganization

1. See T. Roberts, "Motor Vehicular Air Pollution Control in California: A Case Study in Political Unresponsiveness" (Honors thesis, Harvard College, 1969), pp. 37, 89-90; interview with B. Samuel and S. McCausland, 30 July 1971.

2. Interview with D. Jensen, 8 Sept. 1971.

3. *California Motor Vehicle Pollution Control Board Bulletin,* Vol. 5, No. 1, January 1966.

4. Interview with E. Grant, 12 July 1971; interview with M. Sweeney, 23 July 1971; Editorial, *Los Angeles Times,* 31 Jan. 1966.

5. *California Motor Vehicle Pollution Control Board Bulletin,* Vol. 5, No. 4, April 1966; "New Smog Bill Offered by Carrell," *Los Angeles Times,* 17 March 1966.

6. G. Siegel, "The Status of Motor Vehicle Emission Control Public Policy" (mimeo., n.d.), pp. 42-43; *California Motor Vehicle Pollution Control Board Bulletin,* Vol. 6, No. 1, January 1967; ibid., No. 2, February 1967; "Law Sought on Car Smog Unit," *Los Angeles Times,* 20 Jan. 1967.

7. Ch. 1545, § 1, [1967] Cal. Stats. Reg. Sess. 3679, repealing and adding to portions of California Health and Safety Code relating to air pollution. See generally Roberts, "Motor Vehicular Air Pollution Control," p. 38; D. Sokolow, "AB 357: The Passage of California's 'Pure Air' Law in 1968" (Institute of Governmental Affairs, *Environmental Quality Series,* No. 3, University of California, Davis, 1970), p. 4; H. Kennedy, "Some Legal Ramifications of Air Pollution Control and a Review of Current Control of Automotive Emissions," *Arizona Law Review* 10(1968):1, 5-6; Siegel, "Status of Policy," pp. 43-44.

8. Interview with L. Breslow, 2 Aug. 1971.

9. Interview with B. Samuel and S. McCausland.

10. "Reagan Appoints Nine to Air Resources Board," *Los Angeles Times,* 21 Dec. 1967; interview with N. Yost, 27 July 1971; interview with L. Bintz, 19 July 1971.

11. Unless otherwise indicated, the discussion is from J. Davies, *The Politics of Pollution* (New York: Pegasus, 1970), pp. 54-57. For more detailed discussion, see R. Dyck, "Evolution of Federal Air Pollution Control Policy" (Ph.D. diss., University of Pittsburgh, 1971), pp. 76-124.

12. See J. Fromson, "A History of Federal Air Pollution Control," *Ohio State Law Journal* 30(1969):516, 529, 532, footnote 127.

13. See *Proceedings of Third National Conference on Air Pollution* (Washington, D.C., 12-14 Dec. 1966).

14. Fromson, "History of Federal Control," p. 530, footnote 106; Dyck, "Evolution of Policy," p. 102.

15. Fromson, "History of Federal Control," pp. 529-530; J. Esposito, *Vanishing Air* (New York: Grossman Publishers, 1970), pp. 269-280.

16. Esposito, *Vanishing Air,* pp. 287-294.

17. R. Martin and L. Symington, "A Guide to the Air Quality Act of 1967," *Law and Contemporary Problems* 33(1968):239, 240; Esposito, *Vanishing Air,* p. 281.

18. D. Currie, "Motor Vehicle Air Pollution: State Authority and Federal Pre-Emption," *Michigan Law Review* 68(1970):1083, 1089-1090.

19. Roberts, "Motor Vehicular Air Pollution Control," p. 71; interview with A. Wiman, 5 Aug. 1971; A. Wiman, "A Breath of Death" (Transcript of special report on KLAC Radio, Metromedia, Inc., Los Angeles, Calif., 1967), p. 12.

20. Sokolow, "AB 357," pp. 8-9.

21. Quoted in ibid., p. 9.

22. Ibid.

23. "State and County Unite in Smog War," *Los Angeles Times,,* 6 Oct. 1967.

24. Sokolow, "AB 357," p. 9; Roberts, "Motor Vehicular Air Pollution Control," pp. 55-56.

25. Wiman, "A Breath of Death"; Roberts, "Motor Vehicular Air Pollution Control," pp. 55-56.

26. Interview with A. Wiman; Sokolow, "AB 357," p. 10.

27. H. Kennedy and M. Weekes, "Control of Automobile Emissions—California Experience and the Federal Legislation," *Law and Contemporary Problems* 33 (1968):297, 311.

28. Pub. L. No. 90-148, 81 Stat. 485, amending 42 U.S.C. §§ 1857-1857l. On the major provisions, see generally Fromson, "History of Federal Control," pp. 529-536; Kennedy and Weekes, "Control of Automobile Emissions," pp. 310-312; Martin and Symington, "Guide to the Air Quality Act"; J. O'Fallon, "Deficiencies in the Air Quality Act of 1967," *Law and Contemporary Problems* 33(1968):275; Currie, "Motor Vehicle Air Pollution," pp. 1089-1091; M. Greco, "The Clean Air Amendments of 1970: Better Automotive Ideas from Congress," *Environmental Affairs* 1(1971): 384, 387-391.

29. 42 U.S.C. § 1857 (a)(3) (Supp. IV, 1965-1968).

30. 42 U.S.C. §§ 1857f-1(a), 1857f-6(a) (Supp. IV, 1965-1968).

31. Unless otherwise indicated, the discussion is from Sokolow, "AB 357"; Roberts, "Motor Vehicular Air Pollution Control," pp. 39-40, 134-150; interview with W. Heinz, 30 Sept. 1971; interview with B. Samuel and S. McCausland; interview with E. Starkman, 2 Dec. 1971; interview with J. Foran, 12 Nov. 1971; interview with J. Maga, 23 June 1971; interview with G. Taylor, 23 June 1971.

32. Sokolow, "AB 357," p. 4.

33. Technical Advisory Panel on the Pure Air Act of 1968, *Report to the California Assembly Committee on Transportation and Commerce* (1968).

34. Sokolow, "AB 357," p. 20. See Ch. 764, § 1, [1968] Cal. Stats. Reg. Sess. 1463, amending, repealing, and adding to portions of California Health and Safety Code, Vehicle Code, and Government Code relating to air pollution.

35. Kennedy and Weekes, "Control of Automobile Emissions," p. 311; California Air Resources Board, *1968 Annual Report* (1969), p. 18; idem, *1969 Annual Report* (1970), pp. 2, 16.

36. See L. Lees et al., *Smog: A Report to the People* (Los Angeles: Ward Ritchie Press, 1972), p. 36.

37. California Air Resources Board, *1968 Annual Report,* p. 27.

38. California Air Resources Board, *1969 Annual Report,* p. 5.

39. Ibid., p. 3.

40. Ibid., p. 15. See Ch. 1253, § 1, [1969] Cal. Stats. Reg. Sess. 2447, 2448, amending California Health and Safety Code § 39052.

41. California Air Resources Board, *1969 Annual Report,* p. 3.

42. California Air Resources Board, *1968 Annual Report,* pp. 4-5, 15.

43. California Air Resources Board, *1969 Annual Report,* p. 31.

44. See A. Kneese and C. Schultze, *Pollution, Prices, and Public Policy* (Washington, D.C.: The Brookings Institution, 1975), p. 50.

45. See ibid., p. 50.

46. See Esposito, *Vanishing Air*; O'Fallon, "Deficiencies in the Air Quality Act"; E. Angeletti, "Transmogrification: State and Federal Regulation of Automotive Air Pollution," *Natural Resources Journal* 13(1973):448, 456; Greco, "The Clean Air Amendments," pp. 388-391.

47. Interview with E. Grant; interview with W. Heinz; interview with E. Starkman; interview with M. Levee, 8 July 1971; interview with J. Askew, 26 July 1971; interview with A. Batchelder, 23 July 1971; interview with A. Ingersoll, 1 July 1971; Office of the California Legislative Analyst, *Air Pollution Control in California* (1971), pp. 10-15, 56-66.

48. California Air Resources Board, *1968 Annual Report,* p. 27; idem, *1969 Annual Report,* p. 23.

49. California Air Resources Board, *1969 Annual Report,* p. 11.

50. Office of the California Legislative Analyst, *Air Pollution Control in California,* p. 51.

51. Lees et al., *Smog,* p. 34.

52. Interview with J. Maga.

53. California Health and Safety Code § 39176 (1968).

54. P. McDaniel and A. Kaplinsky, "The Use of the Federal Income Tax System to Combat Air and Water Pollution: A Case Study in Tax Expenditures," *Boston College Industrial and Commercial Law Review* 12(1971):351, 352.

55. See J. Krier, "The Pollution Problem and Legal Institutions: A Conceptual Overview," *U.C.L.A. Law Review* 18(1971):429, 467-470.

56. Dyck, "Evolution of Policy," pp. 94-95.

57. See A. Kneese and B. Bower, *Managing Water Quality: Economics, Technology, Institutions* (Baltimore: Johns Hopkins Press, 1968).

58. Dyck, "Evolution of Policy," p. 100; Working Committee on Economic Incentives, Federal Coordinating Committee on the Economic Impact of Pollution Abatement, *Cost Sharing with Industry?* (1967).

59. See, e.g., Kneese and Bower, *Managing Water Quality*; J. Dales, *Pollution, Property and Prices* (Toronto: University of Toronto Press, 1968); H. Wolozin, "The Economics of Air Pollution: Central Problems," *Law and Contemporary Problems* 33(1968):227, 233-237 (and literature cited).

60. L. Ruff, "Price Pollution Out of Existence," *Los Angeles Times,* 7 Dec. 1969. See also idem, "The Economic Common Sense of Pollution," *The Public Interest,* Spring 1970, p. 69.

61. Kneese and Schultze, *Pollution, Prices, and Public Policy,* p. 48.

62. See H. Jacoby et al., "Federal Policy on Automotive Emissions Control" (Harvard University Environmental Systems Program, 1973), p. 10.

63. See Dyck, "Evolution of Policy," p. 200; S. Brubaker, *To Live on Earth* (Baltimore: Johns Hopkins Press, 1972), p. 1.

64. Pub. L. No. 91-190, 83 Stat. 852, codified at 42 U.S.C. §§ 4331-4335.
65. Pub. L. No. 91-604, 84 Stat. 1205, codified at 42 U.S.C. §§ 1857-1858a.

12. The New Federal Program

1. Pub. L. No. 91-604, 84 Stat. 1205, codified at 42 U.S.C. §§ 1857-1858a.
2. See especially F. Grad et al., *The Automobile and the Regulation of Its Impact on the Environment* (Norman: University of Oklahoma Press, 1975); H. Jacoby et al., *Clearing the Air* (Cambridge, Mass.: Ballinger Publishing Company, 1973). See also M. Greco, "The Clean Air Amendments of 1970: Better Automotive Ideas from Congress," *Environmental Affairs* 1(1971):384; E. Angeletti, "Transmogrification: State and Federal Regulation of Automotive Air Pollution," *Natural Resources Journal* 13(1973):448; "Clean Air Amendments of 1970: A Congressional Cosmetic," *Georgetown Law Journal* 61(1972):153; T. Bracken, "Transportation Controls under the Clean Air Act: A Legal Analysis," *Boston College Industrial and Commercial Law Review* 15(1974):749.
3. 42 U.S.C. § 1857(a)(3) (1970).
4. A. Kneese and C. Schultze, *Pollution, Prices, and Public Policy* (Washington, D.C.: The Brookings Institution, 1975), p. 51.
5. Grad et al., *Automobile Regulation*, pp. 328, 332-333.
6. See, e.g., Jacoby et al., *Clearing the Air*, p. 12; H. Molotch and R. Follett, "Air Pollution as a Problem for Sociological Research," in *Air Pollution and the Social Sciences*, ed. P. Downing (New York: Praeger Publishers, 1971), pp. 15, 21.
7. "Congressional Cosmetic," p. 157.
8. See ibid., pp. 157-158.
9. Grad et al., *Automobile Regulation*, p. 328.
10. Senate Committee on Public Works, *National Air Quality Standards Act of 1970*, S. Rep. No. 91-1196, 91st Cong., 2d Sess. 10-11 (1970).
11. Ibid., pp. 10-11.
12. R. Finch, Testimony in Hearings before Subcommittee on Air and Water Pollution of the Senate Committee on Public Works, 91st Cong., 2d Sess., Pt. I, p. 134 (1970).
13. See Grad et al., *Automobile Regulation*, p. 332; Senate Committee on Public Works, *National Air Quality Standards Act of 1970*, pp. 2-3, 24.
14. Finch, Testimony in Hearings, p. 135. See also A. Stern, "National Emission Standards for Stationary Sources," in Hearings before Subcommittee on Air and Water Pollution, Pt. IV, pp. 1550, 1563.
15. Finch, Testimony in Hearings, p. 135.
16. Senate Committee on Public Works, *National Air Quality Standards Act of 1970*, pp. 12-13.
17. Staff of the Senate Commiteee on Commerce, 91st Cong., 1st Sess., *The Search for a Low-Emission Vehicle* 2 (Comm. Print 1969).
18. Ibid., pp. 3-9.
19. J. Esposito, *Vanishing Air* (New York: Grossman Publishers, 1970).
20. Jacoby et al., *Clearing the Air*, p. 11.
21. Ibid.
22. Ibid., pp. 11-12.

23. Grad et al., *Automobile Regulation,* p. 334.
24. Jacoby et al., *Clearing the Air,* pp. 12-13.
25. See, e.g., Kneese and Schultze, *Pollution, Prices, and Public Policy,* p. 53.
26. For more detailed discussion of the legislation, see Grad et al., *Automobile Regulation,* pp. 328-338; Greco, "The Clean Air Amendments," pp. 391-420; J. Stevens, "Air Pollution and the Federal System: Responses to Felt Necessities," *Hastings Law Journal* 22(1971):661, 670-676.
27. *Congressional Record* 116(21 Sept. 1970):16091 (remarks of Senator Muskie).
28. D. Barth et al., "Federal Motor Vehicle Emission Goals for CO, HC and NOX Based on Desired Air Quality Levels," in Hearings before Subcommittee on Air and Water Pollution, Pt. V, pp. 1639-1645.
29. Grad et al., *Automobile Regulation,* p. 333.
30. *Congressional Record* 116(21 Sept. 1970):16095.
31. See generally W. Schwartz, "Mandatory Patent Licensing of Air Pollution Control Technology," *Virginia Law Review* 57(1971):719.
32. See 40 Code of Federal Regulations §§ 50.1 to 50.11 (1974).
33. See, e.g., Jacoby et al., *Clearing the Air,* p. 13; J. Wanniski, "How the Clean Air Rules Were Set," *Wall Street Journal,* 29 May 1973.
34. D. Fisher, "U.S. Aide Wonders if Clean Air Standards Are Too High," *Los Angeles Times,* 29 July 1974 (quoting the EPA's deputy assistant administrator for automotive pollution control).
35. See Los Angeles Air Pollution Control District, *Profile of Air Pollution Control* (1971), p. 241.
36. 40 Code of Federal Regulations § 51.1(n) (1974). See also §§ 51.11, 51.12.
37. See Grad et al., *Automobile Regulation,* pp. 339-340.
38. See, e.g., L. Lees et al., *Smog: A Report to the People* (Los Angeles: Ward Ritchie Press, 1972); Los Angeles Air Pollution Control District, *Profile,* p. i.
39. Los Angeles Air Pollution Control District, *Profile,* p. i.
40. See TRW Transportation and Environmental Operations, *Transportation Control Strategy Development for the Metropolitan Los Angeles Region* (prepared for the Environmental Protection Agency, APTD-1372) (1972), pp. 11, 23, 30, 98.
41. Los Angeles Air Pollution Control District, *Profile,* p. 59.
42. Ibid., p. 42.
43. Ibid.
44. For a summary of the views, see TRW Transportation and Environmental Operations, *Transportation Control,* pp. 35-36. On the effects of growth, see Lees et al., *Smog,* p. 144.
45. Los Angeles Air Pollution Control District, *Profile,* p. 42.
46. Ch. 714, § 1, [1970] Cal. Stats. Reg. Sess. 1339, adding to California Health and Safety Code §§ 39270-39275. See generally W. Simmons and R. Cutting, "A Many-Layered Wonder: Nonvehicular Air Pollution Control Law in California," *Hastings Law Journal* 26(1974):109, 123-124, 129-130.
47. Ch. 451, § 1, [1970] Cal. Stats. Reg. Sess. 900, adding to California Health and Safety Code § 39067.2.
48. Ch. 1507, § 1, [1971] Cal. Stats. Reg. Sess. 2978, adding to California Health and Safety Code and amending California Vehicle Code § 4602.
49. Ch. 1674, § 1, [1971] Cal. Stats. Reg. Sess. 3598, amending, repealing, and adding to portions of California Health and Safety Code relating to the Air Resources Board.

13. Problems of Implementation

1. For some discussion of implementation plans in other areas, see U.S. Environmental Protection Agency, *Transportation Controls to Reduce Automobile Use and Improve Air Quality in Cities* (1974); T. Bracken, "Transportation Controls Under the Clean Air Act: A Legal Analysis," *Boston College Industrial and Commercial Law Review* 15(1974):749; TRW Transportation and Environmental Operations, *Socio-Economic Impacts of the Proposed State Transportation Control Plans: An Overview* (prepared for the Environmental Protection Agency) (1973).

2. J. Dreyfuss, "Smog Board Chief Urges Gas Tax to Fund Transit," *Los Angeles Times,* 29 June 1971.

3. J. Revis, Institute of Public Administration, "Memorandum to G. Hawthorn, Environmental Protection Agency, on Evaluating Transportation Controls to Reduce Motor Vehicle Emissions in Major Metropolitan Areas—Los Angeles Reconnaissance" (5 April 1972), pp. 4-6.

4. Quoted in J. Dreyfuss, "State, U.S. Agree on Clean Air Plan: It Won't Work," *Los Angeles Times,* 16 Jan. 1972.

5. J. Dreyfuss, "State's Clean Air Proposal OKd [by the Air Resources Board] but Is Called Unworkable," *Los Angeles Times,* 20 Jan. 1972.

6. Letter from R. Reagan, governor of California, to W. Ruckelshaus, administrator, Environmental Protection Agency, 21 Feb. 1972.

7. State of California, Resources Agency, Air Resources Board, *Implementation Plan for Achieving and Maintaining the National Ambient Air Quality Standards* (30 Jan. 1972), Pt. I, pp. 64-70 (hereafter cited as *Implementation Plan*). For a summary, see California Air Resources Board, *1972 Annual Report* (1973), pp. 24-30.

8. See J. Dreyfuss, "Smog Device Deadline for 1955-65 Cars Due," *Los Angeles Times,* 20 April 1972; G. Getze, "Smog Device Law for '55-'65 Autos Goes Into Effect Today," *Los Angeles Times,* 1 Sept. 1972; *Implementation Plan,* Pt. I, p. 69.

9. *Implementation Plan,* Pt. I, pp. 70-74; California Air Resources Board, *1972 Annual Report,* p. 29.

10. *Implementation Plan,* Pt. I, pp. 75-83, Pt. VI, pp. 122-124; California Air Resources Board, *1972 Annual Report,* pp. 29-30.

11. Dreyfuss, "State, U.S. Agree."

12. 37 *Federal Register* 10842 (1972). See also 38 *Federal Register* 2194 (1973); U.S. Environmental Protection Agency, *Transportation Controls,* pp. 22-23; Bracken, "Legal Analysis," p. 752; F. Grad et al., *The Automobile and the Regulation of Its Impact on the Environment* (Norman: University of Oklahoma Press, 1975), p. 366.

13. Much of the discussion in this and the next section is based on the personal participation of Dr. Eugene Leong in the planning process. See generally his study, "Air Pollution Control in California from 1970 to 1974: Some Comments on the Implementation Planning Process" (Doctor of Environmental Science and Engineering thesis, University of California, Los Angeles, 1974).

14. Institute of Public Administration and Tecknekron, Inc. with TRW, Inc., *Evaluating Transportation Controls to Reduce Motor Vehicle Emissions in Major Metropolitan Areas* (1972).

15. GCA Corp. and TRW, Inc., *Transportation Controls to Reduce Motor Vehicle Emissions in Major Metropolitan Areas* (prepared for the Environmental Protection Agency, APTD-1462) (1972).

16. TRW Transportation and Environmental Operations, *Transportation Control Strategy Development for the Metropolitan Los Angeles Region* (prepared for the Environmental Protection Agency, APTD-1372) (1972).

17. See California Department of Public Works, Division of Highways, Urban Planning Department, *Can Vehicle Travel Be Reduced 20 Percent in the South Coast Air Basin?* (1973).

18. Revis, "Memorandum," p. 4.

19. See, e.g., "The Local Agencies Plan for the Los Angeles Air Basin," *Clean Air,* 30 Aug. 1973; Southern California Association of Governments, *Short Range Plan* (Interim Report, 7 March 1974).

20. Revis, "Memorandum," p. 21.

21. Ibid.

22. Quoted in "U.S. Plan to Implement State Air Standards Due by Jan. 15," *Los Angeles Times,* 7 Nov. 1972.

23. J. Dreyfuss, "U.S. Clean Air Plan Proposes Gasoline Rationing in Southland," *Los Angeles Times,* 8 Dec. 1972.

24. TRW Transportation and Environmental Operations, *Transportation Control Strategy,* especially pp. 168–170.

25. Quoted in J. Dreyfuss, "Smog Scientists Urge Taxing Car Users on Mileage," *Los Angeles Times,* 18 Oct. 1972. See also L. Lees et al., *Smog: A Report to the People* (Los Angeles: Ward Ritchie Press, 1972); statement of L. Lees at Environmental Protection Agency Los Angeles Hearings, 6 March 1973.

26. Quoted in J. Dreyfuss, "State, U.S. Agree."

27. Ibid.

28. J. Dreyfuss, "Gasoline Rationing Plan Seen as Move to Pressure Congress," *Los Angeles Times,* 16 Jan. 1973. See also 38 *Federal Register* 2194 (1973).

29. 38 *Federal Register* 2194, 2198 (1973).

30. 38 *Federal Register* 2195, 2197, 2199 (1973).

31. 38 *Federal Register* 2195 (1973).

32. Dreyfuss, "Gasoline Rationing Plan Seen as Move."

33. Letter to the Editor, in section entitled "Gas Rationing No Solution to L.A. Pollution Problems," *Los Angeles Times,* 6 Jan. 1973.

34. Quoted in J. Dreyfuss, "Stringent Control Over Car Use on Smoggy Days Only Proposed," *Los Angeles Times,* 6 March 1973; idem, "Public Will Sacrifice to Fight Smog, Panel Told," *Los Angeles Times,* 7 March 1973.

35. L. Dye, "Failure of State Blamed for Gas Rationing Plan," *Los Angeles Times,* 19 Jan. 1973.

36. Dreyfuss, "Public Will Sacrifice."

37. Dreyfuss, "Stringent Control Proposed."

38. Dye, "Failure of State."

39. Editorial, "Time Is Running Out," *Los Angeles Times,* 21 Jan. 1973.

40. P. Houston, "Delay in L.A. Clean Air Program Sought," *Los Angeles Times,* 24 May 1973.

41. See Leong, "Air Pollution Control in California," pp. 23, 31–34.

42. See, e.g., E. Schuck and R. Papetti, "Examination of the Photochemical Air Pollution Problem in the Southern California Area" (mimeo., 23 May 1973), pp. 2–3.

43. Houston, "Delay in Program Sought."

44. 38 *Federal Register* 17683–84 (1973).

45. 38 *Federal Register* 17683–85 (1973).

46. Quoted in D. Fisher, "U.S. Proposal Could Ban Autos in L.A.," *Los Angeles Times,* 16 June 1973.

47. Quoted in ibid.

48. Bureau of National Affairs, *Environment Reporter, Current Developments,* Vol. 4, No. 8, 22 June 1973, pp. 266, 268.

49. Quoted in R. Hebert, "Some of EPA Proposals Already in Use in L.A.," *Los Angeles Times,* 16 June 1973.

50. State of California, Resources Agency, Air Resources Board, *Proposed Revision to Part I of the State of California Implementation Plan for Achieving and Maintaining the National Ambient Air Quality Standards* (18 April 1973).

51. State of California, Resources Agency, Air Resources Board, *The State of California Implementation Plan for Achieving and Maintaining the National Ambient Air Quality Standards—Revision* 3 (21 June 1973).

52. Letter from A. Haagen-Smit, chairman, California Air Resources Board, to R. Fri, acting administrator, Environmental Protection Agency, 9 July 1973.

53. Letter from T. Bradley, mayor, city of Los Angeles, to P. De Falco, administrator of EPA Region IX, 30 July 1973.

54. See, e.g., L. Pryor, " '77 Clean Air Deadline Spurs Southland Action," *Los Angeles Times,* 28 Oct. 1974.

55. 38 *Federal Register* 17683 (1973).

56. E. Younger, California attorney general, "Confidential Memorandum to J. Maga, executive officer, California Air Resources Board" (18 July 1973), pp. 1-2, 17-19.

57. See, e.g., 38 *Federal Register* 31232 (1973); Leong, "Air Pollution Control in California," pp. 21-25.

58. A. Haagen-Smit, "Statement to Be Made at the EPA Hearing on Proposed Transportation and Other Control Measures for the South Coast Air Basin" (mimeo., 9 Aug. 1973), pp. 1-2.

59. 38 *Federal Register* 31232 (1973).

60. 38 *Federal Register* 31232-31255 (1973). See also L. Pryor, "New Smog Plan Called Turning Point," *Los Angeles Times,* 16 Oct. 1973; R. Hebert, "Parking Charges, Bus and Car Pool Aid Ordered in L.A. Area," *Los Angeles Times,* 16 Oct. 1973.

61. 38 *Federal Register* 31233, 31236-31237 (1973).

62. Quoted in Pryor, "New Smog Plan Called Turning Point."

63. 38 *Federal Register* 31233 (1973).

64. 38 *Federal Register* 34124 (1973).

65. 38 *Federal Register* 1848-1849 (1974).

66. Letter from J. Maga, executive officer, California Air Resources Board, to R. Train, administrator, Environmental Protection Agency, 23 Oct. 1973.

67. 38 *Federal Register* 31244 (1973).

68. W. Simmons, staff counsel, California Air Resources Board, "Memorandum to A. Haagen-Smit, chairman, California Air Resources Board, Lawsuit v. EPA" (16 Nov. 1973), p. 3.

69. Letter from A. Haagen-Smit, chairman, California Air Resources Board, to R. Train, administrator, Environmental Protection Agency, 30 Nov. 1973. See California Air Resources Board v. Environmental Protection Agency (No. 73-3307, 9th Cir., filed 29 Nov. 1973).

70. Interview with K. Macomber, 13 June 1975.

71. Bureau of National Affairs, *Environment Reporter, Current Developments,* Vol. 5, No. 27, 1 Nov. 1974, p. 1053.

72. Quoted in ibid., No. 34, 20 Dec. 1974, pp. 1302-1303.

73. Quoted in L. Pryor, "ARB Parts with U.S. on Smog Emergency Rules," *Los Angeles Times,* 16 May 1975.

74. W. Rood, "U.S. Warns State on Smog Emergencies," *Los Angeles Times,* 20 May 1975.

75. Brown v. Environmental Protection Agency, 521 F.2d 827 (9th Cir. 1975), cert. granted, 44 U.S.L.W. 3681 (1976).

76. 521 F.2d 832.

77. 521 F.2d 832, 838. See also State of Arizona v. Environmental Protection Agency, 521 F.2d 825 (9th Cir. 1975), cert. granted, 44 U.S.L.W. 3681 (1976). Compare Friends of the Earth v. Carey, 9 E.R.C. 1641 (2d Cir. 1977) (permitting citizen suit to enforce transportation control plan promulgated not by the EPA but by the state of New York).

78. Bureau of National Affairs, *Environment Reporter, Current Developments,* Vol. 6, No. 28, 7 Nov. 1975, pp. 1225-1226.

79. Ibid.

80. For a more detailed discussion of the developments summarized in this section, see Grad et al., *Automobile Regulation,* pp. 338-364.

81. International Harvester Co. v. Ruckelshaus, 478 F.2d 615 (D.C. Cir. 1973).

82. See Grad et al., *Automobile Regulation,* pp. 340-360.

83. See ibid., pp. 360-364.

84. Bureau of National Affairs, *Environment Reporter, Current Developments,* Vol. 5, No. 23, 4 Oct. 1974, p. 850; interview with K. Macomber.

85. See D. Fisher, "Auto Makers to Be Given Year's Delay on Emissions," *Los Angeles Times,* 28 Feb. 1975; idem, "State under EPA Pressure to Roll Back Emission Rules," *Los Angeles Times,* 6 March 1975.

86. Fisher, "Auto Makers to Be Given Year's Delay." See also idem, "State under EPA Pressure to Roll Back Emission Rules."

87. See Fisher, "Auto Makers to Be Given Year's Delay"; idem, "State under EPA Pressure to Roll Back Emission Rules."

88. Fisher, "Auto Makers to Be Given Year's Delay."

89. Quoted in Fisher, "State under EPA Pressure to Roll Back Emission Rules"; idem, "State Air Unit Challenges EPA," *Los Angeles Times,* 7 March 1975.

90. Fisher, "State under EPA Pressure to Roll Back Emission Rules."

91. Quoted in Fisher, "State Air Unit Challenges EPA."

92. Bureau of National Affairs, *Environment Reporter, Current Developments,* Vol. 5, No. 47, 21 March 1975, p. 1828.

93. Interview with K. Macomber.

94. Bureau of National Affairs, *Environment Reporter, Current Developments,* Vol. 5, No. 50, 11 April 1975, p. 1952.

95. Ibid., Vol. 6, No. 3, 16 May 1975, p. 136.

96. Quoted in ibid., Vol. 5, No. 51, 18 April 1975, p. 1990.

97. National Research Council, *News Report,* Dec. 1973, p. 1.

98. The discussion is taken from "House Interstate and Foreign Commerce Committee Print on Clean Air Act Amendment Proposals," in Bureau of National Affairs, *Environment Reporter, Current Developments,* Vol. 5, No. 49, 4 April 1975, pp. 1929-1935; "Letter from Environmental Protection Agency Administrator to

Senate Public Works Committee Chairman Supporting Proposed Amendments to the Clean Air Act," in ibid., No. 41, 7 Feb. 1975, pp. 1570–1572.

99. "House Interstate and Foreign Commerce Committee Print," p. 1930.

100. Ibid., p. 1934.

101. Ibid.

102. Ibid., p. 1929.

103. California Health and Safety Code §§ 39177.1(b)(2) and (3) (1971).

104. L. Dye, "Smog Device: Engineers Take a Second Look," *Los Angeles Times,* 28 Dec. 1971.

105. California Health and Safety Code § 39177.3 (1971).

106. Quoted in Dye, "Smog Device."

107. Quoted in J. Dreyfuss, "Smog Control Devices Ordered for 1966–70 Autos by Next February," *Los Angeles Times,* 24 Aug. 1972.

108. California Vehicle Code § 4602(b) (1971).

109. D. Fisher, "Smog Device Law Hits Snag; Deadline to Be Extended a Year," *Los Angeles Times,* 27 Aug. 1972.

110. D. Fisher, "Deadline on Smog Devices for '66–'70 Cars Delayed by Board," *Los Angeles Times,* 28 Sept. 1972.

111. See "State Board OKs Third Smog Control Device for Older Cars," *Los Angeles Times,* 1 Dec. 1972; D. Fisher, "Low-Cost Smog Kit for 1966 to 1970 Cars OKd," *Los Angeles Times,* 19 April 1973; idem, "Board Revives 4 Smog Devices for Used Cars," *Los Angeles Times,* 19 July 1973; idem, "Smog Device Rule Reaffirmed; State Lists Exempted Vehicles," *Los Angeles Times,* 31 Oct. 1973.

112. Quoted in Fisher, "Low-Cost Smog Kit."

113. Fisher, "Smog Device Rule Reaffirmed."

114. See California Health and Safety Code § 39030 (1971). See also § 39021 and its accompanying historical note.

115. Quoted in D. Fisher and R. Fairbanks, "Reagan Replaces 4 on State Air Board," *Los Angeles Times,* 15 Dec. 1973.

116. See Clean Air Constituency v. California State Air Resources Board, 11 Cal. 3d 801, 523 P.2d 617, 114 Cal. Rptr. 577 (1974).

117. See Bureau of National Affairs, *Environment Reporter, Current Developments,* Vol. 4, No. 38, 18 Jan. 1974, p. 1553; M. Oliver, "Year's Delay on Car Smog Devices Upheld By Court," *Los Angeles Times,* 15 Feb. 1974.

118. Quoted in Oliver, "Year's Delay."

119. M. Oliver, "Smog Device Delay for Older Cars Challenged in High Court," *Los Angeles Times,* 14 June 1974.

120. Clean Air Constituency v. California State Air Resources Board.

121. L. Pryor, "Air Resources Board Orders Smog Controls for '66–'70 Cars," *Los Angeles Times,* 11 July 1974.

122. Ch. 670, § 1, [1974] Cal. Stats. Advance Service 749, amending California Health and Safety Code § 39177.1 and Vehicle Code § 4602.

123. M. Oliver, "Court Backs Law on Auto Smog Devices," *Los Angeles Times,* 21 Jan. 1975.

124. Ibid.

125. Quoted in J. Gillam, "Panel Rejects Repeal of Smog Device Rule," *Los Angeles Times,* 27 Feb. 1975.

126. K. Burke, "CHP Told to Start Citing Cars Lacking Smog Device," *Los Angeles Times,* 2 March 1975.

127. R. Zeman, "Lead Peril in Auto Smog Units Charged," *Los Angeles Times,* 7 March 1975.

128. J. Gillam, "Bill to Repeal Smog Device Law Fails Again," *Los Angeles Times,* 13 March 1975.

129. M. Oliver, "Judge Refuses to Ban Tickets on Smog Units," *Los Angeles Times,* 18 March 1975.

130. G. Clark, "Writ Restrains CHP From Issuance of NOX Tickets," *Los Angeles Times,* 20 March 1975.

131. Quoted in "Antismog Repeal to Be Reviewed," *Los Angeles Times,* 19 March 1975.

132. J. Gillam, "Assembly Rejects Repeal of NOX Smog Unit Law," *Los Angeles Times,* 8 April 1975.

133. J. Gillam, "Smog Device Foes Keep Heat On for Next Vote," *Los Angeles Times,* 9 April 1975.

134. Quoted in J. Gillam, "NOX Device Repeal OKd by Assembly," *Los Angeles Times,* 11 April 1975.

135. J. Gillam, "NOX Repeal Bill OKd by Senate, Sent to Brown," *Los Angeles Times,* 15 April 1975.

136. Ibid.

137. Quoted in K. Reich, "Brown Signs Bill Killing Controversial NOX Rules," *Los Angeles Times,* 24 April 1975. See also Gillam, "NOX Repeal Bill OKd by Senate."

138. Bureau of National Affairs, *Environment Reporter, Current Developments,* Vol. 6, No. 1, 2 May 1975, p. 11.

139. R. Fairbanks, "Mettle of Lawmakers Tested in Smog Dispute," *Los Angeles Times,* 20 March 1975.

140. Gillam, "NOX Repeal Bill OKd by Senate."

14. Some Themes in the Policy Process

1. J. Hurst, *Law and Economic Growth* (Cambridge: Belknap Press of Harvard University, 1964), p. xi.

2. California Department of Public Health, *Clean Air for California* (Initial Report, 1955), p. 47.

3. California Assembly Committee on Air and Water Pollution, *Final Summary Report* (n.d., c. 1952), p. 25 (hereafter cited as *Final Report*).

4. W. Simmons and R. Cutting, "A Many-Layered Wonder: Nonvehicular Air Pollution Control Law in California," *Hastings Law Journal* 26(1974):109, 123.

5. "The Bay Area Too," *Riverside* [*Calif.*] *Daily Press,* 14 Aug. 1974.

6. See generally F. Fredrick and L. Lowry, "Legislative History and Analysis of the Bay Area Air Pollution Control District" (mimeo., n.d.), pp. 5-19.

7. See generally J. Davies, *The Politics of Pollution* (New York: Pegasus, 1970).

8. H.R. Rep. No. 968, 84th Cong., 1st Sess. 4 (1955).

9. L. McCabe, "Technical Aspects—the 1950 Assessment," in *Proceedings of National Conference on Air Pollution* (Washington, D.C., 18-20 Nov. 1958), pp. 25, 27.

10. H. Kennedy, "Levels of Responsibility for the Administration of Air-Pollution-Control Programs," in ibid., pp. 389, 397.

11. T. Kuchel, "Public Interest Demands Clean Air," in ibid., pp. 15, 17.

12. National Academy of Sciences, *Technology: Processes of Assessment and Choice* (1969), pp. 33-34.

13. See, e.g., Hurst, *Law and Economic Growth,* pp. 213, 224.

14. B. Barger, "Drastic Refinery Controls Urged by Governor's Group," *Riverside [Calif.] Daily Press,* 7 Dec. 1953.

15. "Bill that Created L.A. County Air Pollution Control District Now Recognized as Model for Entire Nation," *Post Advocate* (Alhambra, Calif.), 8 Aug. 1957.

16. Subcommittee of California Assembly Interim Committee on Governmental Efficiency and Economy, *Study and Analysis of the Facts Pertaining to Air Pollution Control in Los Angeles County* (1953), p. 8.

17. H. Kennedy, *The History, Legal and Administrative Aspects of Air Pollution Control in Los Angeles County* (Los Angeles County Board of Supervisors, 1954), pp. 14-15.

18. L. McCabe, "National Trends in Air Pollution," in *Proceedings of First National Air Pollution Symposium* (Los Angeles, Calif., 1949), pp. 50, 51-52.

19. Statement of S. Griswold (mimeo., n.d., c. 1955).

20. Letter from D. Chabek, News Dept., Ford Motor Co., to K. Hahn, Los Angeles County Board of Supervisors, March 26, 1953, in *Smog—A Factual Record of Correspondence between Kenneth Hahn and Automobile Companies,* 6th ed. (Los Angeles, Calif., 1970), p. 4.

21. Air Pollution Foundation, *1954 President's Report* (Los Angeles, Calif., 1954), p. 8.

22. J. Maga and G. Hass, "The Development of Motor Vehicle Emission Standards in California" (Paper presented at 53d Annual Meeting of the Air Pollution Control Association, Cincinnati, Ohio, 25 May 1960), p. 1.

23. California Department of Public Health, *A Progress Report of California's Fight Against Air Pollution* (Third Report, 1957), p. 27.

24. California Department of Public Health, *Technical Report: California Standards for Ambient Air Quality and Motor Vehicle Exhaust* (n.d., c. 1960), p. 14.

25. Ibid., p. 27.

26. "Legislature Approves Car Smog-Cutter Bill," and "Dilworth Slaps Exhaust Action," both in *Riverside [Calif.] Daily Press,* 6 May 1960.

27. B. Barger, "Smog Scientists Join Study Forces," *Riverside [Calif.] Daily Press,* 22 Oct. 1953; *Final Report,* pp. 10-11.

28. J. Esposito, *Vanishing Air* (New York: Grossman Publishers, 1970), p. 281.

29. Air Pollution Foundation, *1954 President's Report,* p. 18.

30. "Panel Discussion," in *Proceedings of Second Southern California Conference on Elimination of Air Pollution* (Los Angeles, Calif., 14 Nov. 1956), pp. 147-148.

31. California Department of Public Health, *Clean Air for California* (Initial Report), p. 48.

32. B. Zimmerman, "Smog Session Brings Discord," *Riverside [Calif.] Daily Press,* 4 June 1954.

33. California Department of Public Health, *Clean Air for California* (Initial Report), p. 47.

34. R. Kehoe, "Air Pollution and Community Health," in *Proceedings of First National Air Pollution Symposium,* p. 115.

35. U.S. Department of Health, Education, and Welfare, *Motor Vehicles, Air Pollution, and Health,* H.R. Doc. 489, 87th Cong., 2d Sess. 62 (1962).

36. E. Cassell, "The Health Effects of Air Pollution and Their Implications for Control," *Law and Contemporary Problems* 33(1968):197, 215.

37. See S. Brubaker, *To Live on Earth* (Baltimore: Johns Hopkins Press, 1972), pp. 14-15.

38. See, e.g., R. Lanzillotti and R. Blair, "Some Economic and Legal Aspects of the Pollution Problem—the Automobile: A Case in Point," *University of Florida Law Review* 24(1972):398, 407; J. Krier, "Environmental Watchdogs: Some Lessons from a 'Study' Council," *Stanford Law Review* 23(1971):623, 662-665.

39. Davies, *The Politics of Pollution*, p. 34.

40. See, e.g., Brubaker, *To Live on Earth*, p. 6; R. Dyck, "Evolution of Federal Air Pollution Control Policy" (Ph.D. diss., University of Pittsburgh, 1971), pp. 217, 226-227, 247.

41. C. Haar, *Land-Use Planning* (Boston: Little, Brown and Co., 1959), pp. 130-131.

42. Dyck, "Evolution of Federal Policy," p. 19; Davies, *The Politics of Pollution*, p. 37.

43. Davies, *The Politics of Pollution*, pp. 30, 108-109.

44. See *Final Report*, p. 38.

45. See W. Wise, *Killer Smog* (New York: Ballantine Books, 1968); Tokyo Metropolitan Government, *Tokyo Fights Pollution* (Tokyo, 1971), especially p. 28 (water pollution).

46. Wise, *Killer Smog*, pp. 36-37, 40-41, 58-59.

47. Davies, *The Politics of Pollution*, p. 37. See also ibid., p. 51.

48. See, e.g., N. Jacoby, "The Environmental Crisis," *Center Magazine*, Nov.-Dec. 1970, p. 37; Davies, *The Politics of Pollution*, pp. 21-23; Brubaker, *To Live on Earth*, pp. 9-10.

49. Brubaker, *To Live on Earth*, p. 10.

50. McCabe, "National Trends in Air Pollution," p. 50.

51. *Final Report*, p. 44.

52. Maga and Hass, "Development of Standards," p. 3.

53. See Krier, "Environmental Watchdogs," p. 664.

54. Interview with C. Biddle, 31 Aug. 1971.

55. H. Molotch and R. Follett, "Air Pollution as a Problem for Sociological Research," in *Air Pollution and the Social Sciences*, ed. P. Downing (New York: Praeger Publishers, 1971), pp. 15, 30. See also J. Davies, "How Does the Agenda Get Set?" (mimeo., 1974).

56. S. Griswold, "Air Pollution in California—The Past" (mimeo., n.d.), pp. 3, 5.

57. California Department of Public Health, *Clean Air for California* (Initial Report), p. 38.

58. H. Hart, "Crisis, Community, and Consent in Water Politics," *Law and Contemporary Problems* 22(1957):510, 525.

59. See R. Heilbroner, "The Arithmetic of Pollution," *Economic Problems Newsletter*, Spring 1970, p. 3, and the discussion at p. 25.

60. Molotch and Follett, "Sociological Research," pp. 23-24. See also Davies, *The Politics of Pollution*, p. 80.

61. See, e.g., J. Ferejohn and M. Fiorina, "The Paradox of Not Voting: A Decision Theoretic Analysis," *American Political Science Review* 68(1974):525; M. Fiorina, "The Voting Decision: Investment and Consumption Aspects," *Social Science Working Paper No. 46* (California Institute of Technology, 1974); R.

Posner, *Economic Analysis of Law* (Boston: Little, Brown and Co., 1972), p. 273.

62. W. Oates and W. Baumol, "The Instruments for Environmental Policy" (mimeo., 1972).

63. T. Schelling, *The Strategy of Conflict* (Cambridge: Harvard University Press, 1960), p. 90.

64. L. Kavaler, "How Los Angeles Women Are Fighting Smog—and Winning," *Family Circle*, Sept. 1968, pp. 55, 99.

65. See T. Roberts, "Motor Vehicular Air Pollution Control in California: A Case Study in Political Unresponsiveness" (Honors thesis, Harvard College, 1969), pp. 47-51.

66. Oates and Baumol, "Instruments."

67. Kavaler, "Los Angeles Women," p. 99.

68. M. Boddy, "Pollution and the Public," in *Proceedings of Second National Air Pollution Symposium* (Los Angeles, Calif., 1952), p. 90.

69. Ibid.

70. See E. Ainsworth, "War on Smog Intensified; New Factors Speed Fight," *Los Angeles Times*, 2 Dec. 1946; Boddy, "Pollution and the Public," p. 90. Mr. Boddy was at the time publisher of the *Los Angeles Daily News*.

71. Boddy, "Pollution and the Public," p. 90.

72. Roberts, "Motor Vehicular Air Pollution Control," pp. 52-53.

73. See *Riverside* [*Calif.*] *Daily Press*, 31 Aug. 1955; ibid., 26 Sept. 1956.

74. Roberts, "Motor Vehicular Air Pollution Control," p. 53.

75. Cf. R. Posner, "Theories of Economic Regulation," *Bell Journal of Economics and Management Science* 5(1974):335.

76. Hart, "Crisis, Community, and Consent," p. 525.

77. See, e.g., D. Brezina, "The Role of Crusader-Triggered Controversy in Technology Assessment: An Analysis of the Mass Media Response to *Silent Spring* and *Unsafe at Any Speed*" (Staff Discussion Paper No. 203, Program of Policy Studies in Science and Technology, George Washington University, 1968), p. 40.

78. A. Weinberg, "Can Technology Replace Social Engineering?" *Bulletin of the Atomic Scientists*, Dec. 1966, pp. 4 and 5. See also U.S. Department of Commerce, *The Automobile and Air Pollution*, Part 2 (1967), p. 118.

79. J. Hurst, *Law and Social Process in United States History* (Ann Arbor: University of Michigan Press, 1960), p. 31.

80. Weinberg, "Can Technology Replace Social Engineering," p. 5.

81. Hurst, *Law and Social Process*, p. 46.

82. U.S. Department of Commerce, *The Automobile and Air Pollution*, Part 2, p. 132. See also ibid., p. 118.

83. A. Will, "Abatement Progress under Existing Air Pollution Laws" (Address to Statewide Conference on Air Pollution Legislation, California Chamber of Commerce, San Francisco, Calif., 21 Feb. 1955), p. 4.

84. M. Neiburger, "Weather Modification and Smog," *Science* 126(1957):637.

85. Will, "Abatement Progress," p. 4.

86. See Air Pollution Foundation, *1956 President's Report* (Los Angeles, Calif., 1956), p. 27; idem, *1957 Annual Report* (Los Angeles, Calif., 1957), p. 17; Kennedy, *Los Angeles History*, p. 82.

87. *Final Report*, pp. 10-11.

88. B. Barger, "Shaw Promises Anti-Smog Bill," *Riverside* [*Calif.*] *Daily Press*, 8 Dec. 1953.

89. Barger, "Drastic Refinery Controls Urged by Governor's Group."

90. Letter from K. Hahn to L. Colbert, pres. Chrysler Corp., 14 Dec. 1953, in *Smog—A Factual Record,* pp. 7, 8.

91. S. Griswold, "The Smog Problem in Los Angeles County," in *Proceedings of Southern California Conference on Elimination of Air Pollution* (Los Angeles, Calif., 10 Nov. 1955), pp. 2, 16.

92. "APCD Completes Job—All Except Auto Exhaust," *Montebello* [*Calif.*] *News,* 7 Nov. 1957.

93. Kennedy, "Levels of Responsibility," pp. 389-390.

94. Letter from J. Campbell, administrative director, Engineering, General Motors Corp., to K. Hahn, March 26, 1953, in *Smog—A Factual Record,* pp. 6, 7; "Panel Discussion," pp. 11, 19-20.

95. See "Panel Discussion," p. 19; Griswold, "The Smog Problem," p. 17; idem, "Reflections and Projections on Controlling the Motor Vehicle" (Paper presented at Annual Meeting of the Air Pollution Control Association, Houston, Texas, 21-25 June 1964), p. 5.

96. Oates and Baumol, "Instruments."

97. Hurst, *Law and Social Process,* p. 293.

98. S. Griswold, Testimony in Hearings before Special Subcommittee on Air and Water Pollution of the Senate Committee on Public Works, 88th Cong., 2d Sess. 37, 44 (1964); Air Pollution Foundation, *1954 President's Report,* p. 18; "Panel Discussion," pp. 93, 114.

99. C. Gruber, "What Can Be Done About Air Pollution?" in *Proceedings of National Conference on Air Pollution,* pp. 44, 46-47.

100. *Final Report,* p. 36.

101. See L. DuBridge, "Summation of Conference," in *Proceedings of Southern California Conference on Elimination of Air Pollution,* pp. 130, 132.

102. Will, "Abatement Progress," p. 2; DuBridge, "Summation," p. 132.

103. Will, "Abatement Progress," p. 6.

104. Kennedy, *Los Angeles History,* p. 22 (emphasis added).

105. DuBridge, "Summation," p. 132; Air Pollution Foundation, *1958 Annual Report* (Los Angeles, Calif., 1958), pp. 11-12. See also "Man Who Organized the First Campaign Gives His Answers," *Los Angeles Times,* 19 Nov. 1953; Kennedy, *Los Angeles History,* p. 14.

106. *Final Report,* p. 25.

107. DuBridge, "Summation," p. 133.

108. *Final Report,* p. 40.

109. U.S. Department of Health, Education, and Welfare, *Motor Vehicles, Air Pollution, and Health,* p. 23.

110. Griswold, Testimony in Hearings, p. 50.

111. J. Ludwig, "Air Pollution Control Technology: Research and Development on New and Improved Systems," *Law and Contemporary Problems* 22(1968):217.

112. See Stanford Research Institute, *The Smog Problem in Los Angeles County* (Menlo Park, Calif., Second Interim Report, 1949), p. 9; Barger, "Shaw Promises Anti-Smog Bill."

113. See California Department of Public Health, *Clean Air for California* (Second Report, 1956), p. 23.

114. Air Pollution Foundation, *1956 President's Report,* p. 27.

115. See A. Carlin and G. Kocher, *Environmental Problems: Their Causes,*

Cures, and Evolution, Using Southern California Smog as an Example (Rand Corp., Santa Monica, Calif., Report Number R-640-CC/RC, 1971), pp. 64–67, 90.

116. Ibid., p. 66.

117. Ibid., p. 92.

118. Ibid., pp. 63–64, 90, 93; Davies, *The Politics of Pollution,* pp. 37, 51.

119. L. Friedman and J. Ladinsky, "Social Change and the Law of Industrial Accidents," *Columbia Law Review* 67(1967):50, 77.

120. Pub. L. No. 92-500, 86 Stat. 816, codified at 33 U.S.C. §§ 1251 et seq. See B. Ackerman et al., *The Uncertain Search for Environmental Quality* (New York: Free Press, 1974), pp. 319–330.

121. Bureau of National Affairs, *Environment Reporter, Current Developments,* Vol. 6, No. 2, 9 May 1975, pp. 67–68.

122. Senate Committee on Public Works, *National Air Quality Standards Act of 1970,* S. Rep. No. 91-1196, 91st Cong., 2d Sess. 13 (1970).

123. See, e.g., Jacoby, "The Environmental Crisis," p. 37; Brubaker, *To Live on Earth,* especially chapters 1 and 2.

124. See, e.g., E. Dale, "The Economics of Pollution," *New York Times Magazine,* 19 April 1970, pp. 27, 28, 40–41.

125. Resources for the Future, *Resources,* Jan. 1972, p. 10.

126. See, e.g., ibid., pp. 9–10; J. Burby, "White House Plans Push for Sulfur Tax Despite Strong Industry Opposition," *National Journal,* 28 Oct. 1972, p. 1663; Bureau of National Affairs, *Environment Reporter, Current Developments,* Vol. 4, No. 22, 28 Sept. 1973, p. 857; ibid., Vol. 6, No. 1, 2 May 1975, p. 3; ibid., No. 2, 9 May 1975, pp. 67–68.

127. Resources for the Future, *Resources,* p. 19.

128. Burby, "White House Plans Push," p. 1668.

129. J. Burby, "White House Activists Debate Form of Sulfur Tax; Industry Shuns Both," *National Journal,* 21 Oct. 1972, p. 1643.

130. See ibid., p. 1646; Bureau of National Affairs, *Environment Reporter, Current Developments,* Vol. 4, No. 22, 28 Sept. 1973, p. 857.

131. See J. Dales, *Pollution, Property and Prices* (Toronto: University of Toronto Press, 1968), p. 85.

132. See, e.g., W. Baumol, "On Taxation and the Control of Externalities," *American Economic Review* 62(1972):307, 319.

133. See, e.g., O. Davis and M. Kamien, "Externalities, Information and Alternative Collective Action," in Joint Economic Committee, 91st Cong., 1st Sess., *The Analysis and Evaluation of Public Expenditures: The PPB System* 67, 83 (Joint Comm. Print 1969).

134. Jacoby, "The Environmental Crisis," pp. 37–38.

15. Some Comments on Present Policy

1. See generally H. Jacoby et al., *Clearing the Air* (Cambridge, Mass.: Ballinger Publishing Company, 1973); F. Grad et al., *The Automobile and the Regulation of Its Impact on the Environment* (Norman: University of Oklahoma Press, 1975); U.S. Office of Science and Technology, *Cumulative Regulatory Effects on the Cost of Automotive Transportation* (RECAT) (1972); Report by the Coordinating

Committee on Air Quality Studies, National Academy of Sciences and National Academy of Engineering, 93d Cong., 2d Sess., *Air Quality and Automobile Emission Control* (Comm. Print 1974).

2. 42 U.S.C. §§ 1857c-4(b)(1) and (2) (1970).

3. See generally A. Teller, "Air Pollution Abatement: Economic Rationality and Reality," *Daedalus* 96(1967):1082.

4. Senate Committee on Public Works, *National Air Quality Standards Act of 1970,* S. Rep. No. 91-1196, 91st Cong., 2d Sess. 11 (1970).

5. Ibid., p. 10.

6. R. Finch, Testimony in Hearings before Subcommittee on Air and Water Pollution of the Senate Committee on Public Works, 91st Cong., 2d Sess., Pt. I, p. 134 (1970).

7. J. Maga, "Air Quality Criteria and Standards," in *Proceedings of Third National Conference on Air Pollution* (Washington, D.C., 12-14 Dec. 1966), pp. 469, 470-471.

8. R. Finch, Testimony in Hearings, p. 134.

9. See, respectively, Report by the Coordinating Committee on Air Quality Studies, *Air Quality,* especially Vol. 1, pp. 2-3; U.S. Office of Science and Technology, *Cumulative Effects,* especially p. xi; B. Goeller et al., *Strategy Alternatives for Oxidant Control in the Los Angeles Region* (Rand Corp., Santa Monica, Calif., Report Number R-1368-EPA, 1973), especially p. xii.

10. Goeller et al., *Strategy Alternatives,* p. xiii.

11. See, e.g., Bureau of National Affairs, *Environment Reporter, Current Developments,* Vol. 4, No. 14, 3 Aug. 1973, pp. 569-570, and the discussion in footnote d, p. 311.

12. Senate Committee on Public Works, *National Air Quality Standards Act of 1970,* p. 10.

13. See Senate Committee on Public Works, *Clean Air Amendments of 1976,* S. Rep. No. 94-717, 94th Cong., 2d Sess. 5-6, 29-31 (1976) (hereafter cited as *Senate Report*); House Committee on Interstate and Foreign Commerce, *Clean Air Act Amendments of 1976,* H.R. Rep. No. 94-1175, 94th Cong., 2d Sess. 189-191 (1976) (hereafter cited as *House Report*).

14. *House Report,* pp. 190-191.

15. See, e.g., *Senate Report,* pp. 5-6, 29-31.

16. *House Report,* p. 22; *Senate Report,* p. 6.

17. Cf. Jacoby et al., *Clearing the Air,* p. 5; B. Ackerman et al., *The Uncertain Search for Environmental Quality* (New York: Free Press, 1974), pp. 320-321.

18. See G. Calabresi, "Transaction Costs, Resource Allocation, and Liability Rules," *Journal of Law and Economics* 11(1968):67.

19. See, e.g., "Confusion, Long Lines Mark Beginning of Odd-Even Plan," *Los Angeles Times,* 2 March 1974; "Gasoline Plan Fails to Eliminate Long Lines," *Los Angeles Times,* 3 March 1974.

20. See "Waiting in Line?—Gasoline Burns, Too," *Los Angeles Times,* 8 Feb. 1974.

21. See generally J. Dales, *Pollution, Property and Prices* (Toronto: University of Toronto Press, 1968).

22. Bureau of National Affairs, *Environment Reporter, Current Developments,* Vol. 5, No. 7, 14 June 1974, p. 185.

23. See Grad et al., *Automobile Regulation,* pp. 8-9, 325.

24. For a summary, see D. Currie, "Motor Vehicle Air Pollution: State Authority and Federal Pre-Emption," *Michigan Law Review* 68(1970):1090-1091. See also the discussion at pp. 174-175, 181.

25. See U.S. Council on Environmental Quality, *Environmental Quality—1976* (7th Annual Report, 1976), pp. 213-214.

26. See, respectively, L. Pryor, "Air Quality in Basin Deteriorated in '74," *Los Angeles Times,* 3 Feb. 1975; R. Kovitz, "West L.A. Smog Tears Flow More Freely in 1976," *Los Angeles Times,* 16 Jan. 1977.

27. Cf. A. Stern, "Strengthening the Clean Air Act," *Journal of the Air Pollution Control Association* 23(1973):1021.

INDEX